Sustainable Development Goals in Southeast Asia and ASEAN

Political Ecology in the Asia Pacific Region

Editor-in-Chief

Ronald L. Holzhacker (*University of Groningen*)

VOLUME 1

The titles published in this series are listed at *brill.com/peap*

Sustainable Development Goals in Southeast Asia and ASEAN

National and Regional Approaches

Edited by

Ronald Holzhacker
Dafri Agussalim

BRILL

LEIDEN | BOSTON

The Library of Congress Cataloging-in-Publication Data is available online at http://catalog.loc.gov

Typeface for the Latin, Greek, and Cyrillic scripts: "Brill". See and download: brill.com/brill-typeface.

ISSN 2589-6636
ISBN 978-90-04-37823-0 (hardback)
ISBN 978-90-04-39194-9 (e-book)

This book is printed on acid-free paper and produced in a sustainable manner.

Contents

PART 3
SDGs and Progress on the Social Agenda in Middle-Income and Developing Countries in Southeast Asia

PART 4
SDGs and the New Urban Agenda, Cities, and Transport

PART 5
SDGs and the Environment, Clean Air and Water for All

PART 6
SDGs and the Economic Agenda for Inclusive Economic Growth and Decent Work for All

PART 7
SDGs, Agriculture, and Community Development through Partnerships

Figures and Tables

Figures

Tables

About the Editors

Ronald Holzhacker
University of Groningen, The Netherlands, is Professor of Comparative Multilevel Governance and Regional Structure in the Faculty of Spatial Science, Department of Spatial Planning and Environment, and the Faculty of Arts, Department of International Relations and International Organization. He holds a PhD from the University of Michigan in political science, and a JD from the University of Minnesota Law School. He is broadly interested in questions of governance, human rights, and the interaction between civil society organizations and institutions in political systems. He is founding Director of the Groningen Research Centre for Southeast Asia and ASEAN (SEA ASEAN) located in Groningen and Universitas Gadjah Mada, Yogyakarta, Indonesia.

He leads an inter-disciplinary team of scholars and a group of 20 PhD researchers engaged in theoretically driven comparative research focused on governance, societal impact, and sustainable society in Southeast Asia. He has published in such journals as *Law & Policy, Comparative European Politics, Journal of European Integration, European Union Politics, Nations and Nationalism, Party Politics, Journal of Legislative Studies, The Asia Pacific Journal of Public Administration* and *The Journal of South East Asian Human Rights*. He is co-editor of numerous books, most recently *Decentralization and Governance in Indonesia* (New York: Springer, 2016).

Dafri Agussalim
Universitas Gadjah Mada, Yogyakarta, Indonesia, is Senior Lecturer at the Department of International Relations and Executive Director of the ASEAN Study Center, Faculty of Social and Political Sciences. He is also a Visiting Fellow at the Indonesian Military Academy and some private universities in Indonesia, as well as researcher at the Security and Peace Study Center, UGM; Human Rights Study Center, Universitas Islam Indonesia (UII); and International Institute for International Studies, UGM. He obtained his Master of Arts in International Relations from the Australian National University and PhD from Universitas Gadjah Mada in 2015. His fields of interest are Indonesian foreign policy, ASEAN, human rights and international security. He has published books, articles, book chapters, conference papers, commentaries and reports on issues relevant to politics and security, as well as human rights.

About the Authors

Thuta Aung

has represented Myanmar at ASEAN and South Asia Regional fora in the area of inclusive business. He continues his academic career which started in the UK by regularly giving lectures as a Senior Fellow at the Myanmar Institute of Strategic and International Studies and Central Institutes of Civil Service (Hpaung Gyi and Zeepin Gyi) on leadership and coordination issues. He is also actively involved in start-up development, having set up the first PPP-operated Incubation Centre with the Department of SME Development. With his artist wife Shwe Thiri Khit, Thuta is part of an initiative to build the fashion brand AmaraKhit. Selected into the ASEAN Young Business Leaders Initiative of the Asia New Zealand Foundation and subsequently selected as a Member of the Leadership Network of the Foundation, Thuta is also a Fellow of the Royal Society of Arts and Associate of the Corbett Centre for Maritime Policy Studies. Thuta has recently been appointed by the Myanmar Minister of Commerce to serve as a Commissioner on the Myanmar Competition Commission.

Aliyyah Nuha Faiqah Azman Firdaus

is a project officer at the Asia–Europe Institute and Centre for ASEAN Regionalism at the University of Malaya, Kuala Lumpur. She specializes in applied linguistics and environmental discourse. She has been involved in capacity-building projects in Cambodia and Laos.

Titus C. Chen

is an Associate Professor of International Relations at the Institute of Political Science of National Sun Yat-sen University, Kaohsiung, Taiwan. He earned his PhD in Political Science from the University of California, Irvine, in 2008. Dr Chen specializes in international relations theory, Asia-Pacific regionalism, East Asia international relations, and Chinese foreign policy.

Tran Dinh Lam

is Director of the Center for Vietnamese and Southeast Asian Studies, Vietnam National University of Social Sciences and Humanities, Ho Chi Minh City. He achieved his PhD at Innsbruck University, Austria in 1998. His publications include, *The Economic, Cultural and Social Life of Bahnar People Sustainable Development* (2011) which was awarded The Excellent Research Project by the President of VNU; *Intellectual Property Rights in Developing Countries* (2012); and *Small and Medium-sized Enterprises: The Way to Success* (2014). His research interests focus on economic development, ethnic minorities, culture and economy, copyright, sustainable development of environment and economy. His research projects have been funded by the Japan Foundation, Erasmus Mundus, and the Japan Society for the Promotion of Science. Dr Tran Dinh Lam is also the Vice President of VAPEC, HCMC.

Maharani Hapsari
is lecturer at the Department of International Relations, Faculty of Social and Political Science, Universitas Gadjah Mada, Yogyakarta, Indonesia. She obtained her Master's Degree in 2010 and Doctoral Degree in 2012 from Nagoya University, Japan, specializing in International Development Studies. Since returning to Indonesia, she has been researching the politics of global environmental change and political economy of development. Her most recent publication is a chapter on *Constructing Palm Oil Justice Movements in Indonesia: Citizenship and Collective Identity*, part of an edited volume on *The Politics of Citizenship in Indonesia*, edited by Eric Hiariej and Kristian Stokke (2017).

Noor Hasharina Hassan
is a researcher and lecturer in the Geography, Environment and Development Programme at the Faculty of Arts and Social Sciences, the University of Brunei. She is also a researcher at the Institute of Asian Studies under the 'Borneo in Transition' Research Cluster.

Azirah Hashim
is Executive Director of the Asia–Europe Institute and Director of the Centre for ASEAN Regionalism, University of Malaya, Kuala Lumpur. Her research interests include higher education in ASEAN, language contact in Southeast Asia, English as a lingua franca in ASEAN, and language and law. She leads projects on capacity building in higher education in Malaysia, Cambodia, and Laos and educational leadership in ASEAN and is involved in projects on Asia–Europe in her current roles. In 2009, she was awarded the Georg Foster Fellowship for Experienced Researchers and is a fellow of the Alexander von Humboldt Foundation, Germany.

Phanthanousone (Pepe) Khennavong
obtained his PhD Degree in 2015 at the Australian National University, Canberra, Australia. His PhD thesis is titled: 'Aid to Laos in the Twenty-First Century: Engagement and Change'. The thesis highlights the Lao experience of development assistance in relation to three contemporary aid issues: the influence

of aid on governance; the impact of the Poverty Reduction Strategy (PRS) approach, a major aid reform; and the influence of growing Chinese aid. He obtained his Master's Degree in Economics in 2006 from Sydney University, and his Bachelor's Degree in International Trade in 2001 from Victoria University of Technology, Melbourne. Since 2014, Dr Khennavong has been employed by the United Nations Development Programme in Laos as technical advisor embedded in the Ministry of Planning and Investment and Ministry of Foreign Affairs to support the Lao Government to manage a development cooperation forum called the Round Table Process, to integrate the SDGs into the Lao context and to apply results-based management to the national socio-economic development planning process. His research interests are, among others, aid and governance, development planning process, human development and sustainable development, and Chinese aid.

Karina Larasati
is a researcher at the ASEAN Studies Center, Faculty of Social and Political Sciences, Universitas Gadjah Mada, Yogyakarta, Indonesia. She holds a Bachelor of Arts in International Relations from the Faculty of Social and Political Sciences, Universitas Gadjah Mada (2017). She is actively involved in research related to the evaluation of SDGs achievement at the national and local level in Indonesia, the perception of youth on bilateral relations between Indonesia and Australia, and research on migrant workers. Her interests cover a wide range of topics related to the culture and identity of Europe, ethnicity and migration, human rights, refugee and asylum Law, sustainability and development, and the interdisciplinary study of international relations and international law.

Amador IV Peleo
is currently completing studies in Mandarin Chinese at the National Sun Yat-Sen University in Kaohsiung, Taiwan. He was previously an Assistant Professor at the Department of Political Science of the College of Social Sciences and Philosophy at the University of the Philippines, Diliman. He obtained his MA in International Relations from the International University of Japan in 2003. In 2009, he earned his PhD from the School of Political, Social and International Studies of the University of East Anglia, Norwich, UK.

Ulrich Karl Rotthoff
is Assistant Professor at the Asian Center, University of the Philippines, Diliman. He has studied in Osnabrück, Germany, and Manila, Philippines, and received his doctorate from the State University of Osnabrück. Teaching in the

Philippine and in the Asian Studies program, his research focuses on human rights and rural development issues.

Bakti (Bobi) Setiawan
is Professor in urban planning at the Department of Architecture and Planning, Universitas Gadjah Mada, Yogyakarta, Indonesia. He completed his Master's in urban and regional planning at the University of Waterloo in 1992 and his doctoral degree in urban and community planning at the University of British Columbia, Canada, in 1998. Currently he serves as the Director of the Graduate Program in Urban and Regional Planning, Universitas Gadjah Mada. He teaches planning theory and planning process and his research interest covers issues such as informal housing, land management, urban politics, and environmental management in general.

Kimsan Soy
currently serves as Director of the Center for the Study of Humanitarian Law (CSHL), based at the Royal University of Law and Economics in Cambodia (RULE). Kimsan joined CSHL as one of its researchers in August 2014, before becoming its Director in September 2016. He holds a Master of Laws degree, focusing on public international law, from the Transnational Law and Business University and a Master's in Human Rights and Democratization from the University of Sydney. Kimsan's major areas of research include fair trial rights, judicial independence, international criminal justice and business and human rights. Kimsan has provided legal training to legal professionals at the Royal Academy for Judicial Professions, as well as to law students at RULE and provincial Cambodian universities. Before joining the CSHL, he also worked at the Asian International Justice Initiative (AIJI) as a trial monitor for two years on Case 002/01 of the Extraordinary Chambers in the Courts of Cambodia.

Wendy Tan
is Associate Professor at the Department of Civil Engineering Western Norway University of Applied Sciences, and senior researcher at Wageningen University and Research, the Netherlands. She is also the Associate Director of SEA ASEAN, the Groningen Research Centre for Southeast Asia and ASEAN at the University of Groningen. Her research focuses on the implementation of land use and transportation integration, mobility issues, and institutional perspectives in planning processes in Europe, America and Southeast Asia. From 2010–2014, she was a board member of the Megacities Foundation. She received the Georges Allaert Prize in 2014 for her research contributions to mobility and spatial planning in our society.

Julio C. Teehankee
is Full Professor and Research Fellow of Political Science and International Studies at De La Salle University, Manila, the Philippines, where he served as Dean of the College of Liberal Arts from 2013 to 2017. He is the Executive Secretary of the Asian Political and International Studies Association (APISA), the regional professional organization of scholars in political science, international relations, and allied disciplines. He is also currently the President of the Philippine Political Science Association (PPSA).

Saikaew Thipakorn
is currently researcher at the Institute of Asian Studies, Chulalongkorn University, Bangkok, Thailand. She earned her PhD in Thai Studies from the Department of Arts, Chulalongkorn University. Dr Saikaew Thipakorn's background is in international relations, with a specialization on Japan–Thailand relations. Her current research is on different aspects of cultural relations between Japan and ASEAN countries. She is also guest lecturer in many academic institutions in Thailand.

Dio H. Tobing
is a Research Fellow at the ASEAN Studies Center, Universitas Gadjah Mada, Yogyakarta, Indonesia. His research interests focus on the topics of international peace and security, human rights and non-traditional security, and international law. He formerly served as the research manager at the ASEAN Studies Center, UGM (2017).

He is currently serving as a Senior Policy Advisor for Indonesia's foreign policy, ASEAN affairs, and Timor Leste to the Embassy of the Kingdom of the Netherlands in Jakarta. He holds a BA in International Relations from Universitas Gadjah Mada (2016) and a LLM in Public International Law from Leiden University, the Netherlands (2018).

Laksono Trisnantoro
is a Professor in Health Policy and Administration and a senior researcher in the Faculty of Medicine, Universitas Gadjah Mada, Yogyakarta, Indonesia. He obtained his MD from Universitas Gadjah Mada in 1987, and Master of Science in Health Economics from the Department of Economics, University of York, UK. His PhD is from the London School of Hygiene and Tropical Medicine. In 2001 he spent a year at the Department of Social Medicine, Harvard Medical School, Boston, USA. He is currently the Director of Graduate Programme in Hospital Management at Universitas Gadjah Mada and Chief Editor of the *Indonesian Journal of Health Policy*.

Ahmad Rizky M. Umar
is a PhD Candidate of the school of Political Sciences and International Studies, University of Queensland, Australia, and a research fellow at the Earth System and Governance Project. He was previously a researcher at the ASEAN Studies Center, Universitas Gadjah Mada, Yogyakarta, Indonesia. His research interests cover a wide range of areas related to ASEAN, Indonesia's foreign policy, global environmental governance, and international relations theory.

Helena Varkkey
is a Senior Lecturer at the Department of International and Strategic Studies, University of Malaya in Kuala Lumpur, Malaysia. She received her Doctorate from the University of Sydney in 2012. She has been interested in sustainable development throughout her academic career. Her interest in the field has evolved to a focus on transboundary pollution in Southeast Asia, particularly pertaining to the role of patronage in agribusiness, especially the oil palm industry, and its link to forest fires and haze in the region. Her findings have been published as a book in 2016 as part of the Routledge Malaysian Studies Series. She continues to undertake research in this field.

Gabriel Y. V. Yong
is a researcher and a lecturer in the Geography, Environment and Development Programme at Faculty of Arts and Social Sciences, Universiti Brunei Darussalam.

PART 1

Institutions and Governance for the Sustainable Development Goals

CHAPTER 1

Introduction: Sustainable Development Goals in Southeast Asia and ASEAN

Ronald Holzhacker

1 Introduction

This volume brings together an impressive group of scholars from leading universities in each of the 10 ASEAN states. The editors requested that each author reflect on how their own research interests relate to the SDGs. We thus explore the issues in developing (Cambodia, Laos, Myanmar, Vietnam), middle-income (Indonesia, Malaysia, Philippines, Thailand) and developed countries (Brunei Darussalam, Singapore) in the region. These marked differences in development levels across Southeast Asia mean that the political and policy challenges in implementing the SDGs are very diverse and that regional plans must be sufficiently flexible to take this diversity into account. The perspectives on governance and the SDGs presented here emerge from the fields of political science, international relations, geography, economics, law, health, and the natural sciences.

The contributions to this volume on the Sustainable Development Goals are organized into seven sections, beginning with broad institutional features and proceeding through the various goal areas. The first section introduces the institutions and governance involved in the SDGs, at the international, regional, national, and local levels. The second section focuses on accountability to citizens and human rights to ensure progress on the SDGs in Southeast Asia. The third section discusses the SDGs and progress on the social agenda in both the developing and middle-income countries in Southeast Asia. The fourth section looks at the new urban agenda, with a concern for cities, housing, and urban transport, including issues in the developed countries in the region. The fifth section focuses on the environment, and clean air and water for all. The sixth section focuses on the economic agenda for inclusive growth and decent work for all. Finally, the seventh section focuses on the SDGs and agriculture and community development through international partnerships.

The scholars involved were able to come together and actively discuss their work with each other. We had wide-ranging discussions about national policies and about the challenges of getting multiple levels of governance

 | DOI:10.1163/9789004391949_002

within the Southeast Asian states engaged and focused on implementation of the SDGs. Bringing together scholars whose work focused on different elements of the SDGs — social and economic spheres, the environment, climate change — spurred discussions about the possible synergies, but also about the perceived trade-offs between the goals which policy makers and governments face. A working conference was held on 3–5 October 2017 at Gadjah Mada University (UGM), Yogyakarta, Indonesia; this served as the initial stage of the book project and succeeded in generating insightful ideas and feedback for all the participants. We discussed and debated the relevance and importance of achieving the SDGs through national and regional perspectives, which led to a stimulating dialogue that continued outside of the room and beyond the conference. Participants not only presented their own research but engaged actively in providing critical comments and building collaboratively on each other's work.

The conference at UGM with invited scholars from across the region was inspired by an earlier event hosted at UGM, at the Open Science Meeting on Scientific Cooperation Indonesia–Netherlands, entitled 'Toward Resilient Society'. This was organized by the Indonesian Academy of Sciences (AIPI), the Royal Netherlands Academy of Arts and Sciences (KNAW), and the Netherlands Organization for Scientific Research (NWO). It is an example of a convergence of the academic agendas of European and Southeast Asia research institutions around the SDGs.

It is important that scholarly work related to the SDGs be shared with a wide range of government officials, policy makers, civil society organizations, and other scholars. For this reason, we have designed an integral conference to launch this volume and to promote a dialogue with a broad audience to increase the potential for societal impact and learning. We will bring together the authors involved in this project from across Southeast Asia, together with invited PhD candidates from the region, for a conference with stakeholders to be held in Brussels in October 2018. We have invited keynote speakers from ASEAN, from the United Nations Economic and Social Commission for Asia and the Pacific (UN ESCAP), the EU External Action Service and the European Commission, Directorate-General for International Cooperation and Development. In the evenings, discussions will continue at selected embassies to which we have been invited by the Ambassadors, including the Permanent Representative to the EU from the Kingdom of the Netherlands, the Embassy of the Republic of Indonesia, and the Embassy of the Republic of Singapore. In 2018, Singapore fulfils the role of the rotating chairmanship of ASEAN, with a focus on a resilient and innovative ASEAN (Singapore Foreign Ministry, 2018). Thus, not only in the creation of this book, but also through a process of dissemination and discussion, the authors of this volume wish to contribute to the achievement of the SDGs.

2 Themes of the Book

The authors in this volume were already engaged in a wide variety of research areas when we approached them with a request to extend their expertise to consider their research in light of the SDGs. Before our invitation, they may not have consciously considered how the SDGs relate to their field of interest. Our contribution here is to encourage the broader university research community across Southeast Asia to engage with the SDGs, moving the discussion away from the exclusive domain of the UN and national policy makers. We proposed three broad themes for the book, related to governance and the challenges which may emerge. At our request, the authors have attempted to include one or more of these themes in their chapters, and to specify how some of these governance challenges have become apparent in their policy areas.

Challenge 1: The shift from the international to the national
- Moving from internationally agreed goals and targets to national institutional arrangements (the nature of the challenge varies by state and by the policy sector involved);
- Coordinating the multiple national ministries or agencies that may be involved in a given goal or target;
- Recognizing that overarching approaches to a set of the SDGs may be more effective than the separate policy pursuit of a single SDG or target.

Challenge 2: Multi-level challenges of implementation
- The necessity of including provincial and local governments in the planning and implementation of SDGs within governing structures which may vary in terms of the degree of state decentralization;
- ASEAN as a regional actor may be engaged and promote learning about best practices in the implementation of the SDGs across Southeast Asia. This learning should not only be between national levels of government across the states, but may also be at the local level.

Challenge 3: Involvement by stakeholders and civil society
- Strengthening engagement by stakeholders, businesses, civil society organizations and citizens involved in consultation and implementation of the SDGs.

In this introductory chapter, these three themes will be explored in greater detail. We will first briefly discuss the development of the institutional idea of sustainable development and the SDGs within the world community, and then examine how other countries and regions active in the international system have assisted Southeast Asia and engaged with ASEAN on the SDGs, focusing especially on the European Union (EU), but also briefly on the United States, China, Japan, and Australia. The final section of the introduction provides a brief overview of the other contributions to this book.

3 The Sustainable Development Goals

The global community and the UN, with a broad mission of global peace and security, formulated the 2030 Agenda for Sustainable Development (Agenda 2030) with the aim of ending poverty, ensuring prosperity for all, and protecting the planet. The internationally agreed Agenda 2030, covering the period 2015 to 2030, is designed to guide states towards global sustainable development through the Sustainable Development Goals (UN, 2015). This agreement followed the previous focus of the world community on the Millennium Development Goals (MDGs). A key change from the MDGs to the SDGs is that whereas the MDGs were targeted toward developing countries, the SDGs are global goals for all countries, and include targets like greater economic equality, sustainable consumption, and climate change related goals (see Figure 1.1). This places obligations and responsibilities on all states of the international community, both within their own country and in partnership with other nations. Adopted by the member states on 25 September 2015, during the United Nations Sustainable Development Summit in New York, the 17 SDGs face considerable challenges. Here we aim to provide an understanding of issues surrounding the implications of the policy goals, policy integration across the goals and with domestic agendas, and multi-level implementation.

Attempts to achieve global sustainable development should be seen within the longer trajectory of the idea of development. The term 'sustainable development' was first introduced as an outcome of the World Charter for Nature in 1982 (Hák, Janousková, & Moldan, 2016; UN, 1982). Five years later, the necessity of sustainable development was addressed in the now well-known Brundtland Report (WCED, 1987). However, it was not until 1992, with the UN's Agenda 21, that a clear action document was formulated for countries to engage in sustainable development (UN, 1992). This document is often seen as the first successful attempt to reconcile economic growth and the protection of the planet. During the World Summit on Social Development in Copenhagen in 1995, the need emerged for a third component — a social pillar — to be added to the definition of sustainable development. Even though the social

FIGURE 1.1 The sustainable development goals
SOURCE: UNITED NATIONS FOUNDATION (2018)

pillar was endorsed by the World Summit on Sustainable Development in Johannesburg in 2002 (UN, 2002), it was not until a decade later that this third pillar was fully embraced in the report 'The Future We Want' (UN, 2012). Thus it was only recently that the social aspect was fully integrated into the idea of sustainable development, signalling a shift from the earlier focus on economic development.

The precursor to the SDGs, the MDGs were presented by the UN in 2000 as preparation for the Millennium Summit in New York (UN, 2000). Since then, scholars and activists have argued for more participation and dialogue in sustainable development (Fukuda-Parr, 2008). The lack of participation undermined the sense of national ownership of the MDGs in the member states, which led not only to uneven progress between countries, but also to issues related to measurement, analysis, and overall justification (Fehling, Nelson, & Venkatapuram, 2013; Fukuda-Parr, 2008; Haines & Cassels, 2000). When, in 2015, the SDGs succeeded the MDGs, the social pillar officially became part of sustainable development for the first time (Biermann, Kanie, & Kim, 2017). In 'Transforming our World: The 2030 Agenda for Sustainable Development' (UN, 2015), the act of balancing social, economic, and environmental values is declared as the basis for sustainable development. This balancing act refers to an equal co-existence of the three sustainable development pillars in terms of priority, policy integration, and governance.

Various parts of the UN are involved in the agenda surrounding the SDGs. For example, the United Nations Development Programme (UNDP) states that, in the transition from the MDGs to the SDGs, it will support a process of "learning and adapting to support their implementation" (UNDP, 2016, p. 9), focusing on three areas of support collected under the headings of: "1. Sustainable Development Pathways; 2. Inclusive and Effective Democratic Governance; and 3. Resilience Building" (ibid., p. 1). The UNDP notes that "[t]ogether they present a global package of services that can be tailored to each national setting" (ibid., p. 13). UNDP has now produced a wide range of documents focused on implementation of specific goals.

4 Synergies between Human Rights and the SDGs

The creation of the MDGs, and later the SDGs, grew out of a longer-term debate within the international community concerning development and human rights. Many clearly see the adoption of the SDGs as beneficial to the attainment of human rights and a response to the treaty requirements and institutional developments around human rights in the post-WWII period. For example, the Danish Institute for Human Rights (DIHR), which serves as the National Human Rights Institution for Denmark but is also engaged internationally, declares that "more than 90% of the Sustainable Development Goals ... targets are linked to international human rights and labour standards" (DIHR, 2018a).

The overall connection between human rights and SDGs is also important in terms of implementation, and the Danish Institute carefully constructs and communicates an argument about the convergence and complementarity of the approaches. "Human rights offer guidance for the implementation of the 2030 agenda" and likewise the agenda "can contribute substantially to the realisation of human rights" (ibid.). It claims that "drawing on existing human rights mechanisms will ease the reporting burden of states, and enhance coherence, efficiency and accountability" (ibid.), and reaches a forceful conclusion with regard to the unique power of law in this process for human rights and SDGs: "The entitlements and obligations under international human rights law define the roles of rights-holders (individuals and groups to which the rights in question apply) and duty-bearers (state and non-state actors with corresponding obligations to respect, protect and fulfil human rights)" (ibid.).

We are beginning to see some official recognition of, and institutional developments that further explore, these synergies between human rights and SDGs. For example, a small group of states from Europe, the Americas and Africa came together at the 34th session of the Human Rights Council on 31 March 2017 in Geneva. This Group of Friends on Human Rights and the

2030 Agenda for Sustainable Development aimed to facilitate an "action orientated and inclusive conversation, to be followed by concerted and inclusive action, about the optimal role of the human rights pillar" toward fulfilment of the 2030 Agenda (Denmark, 2017, p. 3). The Group later expanded to include countries from the Asia-Pacific region, including Thailand. During a meeting in June 2017, the Group discussed how the UN human rights system, including its monitoring mechanisms, might best support states to pursue synergies between the implementation of international human rights obligations and commitments and the attainment of the SDGs. The states involved discussed, inter alia, the role of the Human Rights Council, and the potential use of the Universal Periodic Review (UPR) to provide monitoring for the SDGs. A further extensive report prepared by the Danish Institute discusses the involvement of national human rights institutions in this synergy (DIHR, 2018b).

Others have offered more sceptical views on possible synergies between SDGs and human rights. For example, Arat and MacDonald (2018) conducted a critical interpretive analysis of the documents related to the Convention on the Elimination of All Forms of Discrimination against Women (CEDAW) and the SDGs, and the "extent to which they may supplement and reinforce each other's provisions and thus help improve women's rights, status, and lives" (ibid., p. 1). They conclude that although the shift in language from anti-discrimination in CEDAW to empowerment in the SDGs implies a stronger commitment, it is embodied in a framework which corresponds to a shift from liberalism to neo-liberalism which "tends to undermine, if not violate, the rights of a vast majority of women" (ibid.).

5 The Involvement of ASEAN as a Regional Actor

ASEAN has announced that it recognizes a constructive synergy between its regional integration plans and the SDGs. For example, at a meeting between ASEAN, UN ESCAP, and the Kingdom of Thailand in 2016, the ASEAN Deputy Secretary-General for Socio-Economic Community declared that "ASEAN recognizes the importance of creating synergy between the mutually-reinforcing ASEAN 2025 Blueprints and the Sustainable Development Goals (SDGs) within the framework of ASEAN–UN cooperation" (ASEAN, 2016). The Deputy described the SDGs as "a shared goal at the multilateral level but also at the region level and this initiative shows that ASEAN is at the forefront of the effort to achieve the mutually-reinforcing goals of the Blueprints and the SDGs" (ibid.).

The following year, the UN ESCAP secretariat for the Fourth Asia-Pacific Forum on Sustainable Development in Bangkok (March 2017), prepared a 'Regional road map for implementing the 2030 Agenda for Sustainable Development in Asia and the Pacific' (UN ESCAP, 2017b). The roadmap contains priority areas for cooperation in implementing the SDGs in the region. At the Fifth Asia-Pacific Forum (March 2018), there was greater specificity and a new focus not only on regional, but also on subregional perspectives, including ASEAN. The note by the secretariat titled 'Regional and subregional perspectives on the transformation towards sustainable and resilient societies' demonstrates a new focus on risk profiles and resilience and climate change (UN ESCAP, 2018a).

The symbolic language and rhetoric of the early ASEAN pronouncements have thus progressed to more concrete steps, for example with initiatives proposed between ASEAN and UN ESCAP to achieve the SDGs in the region (UN ESCAP, 2017c; see also IISD, 2018). In addition to ASEAN's engagement with UN ESCAP, the Government of Thailand is also involved in its role as ASEAN Coordinator in Sustainable Development Cooperation. ASEAN has developed a report on the complementarities involved and a framework for action. In term of priority areas, the report calls for ASEAN and its member states to focus on poverty reduction, infrastructure and connectivity, sustainable management of natural resources, sustainable consumption and production, and resilience (UN ESCAP, 2017c, p. 5).

6 The Sustainable Development Goals and Southeast Asia

We will now take a closer look at the successes achieved and the challenges faced by Southeast Asia in terms of the Sustainable Development Goals. The UN ESCAP office has studied the progress made with the MDGs, as well as laying out the unfinished agenda for the SDGs. Specifically focusing on the subregion of Southeast Asia (within its broader Asia-Pacific mandate), it has identified areas of success, including social goals such as cutting extreme poverty and improving school enrolment, health goals such as efforts toward eradicating tuberculosis and other diseases, and the provision of safe drinking water, and environmental goals such as protecting areas with high biodiversity (UN ESCAP, 2017a). The report notes persistent issues in areas of health and the environment, including infant and maternal mortality, deforestation, and carbon dioxide emissions. In the list of priorities of the unfinished agenda, one sees a broad and complex set of issues, focused not only on, for example, reducing poverty, but also on reducing inequality and achieving broad-based economic progress including 'decent' jobs. Similarly, the focus moving forward is not just on increasing school enrolment per se, but on quality education, while a broader set of environmental goals goes beyond biodiversity protection to include measures to ensure environmental sustainability, and resilience to climate impacts and natural disasters. There are also priorities related to women's empowerment, and measures to promote peace and security. The most important areas that UN ESCAP identifies as major challenges are capacity weaknesses and ineffective implementation strategies (ibid., p. iii). Table 1.1 summarizes these successes and challenges.

TABLE 1.1 View of UN ESCAP on MDGs and SDGs in Southeast Asia

Successes:
- cutting extreme poverty
- improving school enrolment
- eradicating TB and other diseases
- protecting areas with high biodiversity
- provision of safe drinking water

Persistent issues:
- infant and maternal mortality
- cohort survival
- deforestation
- carbon dioxide emissions

Priorities of the unfinished agenda:
- reducing poverty and inequality
- broad-based economic progress with decent jobs
- promotion of healthy lives
- quality education
- environmental sustainability
- resilience to climate impacts and natural disasters
- gender equality and women's empowerment
- promoting peace and security

Major challenges:
- capacity weakness
- implementation challenges

SOURCE: ADAPTED FROM UN ESCAP (2017A)

One of the difficulties of addressing these issues from a regional perspective is the great diversity in the countries within Southeast Asia. In Table 1.2, we have placed the 11 Southeast Asian states into three groups, high, medium, and low, based on their 2015 ranking in the UN's Human Development Index, HDI (UNDP, 2015). All of these countries are also members of ASEAN, except Timor-Leste, which desires accession and has an application pending before the regional body. The HDI ranking is based on three basic dimensions — a long and healthy life, being knowledgeable, and having a decent standard of living — with multiple indicators across the dimensions. One can see from Table 1.2 that Singapore and Brunei Darussalam are both ranked highly, at positions 5 and 30 respectively. Much of Southeast Asia may be seen as medium in the HDI rankings, with two countries at the high end: Malaysia with a ranking of 59 and Thailand at 87. Three countries ranked below that are close together in the ranking, Indonesia at 113, Vietnam at 115, and the Philippines at 116. In the lowest category for HDI in the region, are Timor-Leste at 133, Laos at 138, Cambodia at 143, and Myanmar at 145. (It should be noted that the UN index classifies this final group of countries as 'medium' in terms of human development, because there is an additional group of countries below this labelled 'low' from a world-wide perspective.) These large differences across the Southeast Asia region in terms of HDI — from rank 5 to rank 145 — may hamper opportunities for learning across the states, and may make the involvement of ASEAN in the SDGs more complicated. More encouragingly, it may also mean that South–South learning within the region may occur between countries with a higher human development ranking and those countries with a lower ranking.

While the UN prefers to use the broader HDI to rank countries, the World Bank uses Gross National Income (GNI) per capita to draw up rankings. Within Southeast Asia, the World Bank ranks Malaysia and Thailand as upper middle-income countries, and the following countries as lower middle-income countries: Cambodia, Indonesia, Philippines, Timor-Leste, and Vietnam (World Bank, 2018). Some middle-income countries have expressed special concern about the needs of middle-income states in terms of the international cooperation and support that may be available to help them meet the SDGs. For example, a statement delivered at the High-level Political Forum on Sustainable Development 2017, in New York, by the so-called group of Like-Minded Countries in Support of Middle-Income Countries, declares that middle-income countries face significant challenges to achieve sustainable development. They call for a continued exchange of experiences, improved coordination, and better support, but note that as countries move from low-income to middle-income status, "the international cooperation is significantly reduced" (Like-Minded Countries, 2017). Therefore, the group calls for "an

TABLE 1.2 Human Development Index ranking of countries in Southeast Asia

Human Development Index (HDI rank)	
High	
Singapore	5
Brunei	30
Medium	
Malaysia	59
Thailand	87
Indonesia	113
Vietnam	115
Philippines	116
Low	
Timor-Leste	133
Laos	138
Cambodia	143
Myanmar	145

SOURCE: UNDP (2015)

open dialogue on 'development in transition' and an approach that encourages a new perspective to develop graduation policies that are sequenced, phased and gradual" (ibid.). It notes that such engagement should be "tailored to specific country needs" and that methodologies need to be devised to "better account for the complex and diverse realities of these countries, reflecting the multidimensional nature of development and poverty" (ibid.).

Middle-income countries may also face domestic challenges in meeting the SDGs. While the HDI ranks each country as a whole, huge differences in health, education, and living standards exist within many middle-income countries. There is a great deal of difference, for instance, in living standards between the political and business capitals of Southeast Asia, such as Jakarta and Manila, and the peripheries of those countries. There is considerable political power in the capitals, in both democratic and non-democratic states. This may impact on efforts to achieve SDGs in the poorer and more remote areas. An example of this may be seen in the chapter by Trisnantoro in this volume, in which the author notes the recent commitment of Indonesia to universal health access for all, while pointing out that spending priorities remain focused on tertiary care in the capital over primary health clinics in rural areas.

7 Multi-level Governance and the Sustainable Development Goals

In this section, we return to the three themes mentioned earlier, as we focus on the implementation of the SDGs and the challenges of moving from the internationally agreed goals and targets to national institutional arrangements. We posit that, at times, overarching approaches to a given group of SDGs involving multiple ministries will be more effective than the separate pursuit of a single SDG or target. Secondly, we focus attention on the challenges of multi-level implementation, including the role of provincial and local governments within state structures. Many SDGs require engagement by subnational structures, but resources and coordination from the centre are essential. Finally, we investigate ways in which ASEAN as a regional actor may be engaged and may promote learning about best practices in the implementation of the SDGs across Southeast Asia. This learning should not only be between national levels of government, but also at the district or municipal levels, for example through the promotion of transnational municipal networks such as ICLEI Global (Local Governments for Sustainability), supported by the UN. In this complex, multilevel process, we are interested in ways in which stakeholders — businesses, civil society organizations, and citizens — are involved in consultation and contribute to the effective implementation of the SDGs. In elaborating on these themes, we offer a brief review of some of the existing literature on the implementation of the SDGs.

7.1 *The Shift from the International to the National*

The SDGs are expected to be integrated within national policy frameworks, taking into account national aspirations as well as local contexts (Biermann, Kanie, & Kim, 2017). While the MDGs had a rigid design that was later heavily criticized for their 'one size fits all' approach, the SDGs have a broader, qualitative character. Gupta and Nilsson (2017) argue that the translation of these non-legally binding global aspirations into national policies depends on the capacities of national administrative institutions. Despite the challenges of integrating internationally formulated goals into national policies, it is important that multiple administrative institutions, such as ministries or agencies, should be involved if targets and goals are to be met, because overarching approaches to sets of SDGs are generally considered more effective than single-goal approaches.

Table 1.3 indicates for each Southeast Asian country the national institutional mechanisms for implementation of global goals. The first column lists the institutional mechanisms for the MDGs (2000–15), the second for the longer-term sustainable development (SD) agenda (1992–2015), and the final column

TABLE 1.3 National institutional mechanisms for implementation

Country	MDG (2000–2015)	SD (1992–2015)	SDG (2015–2030)
Brunei Darussalam	Ministry of Development	National Committee on the Environment	Ministry of Development
Cambodia	Ministry of Planning	National Council of Sustainable Development (MOE)	National Team 2030 Agenda/National Council for Sustainable Development (MOE)
Lao PDR	Deputy Prime Minister and Ministry of Foreign Affairs	Ministry of Natural Resources and Environment	National Steering Committee for SDG (MOFA and MPI)
Indonesia	Ministry of National Development Planning (BAPPENAS)	Ministry of Environment	National SDF Coordinating TEAM 4 Working Groups (BAPPENAS)
Malaysia	Prime Minister's Office (Economic Planning Unit)		National SDG Council SDG National Steering Committee 5 Multi-Stakeholder Working Committees (Economic Planning Unit)
Myanmar	Ministry of National Planning and Economic Development	National Commission for Environmental Affairs	Ministry of Foreign Affairs
Philippines	Multi-Stakeholder Committee on International Human Development Commitments (NEDA)	Philippine Council for Sustainable Development (NEDA)	Committee on SDGs proposed to be placed under the NEDA Board (Executive Order is for signature)
Singapore	Ministry of National Development	Inter-Ministerial Committee on Sustainable Development	Sustainable Development Policy Group (MEWR and MND)
Thailand	National Economic and Social Development Board (NESDB) Office	National Environment Board (MOE)	National Committee on Sustainable Development 3 Sub-Committees (NESDB)
Timor Leste	MDGs Secretariat led by the Minister of Finance	Ministry of Economy and Development	SDG Task Force Ministry of Economy & Development
Vietnam	Ministry of Planning and Investments	National Council for Sustainable Development	

SOURCE: UN ESCAP (2017A, P. 25, CITING "COUNTRY GOVERNMENT WEBSITES, PRESENTATIONS, AND REPORTS; INFORMATION FROM STAKEHOLDERS")

lists those institutions assigned or established to coordinate the SDGs, from 2015 until (barring any changes) 2030. For example, in Indonesia the Ministry of National Development Planning (BAPPENAS) was responsible for the implementation of the MDG agenda and the Ministry of the Environment for the SD agenda. For the SDGs, a national sustainable development coordinating team has been created with four working groups within BAPPENAS. In Malaysia, the Economic Planning Unit within the Prime Minister's office was responsible during the MDG and SD periods, whereas now a National SDG Council and a Steering Committee have been established with five multi-stakeholder working committees within the Economic Planning Unit. We may conclude that in general over time, there has been a transition from development-oriented MDG mechanisms, to environment-oriented SD mechanisms, to multi-sector SDG mechanisms. The new institutional mechanisms are often inter-ministerial in structure.

In terms of the research and literature to date, Stafford-Smith et al. (2017) examine the interlinkages between sectors and actors and the relation between countries with regard to the SDGs, focusing on finance, technology, capacity building, trade, policy coherence, partnerships, and monitoring and accountability. They formulate several key recommendations, including the need for co-production of knowledge and trading of information between higher- and lower-income countries to share processes of sustainability. They urge a legal framework that would promote investment by higher-income countries in lower-income countries in the form of 'patient capital' — investments in which the return is not expected immediately but rather over the long term, stretching across decades. They also advocate the encouragement of trade in locally sustainable products and services (ibid.).

The SDGs not only mark a new era in which the social pillar is officially integrated in the concept of sustainable development; they also aim to relocate goal setting to the centre of governance and political attention. Undeniably, goal setting has always been part of multilateral agreements of international cooperation between nations or institutions (Kanie, Bernstein, Biermann, & Haas, 2017; Ruggie, 1996). Bowen et al. (2017) stress that the 17 goals and 169 formulated targets generate inevitable tensions and that closer attention should be paid to the trade-offs involved. Bowen et al. examine key challenges in terms of implementing the SDGs, reflecting critically on the individual SDGs while taking into account the ultimate goal of sustainable development. Studies have also been conducted to develop an understanding of how these different goals influence each other, either positively or negatively, and the possible trade-offs involved (Singh et al., 2017).

It is important to study interactions between the goals, and Draper (2016) calls for a rearrangement of the goals towards a "network set of goals". The development of a network set of goals can be seen as a reaction to the critique that several of the SDGs have weak connections to each other (Boas, Biermann, & Kanie, 2016). Policies that support a strong and positive connection between goals can be expected to have a longer planning horizon and longer-term benefits compared to single-goal approaches, as argued by Nilsson (2017). The integration of the goals into national multi-goal policies requires close interaction and cooperation between the different administrative institutions as well as consultation with other (non-governmental) sectors.

7.2 *Multi-level Governance: National, Regional and Local Levels of Government*

We are witnessing a global tendency to move from centralized steering structures towards more decentralized systems of governance, in both democratic and non-democratic states. Successful policy integration requires more inclusive, multi-goal approaches, while single-goal approaches using 'silo' implementation structures should be avoided (Weitz, Huber-Lee, Nilsson, Davis, & Hoff, 2014). However, many governmental institutions operate independently, without taking into account other stakeholders, sectors or institutions (Nilsson, 2017).

Multi-level governance is defined by Mulholland and Berger (2017) as a linking mechanism between vertical (national–regional–local) and horizontal (economic, environmental, and social issues) policy integration. Multi-level governance hence serves as a mechanism to combine these two concepts to improve the quality of SDG integration and implementation (ibid.). In their report, Mulholland and Berger reflect on how the SDGs are implemented within the member states, through multi-level governance. Their aim is to develop an understanding of whether, and how, nations have integrated the SDGs at all governmental levels. This implies the need to establish clear rules for sharing rather than separating competences. In the 2030 Agenda for Sustainable Development document (UN, 2015), multi-level governance and multi-level implementation are referred to several times. However, Mulholland and Berger (2017) make the point that the need for multi-level governance and integration is not sufficiently stressed within the SDGs themselves.

There is also the issue of the single-goal approach or multiple-goal approach toward the SDGs at various levels of government. For most of the countries, a shift away from the single-goal approach will only be feasible through a reorganization of national administrative and governmental structures (Biermann,

Kanie, & Kim, 2017), but this requires close cooperation between all levels. Biermann et al. carried out a collaborative, multi-year research project involving 30 scholars and a series of workshops that were held in Asia, Europe, and North America, as well as a multi-stakeholder analysis which involved UN officials and diplomats in New York. Biermann et al. examine and assess the evolution, rationale, and prospects of the SDGs, focusing especially on governance issues. They stress the need to break down silo structures and other obstacles between governmental authorities to enable cooperation.

Despite the importance of this vertical integration of policies in national, regional, and local authorities for the SDGs, it is not mentioned explicitly in the Agenda 2030 document. The issue of individual ministries acting alone rather than being integrated is one of the main lessons from the MDGs. Mulholland and Berger (2017) reflect on this in their research into the concept of multi-governance and policy integration in SDGs and acknowledge the risk that the concept of policy integration can easily be ignored in favour of an over-reliance on quantitative goals. Whilst exploring the concept of multi-level approaches, Mulholland and Berger remark that the policy coherence between governmental levels should be an overarching theme rather than a directly measurable goal. The context of every country differs and requires special attention while integrating the SDGs into national policy planning.

For encouraging regional learning about best practices around the implementation of the SDGs in Southeast Asia, ASEAN is considered to be a highly suitable institution (Olsen, Teoh, & Miyazawa, 2015). Providing opportunities for learning to take place between states, but also between subnational and local levels, is imperative. If the goals are properly organized and integrated at the supranational ASEAN level, Olsen et al. expect ASEAN's role to include galvanizing support and providing guidance for the lower levels. Furthermore, many issues such as biodiversity, migration, and climate change are cross-border issues, requiring a regional framework to streamline information and action — this could be another role for ASEAN. Olsen et al. also argue for the formulation of "ambitious national responses" by member states of ASEAN in order to guarantee that the future of ASEAN is not exclusively about economic aspirations but also focuses on long-term regional sustainable development. Sustainable aspirations should form a part of the very foundation for the regional integration of ASEAN countries.

7.3 *Stakeholder and Citizen Participation in the SDGs*

In pursuing the SDGs, strengthening decentralization within a country is considered one of the main means to achieve efficient multi-governance for the

SDGs and to increase local involvement (UN ESCAP, 2017a). The multi-level character of the SDGs indirectly points to the multiplicity of stakeholders involved and the need for collective action (Bowen et al., 2017). We argue that the involvement of stakeholders and civil society in the consultation and implementation of SDGs should be strengthened if the process is to be successful.

Bowen et al. see the lack of cooperation between different stakeholders as a widespread, serious issue. Their research addresses governance challenges in the field of SDGs and reflects on the global discussions of transformations for sustainability. Here, the demand for contextually appropriate and inclusive decisions is clear. These decisions should be made across sectors, levels, and scales, whilst always taking into account national aspirations and local contexts. However, incorporating SDGs in national, regional, and local plans will inevitably entail discussions regarding trade-offs or priorities, and where decision making is involved, the issue of power difference remains. Bowen et al. argue that the major challenge in terms of participation is the deep-rooted power differences between the actors. Similarly, Chapin, Knapp, Brinkman, and Cochran (2016), and Hajer et al. (2015) also discuss the important role of change agents from civil society. Through this role, co-production of solutions, processes, and knowledge can be developed together with the government, which will remain the key player in implementing SDGs.

Bernstein (2017) and Fleming, Wise, Hansen, and Sam (2017) argue that the implementation of the SDGs has already led to more meaningful governance structures and participation. However, several global regions still struggle with the issue of cooperation with private stakeholders and civil society. In the report of UN ESCAP (2017a), the organization reflects upon the challenges and priorities in terms of SDG implementation in the ASEAN region. It suggests that policy plans have matured over the years since the MDGs, but the most influential roles in SDG consultation or policy formulation are still reserved for members of policy- and decision-making ministries. The lack of representation in SDG mechanisms at regional or local levels is an acknowledged problem that ASEAN aims to solve, especially since a large part of the 2030 Agenda is intended to be pursued at the local level (UN ESCAP, 2017b). This should be monitored and evaluated to further improve cooperation between governments and civil society. Naturally, the governance style and levels of participation differ between nations. This starting point is defined by the constitutional settings, culture, traditions, politics, and geography that become visible in the social, economic, and environmental conditions of countries. Since more demands for participation are made at the international level, policy coordination across governments at all levels is essential for determining how to shape inclusive and effective governance styles (UN, 2016a). For the ASEAN member states, cooperating on the ASEAN level offers the potential to share learning points and further improve participation and inclusive decision making.

Given that effective participation remains, for the most part, the prerogative of the policy- and decision-making bodies, involving the public remains a challenge. One case which shows promise is the National Council for Sustainable Development (NCSD) in Thailand (UN ESCAP, 2017a). The NCSD is a non-governmental council connected to the office of the prime minister, which advocates for more integrated sustainable policies by providing a platform for knowledge and practice exchange. While most national councils that are engaged in the SDG mechanism are represented by state actors, the NCSD in Thailand is composed of business associations, non-profit organizations, and experts, at the same level as the government officials. To enable this multi-stakeholder mechanism, the Thai government needed not only to develop such a platform, but also to welcome input from the broader society, which required the (re)building of mutual trust and respect between the government and non-governmental actors.

In other countries, including Indonesia, Vietnam, and Malaysia, public participation is organized through working committees. For Indonesia, it is also important to take into account an inherited paternalistic relationship between government and the public which shapes interactions (Irawanto, Ramsey, & Tweed, 2013). To date, participation has remained at the local level, generally known as 'Musrembang'; this involves a community discussion which is integrated into development planning. Through these instruments, citizens can, in theory, engage in shaping their local living environment, although in practice it has been shown that active engagement of the public in policy formulation and integration still remains a challenge. Nevertheless, such councils and committees can become an important forum for reshaping the discussions towards SDG-oriented planning (Purba, 2011).

8 Engagement of UN ESCAP and UCLG in the SDGs

There are three important publications from the international institutional community that address the SDGs in the region — either the broad Asia-Pacific region or the narrower Southeast Asia subregion — that merit special mention. The first is a joint publication of UN ESCAP with the ADB (Asian Development Bank), and the UNDP titled 'Asia-Pacific Sustainable Development Goals Outlook' (UN ESCAP, ADB, & UNDP, 2017). This 150-page document analyses each of the 17 goals and the accompanying targets, with an overview, a discussion of bright spots, hot spots, and emerging issues, presenting both statistical overviews and focus areas for improvement. The second document — which we have already referred to earlier in this chapter — is exclusively from UN ESCAP, and focuses on the SDGs within the narrower environment of Southeast Asia and its regional institution, ASEAN; it is titled 'Southeast Asia Subregion Challenges and Priorities for SDG Implementation' (UN ESCAP, 2017a). This second volume notes that a major challenge in the region is "capacity weaknesses, indicating that a sharp focus on capacity building is imperative for the achievement of SDGs" (UN ESCAP, 2017a, p. ii). It refers to ASEAN as the subregion's "mechanism that boosts the implementation" of the SDGs, through its own vision and roadmaps which coincide with the global goals (ibid.). It reports, favourably, that in most ASEAN countries the SDGs have been assimilated into national development plans, but on a less positive note, it claims that "capacity gaps" for implementation across Southeast Asia are "vast, multi-level and multi-dimensional" (ibid., p. iii).

The third publication, a booklet produced by United Cities and Local Government (UCLG) — an umbrella organization for cities and local and regional government bodies worldwide — provides information and advice to cities and local authorities regarding the implementation of the SDGs (UCLG, 2015). It notes, for example, that Goal 11 to "Make cities and human settlements inclusive, safe, resilient and sustainable" is a result of a campaign by local governments and their associations and is a step forward in the "recognition of the transformative power of urbanization for development, and of the role of city leaders in driving global change from the bottom up" (UCLG, Preface). It mentions the rapid urbanization occurring in the global South and the accompanying growth in slums and inadequate housing, and notes that city governments are responsible for promoting the use of public transport, increasing accessibility to services for all residents, improving safety, and reducing emissions. The UCLG also points out that it is not only SDG 11 that involves cities and local governments, but rather that "All of the SDGs have targets that are directly or indirectly related to the daily work of local and regional governments. Local governments should not be seen as mere implementers of the agenda. Local governments are policy makers, catalysts of change and the level of government best-placed to link the global goals with local communities" (ibid.).

The UCLG booklet proceeds to discuss each of the 17 SDGs, identifying targets that are relevant for local governments. Some of the targets are broad and relate to linkages between the SDGs. For example, in discussing SDG 1 on ending poverty, it notes that this goal is not only about raising the incomes of the poorest, but also about access to basic services. It highlights four targets that are directly relevant to local authorities, including Target 1.5 — "by 2030, build the resilience of the poor and those in vulnerable situations and reduce their exposure and vulnerability to climate-related extreme events and other economic, social and environmental shocks and disasters" (UCLG, 2015, SDG 1; UN, 2015, p. 15). But some of the targets related to local governments are very specific. In discussing SDG 17 on strengthening the means of implementation and revitalizing global partnership for sustainable development, UCLG notes Target 17.18, to increase the availability of "high-quality, timely and reliable data", and Target 17.19, to support "statistical capacity building" (UCGL, SDG 17). This level of detail is very important because the reliability of local data is an "essential tool to monitor sub-national variations in progress and in targeting resources to make sure that no-one is left behind" (ibid.). In other words, it is vital that nation states do not just focus on a few regions where economic growth and progress toward the SDGs may be easiest, but ensure that efforts are made across the entire breadth of the country.

9 International Cooperation and the Global Goals in Southeast Asia and ASEAN

There has been a broad engagement by the international community to assist in the attainment of the SDGs in Southeast Asia. This reflects the status of many of the countries in the region as either low- or middle-income countries, which have traditionally been beneficiaries of development assistance. This engagement has been true of more geographically distant actors like the EU and the USA, but also more immediate neighbours in the Asia-Pacific such as China, Japan, and Australia. These activities are consistent with SDG 17 (Strengthen the means of implementation and revitalize the Global Partnership for Sustainable Development) and the call for international partnerships to assist and cooperate with countries in meeting the SDGs.

ASEAN and the EU have benefited from a broad collaboration and engagement in achieving the SDGs. This inter-regional cooperation is grounded in a comprehensive framework known as the 'Blue Book 2017 EU–ASEAN Development Cooperation' (European Union, 2016), and the explicit link to the SDGs was reinforced during the first High-level ASEAN–EU Dialogue on Sustainable Development (with a focus 'Towards achieving the Sustainable Development Goals'), which took place in Bangkok in November 2017 (European Commission, 2017). The dialogue partners issued a joint press statement which underscores that promoting sustainable development and the implementation of the 2030 Agenda (i.e., the SDGs), is a "critical priority area for both ASEAN and the European Union" and an integral part of regional integration processes to ensure that they are "people-centered and leave no one behind" (ibid.). The statement notes that the EU has increased development cooperation for ASEAN regional integration to €170 million for the period 2014–2020, as well as providing over €2 billion of funding to ASEAN member states bilaterally over the same period. At the same time, the statement warns that achieving the SDGs will require the mobilization of "domestic and international, public and private financial resources" (ibid.).

One of the topics covered by the November 2017 Dialogue was the contribution of regional integration processes towards "narrowing the development gap" (ibid.). There was a particular focus on "important cross-cutting development catalysts" in efforts to achieve multiple SDGs, including gender equality, promoting a green economy with environmentally sustainable and resilient cities, sustainable consumption and production, and the challenges of climate change (ibid.). The Dialogue also highlighted an open form of decision making in the implementation of the SDGs, noting the importance of transparent and accountable processes involving multiple stakeholders, including the private sector, civil society, and academia, in promoting and localizing SDGs. The next ASEAN–EU Dialogue on Sustainable Development is expected in 2019.

The broader Blue Book on EU–ASEAN relations provides background to the recent High-level Dialogue on SDGs. The document notes that the first contacts between the regional organizations began in 1972, and were formalized in 1977. Subsequent years brought the ASEAN-EEC Cooperation Agreement in 1980, Enhanced Partnership in 2007, and in 2012 the adoption of a more comprehensive Plan of Action 2013–2017. A series of high-level meetings and stakeholders' and citizens' debates in Brussels in 2017 marked the formal 40-year anniversary of the relationship between the EU and ASEAN, and a new plan of action with ASEAN is to be adopted by the Ministers of Foreign Affairs (European Union, 2016).

The Blue Book discusses the EU's strategic approach to achieving sustainable development in Europe and around the world, which was set out in the New European Consensus on Development in 2016. This Consensus attempts to integrate the main orientations of the 2030 Agenda into five focus areas: fostering development and reducing poverty; tackling climate change by promoting resource efficiency and sustainable consumption and production; achieving socially inclusive and sustainable growth; promoting peace, justice, and human rights; and improving the effectiveness and impact of development cooperation through involvement of all stakeholders (ibid., p. 8).

The Blue Book delineates cooperation in three areas (which can be compared to the three-pillar structure of ASEAN). In the area of political and security cooperation, the document highlights maritime security, migration, the establishment of an EU Chemical, Biological, Radiological and Nuclear (CBRN) centre in the region, the promotion of preventative diplomacy and mediation, and a policy dialogue on human rights addressing such issues as women's rights, child protection, and the safety of migrant workers. In the area of economic cooperation, the EU has focused on supporting the harmonization of standards and customs procedures in ASEAN, contained in the Master Plan for ASEAN Connectivity. In the area of socio-cultural cooperation, the EU stresses the need for building closer people-to-people contacts: "people-centred and socially responsible action … aims to build an inclusive and sharing society where livelihoods, well-being and welfare have top priority" (ibid., p. 30). A number of concrete programmes are mentioned, including the EU-Share programme which funds student mobility in higher education; support for farmers' adaptation to climate change by strengthening famers' organizations; and work with the ASEAN Centre for Humanitarian Assistance (AHA) to coordinate early warning systems and emergency responses.

A further step in encouraging dialogue between the EU and ASEAN occurred in Brussels in February 2018, during a conference and public debate on 'EU–ASEAN Relations: The next forty years' (European Union, 2018). The conference was organized by the EU, ASEAN, and the Brussels-based NGO Friends of Europe (Friends of Europe, 2018). This conference built on a formal meeting of ASEAN and EU ministers in Manila in August 2017, at which a multi-faceted plan of action and cooperation was decided upon, focusing on bilateral ties, connectivity, security issues, and negotiations toward an EU–ASEAN free trade agreement, but also climate change and sustainable development (ASEAN, 2017).

Other nations have also identified the SDGs as important to their involvement in development, including the USA. However, the priorities of the latter may change from administration to administration depending on political leadership. For example, the part of the website from the United States Agency for International Development (USAID) which focuses on the SDGs, features a quote from President Obama dated 27 September 2015, asserting that the US is committed to the SDGs and that "our common humanity, compels us to act … we reaffirm that supporting development is not charity, but is instead one of the smartest investments we can make in our future" (USAID, 2018a). However, the language on the website's homepage seems to give less emphasis to partnership and is more nationalistic in tone. The homepage first quotes Mark Green, President Trump's appointment as USAID Administrator, stating "The purpose of foreign aid is to end the need for its existence". This is followed by language from USAID focused on American interests: "USAID is the world's premier international development agency and a catalytic actor driving development results. USAID's work advances U.S. national security and economic prosperity, demonstrates American generosity, and promotes a path to recipient self-reliance and resilience" (USAID, 2018b). Nonetheless, "resilience" is a critical component of the SDGs.

Japan has taken specific measures to fulfil the SDGs domestically, but also contributes more broadly to the global goals. In a document entitled 'The 2030 Agenda for Sustainable Development and Japan's Implementation', the country's Ministry of Foreign Affairs mentions its efforts toward "global health, disaster risk reduction, promotion of quality infrastructure investment, and women's empowerment", adopted as part of its Development Cooperation Charter, which takes the concept of human security as a guiding principle (Japan MOFA, 2017, p. 5). It also undertakes to "support developing countries to formulate national strategies and programs and mainstream the SDGs into them" (ibid.).

China, too, has noted the role that the SDGs play domestically in its own development, and in international development abroad. In September 2015, the President of China, H. E. Xi Jinping, was present and spoke at the United Nations Sustainable Development Summit at which world leaders endorsed the 2030 Agenda for Sustainable Development.

> China attaches great importance to implementation of the 2030 Agenda and believes that the principles of Peaceful Development, Win-win Cooperation, Integration and Coordination, Inclusiveness and Openness, Sovereignty and Voluntary Action, as well as 'Common but Differentiated Responsibilities', should be followed in building a new type of international relations featuring win-win cooperation, establishing all-round partnership, and achieving economic, social and environmental development in a balanced manner.
>
> UN, 2016b

Australia, as a strong regional actor, is closely involved in the implementation of SDGs in Southeast Asia. The Australian government has stated that "The 2030 Agenda is both a domestic and international agenda. It is well aligned with Australia's foreign, security, development and trade interests — especially in promoting regional stability, security and economic prosperity. It also helps Australia in advocating for a strong focus on economic growth and development in the Indo-Pacific region and in promoting gender equality, governance and strengthening tax systems" (Australian Government, 2018).

10 Introduction to the Book Chapters

In this volume, we focus on the challenges and contributions of a multi-level governance approach to the SDGs. We hope in some small way to investigate and analyse these challenges to achieving the SDGs in Southeast Asia through the chapters that follow. The chapters adopt different perspectives: some look at the efforts of a particular country in the region to broadly address the SDGs; some focus on the implementation of a specific SDG across numerous countries; while others zoom in on one SDG or a group of closely related SDGs in one country.

The book is divided into seven sections. This first section focuses on institutions and governance for the SDGs in Southeast Asia, starting with the broad discussion in this introductory chapter. The following two chapters then look at particular countries: Chapter 2 by Dafri Agussalim, Ahmad Rizky M. Umar, Karina Larasati and Dio H. Tobing focuses on Indonesia, and Chapter 3 by Phanthanousone Khennavong on Laos. These cases represent one middle-income country and one of the least developed countries in Southeast Asia.

Agussalim et al. acknowledge that Indonesia has produced a compelling Voluntary National Review policy document that discusses the country's technocratic efforts to implement the SDGs, but argue that it is not clear how Indonesia's broader development framework, at either the national or subnational level, truly supports the policy. The authors argue that this requires a further development strategy to "localize" the SDGs within a development framework at both the provincial and local government levels and to ensure deeper involvement of civil society and businesses in the process. In the final analysis, they conclude that involvement at the national technocratic level of the Indonesian Planning Ministry (BAPPENAS), and within the inter-institutional arrangements of multi-level governance, will not be sufficient to meet the demands of the SDGs. They insist that what is critically needed is involvement at the political level — the elite circles around President Joko Widodo (known as President Jokowi). The president has prioritized development in the rural areas, especially in the eastern part of Indonesia, as one of his main policy agendas. While this is no doubt a very important component of the SDGs, Agussalim et al. argue that the SDGS are broader and must encompass nationwide change, in both rural and urban areas.

In his chapter, Khennavong focuses on the internal political processes within Laos that helped the mainstreaming of the SDGs, and led the government to embrace the goals. Having made a commitment to graduate from the status of Least Developed Country (LDC) by the 2020s to become a middle-income country, the government began to see that the 2030 Agenda would be a useful instrument to realize its own goals and long-term vision. This mainstreaming would also encourage the involvement of various partners and sources of development finance for Laos. In addition, Khennavong argues that the new Prime Minister, Thongloun Sisoulith, who had previously been involved in the national planning process and had a role in donor coordination and partnerships, was looking for ways to "introduce reform of the planning process in a highly change-averse system", by suggesting mainstreaming SDGs as a response to pressure from the international community.

The second section of the book focuses on accountability to citizens and respect for human rights to ensure progress toward the global goals. Many of the SDGs are formulated as human rights pursuant to the 1948 Universal Declaration of Human Rights and subsequent treaties and covenants, such as rights related to health, education, gender equality, and a fair justice system. With the SDGs, however, the international community focuses not just on individual human rights and obligations, but moves towards concrete goals and targets for the attainment of these rights and societal objectives. In this part of the book, SDG 16 on promoting peaceful and inclusive societies, access to justice, and building effective, accountable, and inclusive institutions at all levels comes to the fore. Chapter 4, by Julio Teehankee, discusses the accountability challenges to the SDGs across Southeast Asia. He notes various forms of government accountability, including vertical accountability through formal elections, horizontal accountability through institutional restraints, and the growing role of citizens, civil society, and the media in a digital accountability. He argues that the presence of accountability deficits across Southeast Asia are a challenge for the realization of SDGs, including SDG 16, in the region. Chapter 5, by Ulrich Karl Rotthoff, focuses on the Philippine's case and human rights in the international context. Rotthoff is interested in the role of law and the legal remedies which are available at different levels, from the national to the regional and international. He argues that an emphasis on the principle of non-interference by ASEAN has limited its ability to intervene in the Philippines campaign of extrajudicial killings and other human rights issues.

The third section of the book focuses on progress being made in the SDGs associated with a social agenda in both middle-income and developing countries in the region. Thus, this section focuses on SDGs 1–5, which aim to end hunger and poverty, and achieve good health, quality education, and gender equality. Chapter 6 by Azirah Hashim and Aliyyah Nuha Faiqah Azman Firdaus focuses on capacity building in higher education in the ASEAN region. The importance of this is reflected in the links which can be drawn between SDG 4 on quality education and the achievement of other SDGs. As the UN notes, while there is near-universal primary school enrolment across the Asia-Pacific, "[e]nsuring equal access for marginalized groups and quality dimensions of education will be major challenges, with investment needed to provide lifelong learning opportunities" (UN ESCAP, ADB, & UNDP, 2017, p. 37). Moreover, many of the traditional forms of higher education institutions in the region "lack the ability to innovate and create knowledge — and to find innovative application of what is learned, which is problematic in a dynamic region with a rapidly increasing number of middle-income countries" (ibid., p. 39).

Chapter 7 in this section, by Laksono Trisnantoro, is about healthcare system reform in Indonesia. Again, the 2017 report by UN ESCAP, ADB, and UNDP highlights the importance of healthcare in the region, in terms of both communicable diseases like HIV and tuberculosis, and non-communicable diseases. The report concludes that "[a]ccess to affordable quality health care and medicines is constrained and unequal in many countries" (ibid., p. 31), and argues that "[p]romoting gender equality and the human rights of vulnerable groups can lead to markedly better disease prevention, treatment and care" (ibid., p. 32.) Trisnantoro first notes that Indonesia adopted universal coverage of health services in 2014. The law "aims to reform the healthcare system with the primary objective of improving health equity and access to high-quality care". However, after four years of implementation, the goal of health system reform in terms of increasing equity in the distribution of health services has not yet been achieved. Trisnantoro notes that the existing system lacks financial sustainability; he argues for a more fundamental, radical change, adopting broader principles of healthcare system reform with reference to the objectives of the SDGs. Programmes to encourage healthier lifestyles and improved environmental practices should be introduced, health sector governance in the decentralized political system should be reinforced, and the use of taxpayer funds to finance medical treatment for the better-off in society should be reduced in favour of primary care for a broader spectrum of citizens.

The fourth section of the book focuses on cities, the urban agenda, and transport. Here the two goals of principal relevance are SDG 11 on sustainable cities and communities, and SDG 9 on industry, innovation, and infrastructure. The UN ESCAP, ADB and UNDP (2017) report notes that hundreds of millions in the Asia-Pacific region are "without access to basic infrastructure, including for connectivity" and argues that quality must be improved for greater resilience and environmental sustainability (ibid., p. 67). This section of the book comprises three chapters. In Chapter 8, Bakti Setiawan discusses urban transformation and housing policy in Indonesia, with a focus on political will and capacity building, examining the New Urban Agenda (NUA) as part of a comprehensive effort to implement and localize the SDGs across cities in Indonesia. In assessing the commitment and capacity of implementation, he focuses on such critical factors as "strong political and leadership commitment, clear and robust legal and institutional frameworks, and effective networks among stakeholders".

Hasharina Hassan and Gabriel Yong, in Chapter 9, discuss housing issues in Brunei Darussalam. Hassan and Yong point out that the aspiration of Brunei to provide every resident with basic facilities and shelter pre-dates the MDGs and SDGs, yet closely aligns with the goals. Both the National Development Strategy (Wawasan 2035) and the current National Development Plan identify the provision of "quality and sufficient housing" for a "high quality of life". Hassan and Yong note two main challenges to the provision of national housing: sufficient funding to meet the growing demand, and land scarcity, which is "exacerbated by rapid urban sprawl and migration to the more developed and urbanized centres". Finally, in this section, Wendy Tan (Chapter 10) focuses on the move away from individual automobile transport in cities, towards sustainable mobility and transport in Southeast Asia, focusing on a comparison between the cities of Singapore and Jakarta.

The fifth section of the book focuses on SDGs 14 and 15, which address life in water and on the land, and SDG 6 on clean water and sanitation. SDG 6, on water, has strong links to the social goals, SDGs 1–5. The report by UN ESCAP, ADB, and UNDP (2017) notes that "[t]hreats related to water scarcity, poor water quality and inadequate sanitation will negatively impact the achievement of other SDGs as well [as SDG 6]" (ibid., p. 49). It cites UN Habitat and ADB sources which show that between 80% and 90% of wastewater across the developing countries in the Asia-Pacific is discharged into water bodies without any treatment; in Indonesia only 14% is treated, in the Philippines 10%, and in Vietnam, only 4% (ibid., p. 51). In this volume, Helena Varkkey (Chapter 11) analyses transboundary haze and the relevant SDGs. She finds that ASEAN's non-interference norm has resulted in "resistance towards ideas of the environment and development that originate from outside ASEAN" and that the institution's sectoral, pillar-based structure impedes progress toward the SDGs. Maharani Hapsari in Chapter 12 focuses on emerging spaces of citizenship and the SDGs and governance over water in Indonesia. She works within a framework of "SDGs from below", and elaborates forms of citizenship produced by grassroots communities around the provision of clean water.

In the sixth section of the book, we turn from social goals and the environment to SDGs related to the economy. Here the focus is on inclusive economic growth and decent work for all (SDG 8), and reducing inequalities (SDG 10). The UN ESCAP et al. (2017) report notes that "relatively strong economic growth has not translated into commensurate increases in decent jobs and higher wages for most economies" (ibid., p. 61). In Chapter 13 of this volume,

Tran Dinh Lam examines economic reform and sustainable development in Vietnam. He notes that the country has had successes in economic achievements and poverty reduction compared to others in the region. Yet he also notes that Vietnam's rapid economic development has been accompanied by the rise of powerful interest groups and growing income inequality. He concludes that the ASEAN community is an important bridge for Vietnam in its efforts to learn from others and improve management capability, and its desire to integrate with countries in the region in a wider effort to promote sustainable development.

Next we turn to Kimsan Soy in Chapter 14, who discusses the protection of workers in the construction industry in Cambodia. His research examines labour standards related to the safe workplace under international human rights treaties, relevant ILO conventions, and SDG Goal 8 on decent work — which Cambodia has ratified and adopted. This chapter looks at the effectiveness of monitoring and enforcement mechanisms, particularly within relevant governmental bodies, and then highlights the challenges and the reasons why the right to a safe workplace is not guaranteed, in the context of the political economy of the construction industry and the limited involvement of NGOs and civil society in Cambodia. In Chapter 15, the final contribution to this section, Thuta Aung discusses the role of stakeholders, specifically small and medium enterprises (SMEs), as a force for achieving the SDGs in Myanmar. He argues that small-scale technology appropriately utilized by SMEs will assist in the attainment of the SDGs in the country.

In the seventh and final section of the book, we turn to agriculture and community development through partnerships. Agriculture is important in the context of the SDGs both in terms of meeting minimum human needs for life (SDG 1, no poverty; SDG 2, no hunger), but also in its contribution to SDG 8 for decent work and economic growth in the agricultural sector, and SDG 12, related to responsible consumption and production. The report by UN ESCAP et al. (2017) notes that "agriculture continues to pose an unsustainable burden on the planet's carrying capacity". It also notes with foreboding that "[c]limate change will have a significant impact on crop yields and harvests, disproportionally affecting the nutritional status and livelihoods of impoverished households" (ibid., p. 25). In Chapter 16 of our volume, Titus Chen and Amador IV Peleo discuss the coffee industry and sustainable development in Southeast Asia. They find that communities growing higher-value cash crops, such as the sought-after and relatively expensive Arabica coffee beans, often perform better in terms of the SDGs. Finally, in Chapter 17, Saikaew Thipakorn focuses on partnership arrangements between Japanese agricultural cooperatives and growers of bananas in Thailand, which have led to more stable and profitable export markets, and had wider community benefits impacting many SDGs. This form of partnership between agricultural cooperatives, related to SDG 17 on partnerships for the goals, is an example of an often overlooked and understudied form of partnership in the multi-level, multi-dimensional institutional implementation of the SDGs.

11 Conclusion

The world community has taken a substantial step forward in coming together to adopt the Sustainable Development Goals. For the first time, in a single document and with an accompanying institutional process guided by the United Nations, broad goals related to social, economic, and environmental spheres, as well as climate and peace, have been agreed upon. Now the successful implementation of the SDGs across the world — including within Southeast Asia — by 2030 will critically depend on a multi-governance approach, with regional, national, provincial, and local levels of government involved and with input from citizens. In this conclusion, we wish to lay out our main findings from a review of the existing academic literature, the expansive research and guidance prepared by international bodies such as the UN, and the contributions by the authors of this volume from each of the 10 ASEAN countries.

First, it is clear that there is a role for greater political leadership to be exercised in the diligent and careful pursuit of the SDGs and Agenda 2030. Much of the existing literature and policy briefs so far have focused on the transposition of the international goals into national goals, targets and action plans. What is needed now is more public statements and institutional steering through national ministries and bureaucracies — not only at the very top of the national political systems, by presidents and prime ministers, but also for example by governors of states or provinces, and mayors of cities and local governments. The implementation of the SDGs is being integrated with existing national plans and priorities; through effective leadership it can become more central to the priorities of governments and society.

Second, there is a clear need for greater involvement by civil society organizations (CSOs), non-governmental organizations (NGOs), and citizens in the process of 'localization' of the SDGs. These groups may ask: what do the SDGs mean concretely in our community; how can we apply political pressure to ensure they receive sufficient attention; and how can we be involved in assessing problems, priorities, and implementation of the SDGs in our community? There also needs to be much more effort to encourage businesses to take an active part in meeting Agenda 2030. What kind of laws, regulations, and incentives need to be in place for the private sector to become involved in the social, economic, environmental, and climate change goals?

Third, there is a much more developed network of CSOs, NGOs, and citizens organized around a human rights approach to development than there is for the SDGs. The power of the human rights approach is based on legal treaty status, ideas of universality, potential judicial remedies for individual victims, and international monitoring. The SDG approach, on the other hand, has the virtues of voluntary commitment by governments, ideas of setting and balancing priorities, involvement by a wide range of institutions beyond the judiciary, and international learning and exchange instead of a process of 'naming and shaming'. It is therefore important that existing groups are encouraged to see the distinct advantages of the SDG approach, with a view to creating a new network of civil society organizations pressing for Agenda 2030.

Fourth, the SDGs may function as an impetus for middle-income countries to more consciously focus on the needs and interests of their more remote regions, over and above the powerful political and economic demands made by those in capital cities. This should lead to more balanced development within these countries, with improved living standards and reduced inequality.

Fifth, within Southeast Asia, the SDGs may be a way for ASEAN to gradually deepen its role beyond its existing intergovernmental functions, to embrace more integrative and societally impactful functions. For example, ASEAN could play a greater role in encouraging learning across the member states and between cities and local governments as they all strive to achieve the SDGs. The great diversity of development across the ASEAN member states, noted at the beginning of this chapter (and see Table 1.2), could be used to advantage by sharing examples and practices across states which occupy different ranks on the Human Development Index. In addition, knowledge, capital, and appropriate technology could flow more efficiently within Southeast Asia. Whereas some development is increasingly led by Chinese investment, intra-ASEAN development would improve the internal cohesion of the region. This would also strengthen the sovereignty and independence of Southeast Asia, by balancing powerful countries and blocs both near and far, including China, the USA, and the EU, as well as Japan and Australia.

Finally, we issue a call for governments, funding institutions, and universities to support additional academic research focused on the Sustainable Development Goals. It is vital to move beyond the stage of analysing policy documents, to conduct empirical research in the field about national and local engagement with the SDGs. The world community is justly proud that a single instrument has incorporated so many goals into one place, but academic scholars need to be able to study such questions as: how are national governments and local communities finding synergies and avoiding perceived trade-offs in these lofty goals? More comparative, longitudinal academic studies are needed to analyse the challenges of implementing the SDGs and achieving Agenda 2030. A better, more just, world awaits, and we all need to be part of the journey.

Acknowledgements

I sincerely thank Kirstin Wittig, a graduate Master's Student in the Faculty of Spatial Sciences at the University of Groningen, for her research and assistance in preparation of this chapter. Her research and critical thinking skills are considerable, and we have had many helpful discussions about the approach to take in this research. Components of this chapter were presented at the International Political Science Association (IPSA) conference in Brisbane, Australia on 25 July 2018. I am grateful for the questions and comments received from the audience, and for the opportunity to share ideas with them about the Sustainable Development Goals.

References

Arat, Z. F. K., & MacDonald, E. (2018). From anti-discrimination to empowerment of women: A critical assessment of the CEDAW, MDGs and SDGs. Paper presented at the International Political Science Association (IPSA) meeting, Brisbane, Australia, 21–25 July.

ASEAN (2016). ASEAN taps on Vision 2025 to support SDGs. Retrieved from http://asean.org/asean-taps-on-vision-2025-to-support-sdgs-2/.

ASEAN (2017). Joint communique of the 50th ASEAN foreign ministers' meeting: Partnering for change, engaging the world. Retrieved from http://asean.org/storage/2017/08/Joint-Communique-of-the-50th-AMM_FINAL.pdf.

Australian Government (2018). 2030 Agenda for Sustainable Development. Australian Government, Department of Foreign Affairs and Trade. Retrieved from http://dfat.gov.au/aid/topics/development-issues/2030-agenda/Pages/default.aspx.

Bernstein, S. (2017). The United Nations and the governance of Sustainable Development Goals. In N. Kanie, & F. Biermann (Eds), *Governing through goals: Sustainable Development Goals as governance innovation* (pp. 213–240). Cambridge, MA: MIT Press.

Biermann, F., Kanie, N., & Kim, R. E. (2017). Global governance by goal-setting: The novel approach of the UN Sustainable Development Goals. *Current Opinion in Environmental Sustainability*, 26(27), 26–31.

Boas, I., Biermann, F., & Kanie, N. (2016). Cross-sectoral strategies in global sustainability governance: Towards a nexus approach. *International Environmental Agreements: Politics, Law and Economics*, 16(3), 440–464.

Bowen, K., Cradock-Henry, N. A., Koch, F., Patterson, J., Häyhä, T., Vogt, J., & Barbi, F. (2017). Implementing the "Sustainable Development Goals": Towards addressing three key governance challenges — Collective action, trade-offs, and accountability. *Current Opinion in Environmental Sustainability*, 26(27), 90–96.

Chapin III, F. S., Knapp, C. N., Brinkman, T. J., Bronen, R., & Cochran, P. (2016). Community-empowered adaptation for self-reliance. *Current Opinion in Environmental Sustainability*, 19(9), 67–75.

Denmark (2017). Human rights and the SDGs: Pursuing synergies. Denmark, Permanent Mission to the United Nations Office at Geneva and the Permanent Mission of Chile to the United Nations Office at Geneva. Retrieved from https://www.universal-rights.org/urg-policy-reports/human-rights-sustainable-development-goals-pursuing-synergies/.

DIHR (2018a). Human rights and the SDGs. Copenhagen: Danish Institute for Human Rights. Retrieved from https://www.humanrights.dk/our-work/sustainable-development/human-rights-sdgs.

DIHR (2018b). Human rights and the 2030 Agenda for Sustainable Development: Lessons learned and next steps. Copenhagen: Danish Institute for Human Rights. Retrieved from https://www.humanrights.dk/sites/humanrights.dk/files/media/dokumenter/sdg/hr_and_2030_agenda-web_2018.pdf.

Draper, S. (2016). Systems thinking: Unlocking the Sustainable Development Goals. Retrieved from https://www.forumforthefuture.org/blog/systems-thinking-unlocking-sustainable-development-goals.

European Commission (2017). Inaugural high-level ASEAN–EU dialogue on sustainable development: Towards achieving the Sustainable Development Goals. Press statement, 17 November. European Commission DG International Cooperation and Development. Retrieved from https://ec.europa.eu/europeaid/news-and-events/co-chairs-press-statement-inaugural-high-level-asean-eu-dialogue-sustainable_en.

European Union (2016). Blue book 2017. EU–ASEAN development cooperation in 2016. Jakarta: EU Mission to ASEAN. Retrieved from https://eeas.europa.eu/sites/eeas/files/eu-asean-bluebook2017-hi.pdf.

European Union (2018). Forum on the future of EU–ASEAN relations. Retrieved from https://eeas.europa.eu/delegations/association-southeast-asian-nations-asean_en/43291/Forum%20on%20the%20Future%20of%20EU-ASEAN%20Relations.

Fehling, M., Nelson, B., & Venkatapuram, S. (2013). Limitations of the Millennium Development Goals: A literature review. *Global Public Health*, 8(10), 1109–1122.

Fleming, A., Wise, R., Hansen, H., & Sam, L. (2017). The Sustainable Development Goals: A case study. *Marine Policy*, 86, 94–103.

Friends of Europe (2018) EU-ASEAN relations: The next forty years. Retrieved from https://www.friendsofeurope.org/sites/default/files/2018-04/ASEAN_PI_report.pdf.

Fukuda-Parr, S. (2008). Are the MDGs priority in development strategies and aid programmes? Only a few are! Brasilia: International Poverty Centre.

Gupta, J., & Nilsson, M. (2017). Toward a multi-level action framework for Sustainable Development Goals. In N. Kanie, & F. Biermann (Eds), *Governing through goals: Sustainable Development Goals as governance innovation* (pp. 275–294). Cambridge, MA: MIT Press.

Haines, A., & Cassels, A. (2000). Can the Millennium Development Goals be attained? *BMJ*, 329(7462), 394–397.

Hajer, M., Nilsson, M., Raworth, K., Bakker, P., Berkhout, F., de Boer, Y., Kok, M. (2015). Beyond cockpit-ism: Four insights to enhance the transformative potential of the Sustainable Development Goals. *Sustainability* 7, 1651–1660.

Hák, T., Janousková, S., & Moldan, B. (2016). Sustainable Development Goals: A need for relevant indicators. *Ecological Indicators*, 60, 565–573.

IISD (2018). ASEAN, ESCAP propose 7 initiatives to achieve SDGs in the region. New York: International Institute for Sustainable Development. Retrieved from http://sdg.iisd.org/news/asean-escap-propose-7-initiatives-to-achieve-sdgs-in-the-region/.

Irawanto, D., Ramsey, P. L., & Tweed, D. M. (2013). The paternalistic relationship: Authenticity and credibility as a source of healthy relationships. *The Journal of Values-Based Leadership*, 6(1), article 5.

Japan MOFA (2017). The 2030 Agenda for Sustainable Development and Japan's implementation. Tokyo: Japanese Ministry of Foreign Affairs. Retrieved from http://www.mofa.go.jp/files/000252817.pdf.

Kanie, N., Bernstein, S., Biermann, F., & Haas, P. (2017). Introduction: Global governance through goal setting. In N. Kanie, & F. Biermann (Eds), *Governing through goals: Sustainable Development Goals as governance innovation* (pp. 1–28). Cambridge, MA: MIT Press.

Like-Minded Countries Supporters of Middle-Income Countries (2017). High-level Political Forum on sustainable development. Retrieved from https://sustainabledevelopment.un.org/content/documents/25362middleincomecountries.pdf.

Mulholland, E., & Berger, G. (2017). Multi-level governance and vertical policy integration: Implementation of the 2030 Agenda for Sustainable Development at all levels of government. Vienna: ESDN Office.

Nilsson, M. (2017). Important interactions among the Sustainable Development Goals under review at the High-level Political Forum 2017. Stockholm: Stockholm Environment Institute.

Olsen, S., Teoh, S., & Miyazawa, I. (2015). ASEAN community and the Sustainable Development Goals: Positioning sustainability at the heart of regional integration. In IGES, *Greening integration in Asia: How regional integration can benefit people and the environment* (pp. 59–77). Kanagawa: Institute for Global Environmental Strategies.

Purba, R. E. (2011). Public participation in development planning: A case study of Indonesian Musrenbang. *International Journal of Interdisciplinary Social Sciences*, 5(12), 265–277.

Ruggie, J. G. (1996). *Winning the peace*. New York: Columbia University Press.

Singapore Foreign Ministry (2018). 32nd ASEAN Summit: Transcript of press conference by Prime Minister Lee Hsien Loong. Press release, 28 April. Retrieved from https://www.asean2018.sg/MFA/Newsroom/Press-Releases/Press-Release-Details/20180428_Press_Conference_Speech.

Singh, G. G., Cisneros-Montemayor, A. M., Swartz, W., Cheung, W., Guy, J. A., Kenny, T. A., ... Otak, Y. (2017). A rapid assessment of co-benefits and trade-offs among Sustainable Development Goals. *Marine Policy*, 93, 223–231.

Stafford-Smith, M., Griggs, D., Gaffney, O., Ullah, F., Reyers, B., Kanie, N., ... O'Connell, D. (2017) 'Integration: The key to implementing the Sustainable Development Goals. *Sustainability Science* 12(6): 911–919.

UCLG (2015). The Sustainable Development Goals: What local governments need to know. Barcelona: United Cites and Local Government. Retrieved from https://www.uclg.org/sites/default/files/the_sdgs_what_localgov_need_to_know_0.pdf.

UN (1982). World charter for nature. New York: United Nations. Retrieved from http://www.un.org/documents/ga/res/37/a37r007.htm.

UN (1992). Agenda 21. Rio de Janeiro: United Nations.

UN (2000). Millennium declaration. New York: United Nations.

UN (2002). World summit on sustainable development. Johannesburg: United Nations.

UN (2012). The future we want. New York: United Nations.

UN (2015). Transforming our world: The 2030 Agenda for Sustainable Development. New York: United Nations.

UN (2016a). Global sustainable development report (2016). New York: United Nations.

UN (2016b). Executive summary of China's actions on the implementation of the 2030 Agenda for Sustainable Development. New York: United Nations. Retrieved from https://sustainabledevelopment.un.org/memberstates/china.

UNDP (2015). Human Development Index and its components. New York: United Nations Development Programme. Retrieved from http://hdr.undp.org/en/composite/HDI.

UNDP (2016). UNDP support to the implementation of the 2030 Agenda for Sustainable Development. New York: United Nations Development Programme. Retrieved from http://www.undp.org/content/undp/en/home/librarypage/sustainable-development-goals/strategy-undp-support-to-the-implementation-of-the-2030-agenda.html.

UN ESCAP (2017a). Southeast Asia subregion challenges and priorities for SDG implementation. Bangkok: United Nations Economic and Social Commission for Asia and the Pacific. Retrieved from https://www.unescap.org/sites/default/files/Southeast%20Asia%20Subregion%20Challenges%20and%20Priorities%20for%20SDG%20Implementation%202017.pdf.

UN ESCAP (2017b). Regional road map for implementing the 2030 Agenda for Sustainable Development in Asia and the Pacific. Bangkok: United Nations Economic and Social Commission for Asia and the Pacific. Retrieved from https://www.unescap.org/sites/default/files/publications/SDGs-Regional-Roadmap.pdf.

UN ESCAP (2017c). Complementarities between the ASEAN Community Vision 2025 and the United Nations 2030 Agenda for Sustainable Development: A framework for action. Bangkok: United Nations Economic and Social Commission for Asia and the Pacific. Retrieved from http://asean.org/storage/2017/11/FINAL_Complementarities-Report-no-graphic-on-cover.pdf.

UN ESCAP (2018a). Regional and subregional perspectives on the transformation towards sustainable and resilient societies. Bangkok: United Nations Economic and Social Commission for Asia and the Pacific. Retrieved from http://sdgasiapacific.net/download/SDG_Resilience_Report.pdf.

UN ESCAP (2018b). Asia and the Pacific SDG progress report 2017. Bangkok: United Nations Economic and Social Commission for Asia and the Pacific. Retrieved from https://www.unescap.org/sites/default/files/publications/Asia-Pacific-SDG-Progress-Report-2017.pdf.

UN ESCAP, ADB, & UNDP (2017). Asia Pacific Sustainable Development Goals outlook. United Nations Economic and Social Commission for Asia and the Pacific, Asian Development Bank and United Nations Development Programme. Retrieved from https://www.adb.org/sites/default/files/publication/232871/asia-pacific-sdgoutlook-2017.pdf.

United Nations Foundation (2018) Sustainable Development Goals. Retrieved from http://www.unfoundation.org/features/globalgoals/the-global-goals.html.

USAID (2018a) Taking a holistic view of the Sustainable Development Goals. United States Agency for International Development. Retrieved from https://www.usaid.gov/GlobalGoals.

USAID (2018b). USAID homepage. United States Agency for International Development. Retrieved from https://www.usaid.gov/.

WCED (1987). Brundtland report: Our common future. New York: World Commission on Environment and Development.

Weitz, N., Huber-Lee, A., Nilsson, M., Davis, M., & Hoff, H. (2014). Cross-sectoral integration in the Sustainable Development Goals: A nexus approach. Stockholm: Stockholm Environment Institute.

World Bank (2018) The World Bank in Middle-Income Countries. Retrieved from https://data.worldbank.org/income-level/upper-middle-income and https://data.worldbank.org/income-level/lower-middle-income.

CHAPTER 2

Localizing the Sustainable Development Goals: Assessing Indonesia's Compliance towards the Global Goals

Dafri Agussalim, Ahmad Rizky M. Umar, Karina Larasati and Dio H. Tobing

1 Introduction

This chapter provides an assessment of Indonesia's compliance towards the Sustainable Development Goals (SDGs). More specifically, it aims to analyse whether Indonesia's national development policies have been fully in compliance with the global goals. In 2017, Indonesia submitted its Voluntary National Review (VNR) Report on several of the SDGs. The VNR indicates some successes — if evaluated by the government's measurement and national standard — in achieving several of the SDGs, namely Goals 1, 2, 3, 5, 9, 14, and 17 (see BAPPENAS, 2017). It discusses the ways in which the SDGs are incorporated into national frameworks, mentions three core dimensions of national strategies, and looks at means of implementation (ibid.). It concludes with some steps which it considers necessary to achieve the rest of the SDGs in future.

Whilst the VNR focuses on how to achieve the development goals — which is typical of many government reviews — it is important to examine not only the extent to which the government has complied with the development goals through existing technocratic measurements, but also the extent to which its broader development strategies — as well as its approach towards relevant development stakeholders — support its efforts to achieve the SDGs. This chapter therefore assesses not only 'fulfilment of goals' undertaken by the government, but also the incorporation of the SDGs into the broader development programme, which also involves the multi-level structure of governance, cross-sectoral governmental bodies, and multiple stakeholders (Hooghe and Marks, 2001)

This attempt to analyse Indonesia's efforts at incorporating and locating the SDGs within a multi-level, networked, and cross-sectoral form of governance necessitates an understanding of the goals as a governance strategy to achieve the normative foundations that have been laid out by the SDGs, as a continuation of the Millennium Development Goals (MDGs). This implies a governance

 | DOI:10.1163/9789004391949_003

strategy that should not only be understood in terms of goal-achievement, but also — quoting Kanie and Biermann (2017) — as "goal-setting". As a goal-setting strategy, the SDGs are incorporated by establishing policy priorities, allocating resources to support their implementation, benchmarking, and creating a long-term strategy to achieve the goals (see Young, 2017).

To this end, the chapter draws upon the recent work of Frank Biermann, Rakhyun E. Kim, and Norichika Kanie. These three authors suggest that a further reconceptualization of the SDGs is necessary, arguing that the achievement of the goals goes beyond traditional means such as national reports and reviews. Rather, it demands acknowledgment of the role of networked actors, governance mechanisms, and the multi-level structure of governance in which the SDGs are institutionalized, at either the national or subnational level of government (Biermann, Kanie, & Kim, 2017). Biermann et al. further suggest that the SDGs should be understood in two interconnected ways: (1) as the continuation of the MDGs rather than as distinct development goals; and (2) as a normative foundation to be incorporated into national and subnational levels of government, rather than a detailed roadmap on how to undertake development. The latter point is important given that the wide range of development goals included in the SDGs makes it difficult to achieve *all* goals within a short timeframe.

Based on this specific theoretical framework, it can be asserted that even though Indonesia has set up plans to accomplish the SDGs, its broader development framework is not fully consistent with these plans. This argument is substantiated by an assessment of Indonesia's VNR and its coherence with the broader development framework, namely the National Long-term Development Plan (RPJP) and National Middle-term Development Plan (RPJMN).[1] Based on that assessment, the argument can be further developed that although Indonesia has undertaken a commitment to the SDGs and has incorporated this commitment into development-related policies, as shown by the VNR, the SDGs do not as yet form the core of Indonesia's development policies, due to some inconsistencies with the country's broader development frameworks. This is due to *competing discourses of development* within Indonesia's development framework since the beginning of the period

1 In this chapter, we focus on the national policy framework and assess its coherence. This does not mean that the achievement of the SDGs at the regional or subnational level is not important; however, since the national government has the authority to formulate and implement a policy framework regarding the SDGs, albeit with some variations as a result of the decentralization process in Indonesia, it is worthwhile to assess the SDG process at the national level.

known as *Reformasi* (political reform). Although Indonesia has acknowledged the SDGs and institutionally incorporated them in its national development framework, the role of the goals within the infrastructure-led development regime of the current president, Joko Widodo (known as Jokowi), remains limited. There are still many limitations preventing the goals from becoming the core of Indonesia's national development agenda (Warburton, 2016).

This argument is explored in more depth in three main sections. The first section analyses the SDGs from the perspective of goal-setting, which acknowledges the importance of networked actors and integrated policy frameworks (Biermann et al., 2017; Young, 2017). It traces the history of global goals, the institutional development of the United Nations and its growing concerns over environmental deterioration, as well as the emergence of the sustainable development discourse and its culmination in 2015 in the SDGs. This section also elaborates the approach to SDGs as goal-setting rather than only goal-fulfilling, which will be utilized to assess the SDGs in Indonesia's development framework. The second section examines Indonesia's VNR and assesses its coherence within a broader development framework at the national level. Specifically, this section identifies the goal-setting and specific normative foundations that underpin two development frameworks, namely the 2005–2025 Long-term Development Plan (RPJP), and the 2015–2019 Middle-term Development Plan (RPJMN). The third section of the chapter offers a brief examination of the recent 'new developmentalist turn' during the Jokowi presidency, before putting forward some alternative policy options to ensure Indonesia's compliance with the SDGs in its broader development frameworks. These include possible initiatives to improve the country's progress towards achieving the SDGs by taking into account the multi-level, networked, and cross-sectoral nature of sustainable development.

2 The Origins of the SDGs: a History of Global Goals

The Sustainable Development Goals were conceived at the United Nations Conference on Sustainable Development in 2012, as a set of universal goals designed to expand on the Millennium Development Goals which had been created to confront the environmental, economic, and political challenges facing the world at the end of the 20th century. The roots of the two sets of goals can be traced back to the 1970s and 1980s when, amid increasing concerns about environmental degradation, nations became more aware of the importance of protecting the environment while continuing to follow an economic and social development agenda both nationally and globally.

The concept of sustainable development as an objective policy was born from the World Commission on Environment and Development (WCED), which set up the Brundtland Commission in 1983, tasked to articulate a new vision of development by integrating social and economic development with the need to protect the environment. According to the Brundtland Report, *Our Common Future*, the notion of sustainable development was ultimately agreed upon as a way "to ensure that [humanity] meets the needs of the present without compromising the ability of future generation to meet their own needs" (WCED, 1987, p. 8). The definition of sustainable development contains two key concepts: (1) the concept of "needs", in particular, the essential needs of the world's poor, to which overriding priority should be given; and (2) the idea of limitations imposed by the state of technology and social organization on the environment's ability to meet present and future needs (WCED, 1987).

In 1992, through the United Nations Conference on Environment and Development (UNCED) in Rio de Janeiro, 178 nation states met to rethink an eco-efficient, sustainable development agenda, to tackle the depletion of irreplaceable resources and to stop irreparable damage and pollution being inflicted on the planet. This meeting resulted in a global commitment to a framework of state-led development to promote environmental protection, and to raise awareness of its relevance to other important issues such as human rights, population, social development, gender issues, and human settlements. Eight years later, amid an economic crisis in various parts of the world and a deepening global mistrust of international economic institutions, UN leaders came to the conclusion that a more concerted approach to global partnership was needed to effectively address the global situation at that time — a genuine approach that could address problems in developing countries and actively involve them in the process. In September 2000, following the UN Millennium Summit, eight global development goals were announced, ranging from halving extreme poverty to halting the spread of HIV/AIDS and providing universal primary education, to be achieved by 2015 (UN News Center, 2015). The Millennium Development Goals were seen as a global breakthrough in meeting the needs of the world's poorest. They also provided a uniform language and framework for governments to develop policies, programmes, and relief efforts in the form of aid to help reduce poverty in the world.

Critics have argued that these internationally agreed development goals were unrealistic and overambitious, often overlooking limitations within local capabilities. By focusing on only eight goals and key areas, the MDGs were accused of not addressing the root causes of poverty such as human rights and of neglecting other important issues. As a global development agenda, the MDGs should ideally cover the development needs of all countries, recognizing that the focus of development of each country is different; for developed countries, targets on health and education are appropriate, but in some developing countries, infrastructure such as roads and communications are urgently needed to push the development agenda. Approaching the end of the MDG deadline in 2015, UNICEF reported: "around 1 billion people still live on less than $1.25 a day — the World Bank measure on poverty — and more than 800 million people do not have enough food to eat. Women are still fighting hard for their rights, and millions of women still die in childbirth" (UNICEF, 2015). The eight MDGs seemed to have failed to consider the root causes of poverty, meaning that human rights were neglected and economic development was not addressed as directly as it should be.

Still committed to addressing socio-economic problems and the urgency of global climate change, without jeopardizing future resource availability, the world's leaders met again and embraced a bold, ambitious, and potentially transformational agenda that could be applied to all countries. At the United Nations Conference on Sustainable Development, or the Rio + 20 Conference, held in June 2012, a new set of development commitments was created in the form of the Sustainable Development Goals — 17 global goals and 169 targets — with a view to achieving a better combination of economic development, environmental sustainability, and social inclusion, with differing objectives for societies around the world. The SDGs were adopted by the world's leaders in September 2015. Based on the foundations of well-being, planet, peace, and partnership, these global goals are to be fulfilled before 2030.

2.1 *Sustainable Development Goals as Goal-Setting: an Analytical Framework*

Having outlined the history of the SDGs and their emergence as a global development strategy at the United Nations, this section proposes an analytical framework which can be used to understand the implementation of that strategy in Indonesia. Three competing, but often overlapping, approaches dominate the understanding of the SDGs and their implementation. The first is the normative approach, which sees the SDGs as an institutionalization of norms in the global development strategy (Lafferty, 1996; Sachs, 2012). This normative argument is based on a consideration of the relationships between economic development, environmental deterioration, and global governance. Sachs (2015) draws on the notion of the Anthropocene to argue that the discourses of development need to be transformed by incorporating sustainability. According to Sachs, the issue of environmental deterioration provides sufficient justification to reorient global development; he cites, for example, the rapid rise of the human population by around 75–80 million people per year, with the total population expected to reach 9 billion by the middle of the 21st century, and 10 billion by the end of the century (see also Sachs, 2012). Sachs's argument is a classic example of sustainable development as a new normative framework for global governance, with the aim of tackling environmental deterioration, and justified by the latest trans-disciplinary arguments (van Zeijl-Rozema, Cörvers, Kemp, & Martens, 2008).

Another strand of scholarship, in a similar spirit but using a different methodological framework, argues that the SDGs should be read in conjunction with the emergence of the so-called Earth system science, and should be based on the latest research that could be applied to developed and developing countries (Griggs et al., 2013; see also Le Blanc, 2015; Reid et al., 2010; Steffen et al., 2015). It is in this context that the idea of 'planetary boundaries', in conjunction with broader developments in Earth system science and the Earth System Governance Project, emerged as a guiding concept to understand how the SDGs work (Steffen et al., 2015; see also Biermann 2012). These 'science-based' arguments attempt to reconcile the need for development with the 'progress' in scientific findings related to Earth systems in a trans-disciplinary approach, and address the achievement of the SDGs in that context (Griggs, 2015; Steffen et al., 2015). This approach also reconsiders the idea of sustainable development as a framework that meets the needs of the present while safeguarding Earth's life-support system, on which the welfare of current and future generations depends — an echo of the Brundtland Report (WCED, 1987) and the Rio Declaration (UNCED, 1992).

While the normative and science-based approaches provide a well-argued case for promoting the SDGs, neither can answer comprehensively how the global development framework fits into the established national development frameworks of the UN's member states. In this regard, the third approach points out one limitation of the SDGs as a global development framework: the relative neglect of a multi-structured and multi-layered form of 'development' at the national level, which needs to be engaged in constructing new global governance norms (Geels, 2010, 2012; Pahl-Wostl, 2009; Smith, Voß, & Grin, 2010). Another group of scholars, therefore, has put forward a multi-level approach to understanding the SDGs, which addresses the spatial limitation of the global agenda as a global development framework, and proposes a broader agenda to reconcile local, national, and global development regimes (Geels, 2010; see also Norström et al., 2014). This issue has also been raised by the Earth System Governance Project, by highlighting the issue of scale in guiding the sustainable development agenda (Biermann, 2012). According to this approach, it is important to see the SDGs not only as a set of networked goals, but also as an example of participatory and multi-level governance (Le Blanc, 2015; see also Newig and Fritsch, 2009; Norström et al., 2014).

In order to comprehensively understand the progress of the SDGs, the analysis in this chapter builds upon the third approach, but also considers the merits of the first and second approaches. Following the multi-level approach to the SDGs, we argue that the implementation of the global agenda in Indonesia is best understood as an ongoing 'multi-level developmental process' at the subnational, national, and global levels — a technocratic process aimed at achieving a set of global goals. Acknowledging the multi-level characteristics of development is particularly important in the context of Indonesia, which has undergone a process of decentralization since 1998. This decentralization has, to some extent, turned development into a politicized process, in which development planning is devolved to provincial or district-level governance with only a limited degree of political authority given to the national development agency, BAPPENAS. This multi-level context also creates a space for multi-level and multi-actor engagement in institutionalizing the SDGs, which in turn provides another opportunity to develop 'local adaptive capacity' to guide the implementation of this model of development (Smit & Wandel, 2006). Hence, the implementation of the SDGs is not merely a top-down process; it has some variations in the national and subnational contexts, as well as in the dynamics of incorporating the global goals into national and subnational development planning.

In this context, a new understanding of the SDGs and their dynamic implementation is required. The recent contributions by Biermann, Kanie, and Kim that see the SDGs as goal-setting, in which the development goals are incorporated in a multi-level and multi-layered form of development without neglecting the universal targets, could be used as a starting point. As mentioned earlier, Biermann et al. (2017) argue that the achievement of the SDGs should go beyond traditional means of national reports, as exemplified by the MDGs; they therefore propose a more dynamic approach to accommodate local and subnational variation. This necessitates the acknowledgement of networked actors, governance mechanisms, and the multi-level structure of governance in which the SDGs are institutionalized (ibid.). Sustainable development, in this context, should recognize the complexity of global governance, which consists not only of a system of sovereign states, but also a network of scientists, corporations, and civil society organizations that interact with each other, as well as the environmental context that serves as the basis for global governance.

Biermann et al. (2017) further suggest that the SDGs should be seen in two interconnected ways. First, they should be understood as a continuation of the

MDGs, extending the promises and responding to the limitations of the MDGs in delivering a global development framework. Second, the SDGs should be understood as a normative foundation of global development, justified through scientific research and progress on global environmental change, which necessitates the incorporation of particular environmental standards at the national and subnational levels. However, the form of this incorporation might vary in the light of global inequality and variations in national development strategy. Incorporating the three pillars of sustainable development — economic, social, and environmental — into the national politics of development is thus essential. Rather than proposing detailed measures on how to 'do development', the SDGs invite the UN member states to carry out a Voluntary National Review and to collaboratively evaluate the progress towards sustainable development by 2030.

Taking this new conception of the SDGs as its point of departure, and considering the process of national implementation in Indonesia, the rest of this chapter argues that the progress of SDG implementation needs to be assessed against three important markers: the Voluntary National Review, which is conducted by BAPPENAS and periodically submitted to the United Nations; the National Long-term and Medium-term Planning Strategies; and the government's budget commitments for funding these programmes. Due to space limitations, we will focus on the first two of these, which will be assessed in terms of their coherence at the national level, as well as within the global development framework. In this context, the SDGs are seen as a global normative framework that provides a basis for national implementation. It is important to compare the VNR, which deals directly with the global goals, with the national planning strategies as outlined by BAPPENAS in 2005 (for the Long-term Strategy, RPJP) and in 2015 (for the Medium-term Strategy, RPJMN). Against this backdrop, it is argued that although Indonesia has made a commitment to the SDGs and incorporated this commitment into development-related policies (as demonstrated in the VNR), the SDGs cannot at this point be said to lie at the heart of Indonesia's development policies, due to some inconsistencies with broader development frameworks. This inconsistency will be discussed in the next section.

3 Assessment of the Sustainable Development Goals' Implementation in Indonesia

This section reviews several documents related to the implementation of the SDGs in Indonesia, and analyses the extent to which this implementation is consistent with the broader development framework adopted at the national level. It finds that, although Indonesia has made technocratic progress in implementing the SDGs, and the development framework has at some points mentioned the importance of the global goals for national development, nevertheless the SDGs are not necessarily consistent with national development policies. The first part of the section assesses Indonesia's Voluntary Review of the Sustainable Development Goals' implementation undertaken by National Development Planning Agency (BAPPENAS). The second and third parts then examine Indonesia's Long-term and Medium-term Development Plans, respectively, in relation to the SDGs.

3.1 *Indonesia's Voluntary National Review: an Overview*

Indonesia is one of the few countries in Southeast Asia that is seriously committed to realizing the 2030 global agenda. In 2012, the Indonesian government tasked the Ministry of National Development Planning (BAPPENAS) with implementing this agenda and monitoring the implementation strategy of the SDGs. Indonesia's strong commitment led to the synchronization of its national development planning with the SDG process; as the national focal point, BAPPENAS was responsible for drawing up national guidelines in order to achieve the ambitions of the international community. As a result, 108 of the 169 SDG targets have been incorporated within the country's medium-term development plan (UN Development Group, 2016).

To monitor the implementation of the SDGs, Indonesia participates in the review mechanism formally known as Voluntary National Review. The VNR is undertaken to "conduct regular and inclusive reviews of country-led and state-driven national and sub-national progress, which will serve as the basis for regular review by high-level political forums or ECOSOC of the United Nations" (UN, 2015). In 2017, Indonesia — represented by BAPPENAS — signed the country's Voluntary National Review which was made available to the public at a later stage.

Three important points are highlighted in the Indonesian report: 1) the compatibility of the global agenda with the national development plan; 2) Indonesia's commitment, as shown by the issuing of Presidential Regulation no. 59 in 2017 on the Implementation of the SDGs; and 3) the inclusivity of the process, involving all stakeholders (BAPPENAS, 2017). The VNR also stresses the coherence between the SDGs and the country's national agenda, represented by the National Development Vision, "Nawacita"; the National Long-term Development Plan (RPJP) 2005–2025; and the National Medium-term Development Plan (RPJMN) 2015–2019. The VNR emphasizes the importance of paying attention to critical issues, such as poverty eradication, prosperity realization, and environmental quality improvements. In this endeavour, two key aspects are improving the quality of human resources, and enhancing economic opportunities and sustainable livelihoods (BAPPENAS, 2017). The VNR claims considerable progress on these two aspects. First, Indonesia claims that in 10 years, poverty has been reduced significantly due to the increasing coverage of comprehensive social protection, fulfilment of basic needs, and encouragement towards the improvement of peoples' welfare provided by the National Social Security System (NJSN) and integrated social assistance.

> The Maternal Mortality Rate (MMR) has decreased from 346 in 2010 to 305 in 2015.... Indonesia managed to increase self-sufficiency in food production where rice production reached 75.4 million tons in 2015 and maize 19.6 million tons in the same year. Access to clean water for the poor and vulnerable groups is also found to increase to 61.57% and 61.94% in 2016 while in the same period, access to sanitation has increased from 47.76% to 54.12%. Apart from that the GDP per capita has increased from USD 1,420 in 2006 to USD 3,605 in 2016, where the unemployment rate has declined from 10.28% in 2006 to 5.6% in 2016. In addition to the above improvements, other improvements were also found in the development of road infrastructure, marine protected areas and the involvement of women in parliament and as civil servants by 2015 and the development of innovative partnerships with many sectors.
>
> BAPPENAS, 2017, pp. 91–93

Behind this impressive report on Indonesia's achievements, however, there are still many significant shortcomings in the country's development. In the SDG Index and Dashboard Report on country performance regarding Agenda 2030 and the SDGs, Indonesia was ranked at 98 out of 149 countries in 2016, and 100 out of 157 countries in 2017 (SDSN, 2017). Among ASEAN countries, Indonesia ranks sixth, after Malaysia (ranked 54), Thailand (55), Singapore (61), Vietnam (68), and Philippines (93) (ibid.). To put this into context, it is important to recognize that the SDGs essentially entailed 'too many' goals, causing a diffusion of focus.

One of the limitations of the SDG regime lies in the fact that through VNRs, countries are required to show their compliance to the whole regime, while, in fact, policy incoherence remains. Kanie and Biermann (2017) suggest that there is a weakness in viewing the SDGs as solely aspirational, as this will lead to a degree of utopianism and to only partial or temporary achievement of the goals (Kanie & Biermann, 2017). For example, to meet the SDGs, Indonesia must 'morally' claim a strong commitment to Goal 6 (Ensure access to water and sanitation for all), while the actual commitment at the national policy level remains very low. The case of Jakarta is illustrative: in 1991, a water privatization project was initiated, but by late 2016, only 49.6% of the total population of Jakarta had access to safe drinking water, while the water tariff had increased significantly from Rp. 1.700 per cubic meter at the beginning of the concession to Rp. 7.025 per cubic meter by 2016 (BRPAMDKI, 2016; Zamzamie & Ardhianie, 2015). In 2015, Jakarta's water tariff was the highest in Southeast Asia (Tempo, 2015).

This case demonstrates that, despite Indonesia's claim of coherence between its national development planning and the SDG regime, some of its national development focus is not maintained at a level consistent with the global goals. In accordance with the aim of this chapter — to move beyond a technocratic assessment of Indonesia's SDG implementation record — it is therefore important to analyse the opportunities and challenges of national and local implementation of the SDGs in order to map the extent to which the global goals are at the forefront of Indonesia's national development planning.

3.2 *Policy Coherence 1: RPJP (2005–2025)*

In order to realize the global agenda, it is imperative that the idea and vision of this agenda are disseminated and understood by all levels of society, not only the central government level. The UN's Sustainable Development Solutions Network (SDSN) states that local governments cannot be mere implementers but must be partners in policy making, implementation and evaluation of SDGs in order to support their optimum implementation (SDSN, 2016, p. 15). Local governments must also play a greater and more active role in encouraging the realization of sustainable development by private sectors, as the latter are better informed about local capacities, characteristics, and challenges.

The Indonesian government has made an effort to convert the vision and mission of the SDGs into national and regional development plans. The SDG vison is not a foreign idea forced onto Indonesia; rather, the important mission of the global agenda has the same basic focus as the national development agenda which prioritizes the development of people and the state economy, while maintaining the protection of the environment. The Long-term Development Plan (RPJP) 2005–2025 (BAPPENAS, 2005) entails similar goals to those expressed by the SDGs. What distinguishes the two are the sources and the framework of the goals. The SDGs are the result of the commitment made by the heads of state on 25 September 2015, during the UN Sustainable Development Summit. They are a set of long-term goals to which the international community has committed, without any enforcement or binding effect on the part of the members. Rather, countries comply voluntarily with the SDG regime based on their own perceptions of the goals. In terms of frameworks: the SDGs are more generally formulated and applicable towards all nations, whereas the RPJP is created locally and is therefore more specific, focusing on strategic issues to which attention and priority must be given. Furthermore, government agencies are compelled to work with the RPJP while drafting measures, so that the resulting policies conform with the RPJP.

In order to determine whether there is policy coherence between the two, it is necessary to examine the RPJP more closely. The RPJP covers nine categories of development, namely: politics; defence and security; law and state administration; socio-cultural aspects; human resources; economics; regional–local; infrastructure; and natural resources and living environment. This demonstrates that the RPJP, as a long-term development plan, adopts a similar approach in terms of goal-setting to the SDGs. For example, if we take the global Goal 16 on peace, justice, and strong institutions, this would fall under the umbrella of the RPJP's development goal on defence and security as well as its goal on law and state administration. Another example: global Goal 15 aims to 'protect, restore and promote sustainable use of terrestrial ecosystems, sustainably manage forests, combat desertification, and halt and reverse land degradation and halt biodiversity loss', while similar concerns and indicators are provided under the RPJP development goal on natural resources and living environment.

In addition, the two agendas entail similar visions. In the RPJP this is stated as follows:

> The national vision of long-term development is the creation of healthy, intelligent, productive and noble human beings and an increasingly prosperous society in sustainable development driven by an increasingly developed, self-sustaining and equitable economy throughout the region supported by the provision of adequate infrastructure and robustness of integrity and unity of the nation that is imbued with a formidable character in the Unitary State of the Republic of Indonesia organized with democracy [based on the values of Pancasila] as a guideline in the life of society, nation and state and upholding the rule of law.
>
> BAPPENAS, 2005, p. 17

Although expressed in different words, the vision of the SDGs, as contained in the UN General Assembly resolution, includes very similar aims:

> In these Goals and targets, we are setting out a supremely ambitious and transformational vision. We envisage a world free of poverty, hunger, disease, and want, where all life can thrive.... We envisage a world of

> universal respect for human rights and human dignity, the rule of law, justice, equality and non-discrimination; of respect for race, ethnicity and cultural diversity; and of equal opportunity permitting the full realization of human potential and contributing to shared prosperity.... We envisage a world in which every country enjoys sustained, inclusive and sustainable economic growth and decent work for all. A world in which consumption and production patterns and use of all natural resources — from air to land, from rivers, lakes, and aquifers to oceans and seas — are sustainable. One in which democracy, good governance and the rule of law as well as an enabling environment at national and international levels, are essential for sustainable development, including sustained and inclusive economic growth, social development, environmental protection and the eradication of poverty and hunger.
>
> UN, 2015, pp. 3–4

The RPJP was enacted in 2005, the SDGs were agreed in 2015. This suggests that, in principle, the Indonesian government has been active in the implementation of sustainable development goals since well before the formulation of the SDGs themselves.

3.3 *Policy Coherence 2: RPJMN (2015–2019)*

The second development document that is important in analysing the realization of the SDGs in Indonesia is the Medium-term Development Plan (RPJMN). The RPJMN is a planning document which covers five years; the points included are described in the Government Work Plan (RKP). The RPJMN 2015–2019 is the third phase of the implementation of the 2005–2025 RPJP; it represents the mission and vision of President Jokowi for the period 2015–2019. The RPJMN is seen as a guide for ministries, agencies, and governments in formulating and/or adjusting national or regional development strategies (BAPPENAS, 2014).

Like the RPJP, most of the development plans and strategies set forth in the RPJMN reflect the global agenda of the SDGs. There are at least 94 global indicators listed in the RPJMN. Compared to the earlier conception of global development formulated in the MDGs, it is clear that the SDGs have the potential for far greater achievements, as they emphasize the importance of public stakeholders' participation — an approach that differs from their predecessors. Chapter 3.4 of the RPJMN indicates the government's firm commitment to implement Indonesia's development agenda within the global context of the SDGs (UNDP, 2015). Chapter 6 of the RPJMN includes various key development plans that are compatible with the SDGs. For example, poverty eradication

(SDG 1), halting hunger (SDG 2), health and well-being (SDG 3), improving education (SDG 4), and gender equality (SDG 5) are all or respectively integrated within the the sub-chapters of the RPJMN as Chapter 6.1, 6.3, 6.5, 6.6, 6.7, 6.8, and 6.9 (Ibid.). The global goals focusing on climate change (SDG 13), environmental selection, conservation and sustainable use of marine and forest resources (SDGs 14 and 15) appear as Chapters 6.4 and 6.7 of the RJPMN (ibid.). A more detailed description of convergence between the two documents is presented in Table 2.1.

Although almost all of the vision of the SDGs has been integrated into the RPJMN 2015–2019, some problems nevertheless remain, especially when translating the plan into development policy. One example is the plan to accelerate economic growth to support poverty alleviation through strengthening investment, as outlined in Chapter 6.6.5 of the RPJMN. One of Indonesia's strategies to strengthen the country's economy and improve the trade balance is through investments in palm oil, which the government claims has strategic value in terms of development and poverty alleviation (Indonesian Ministry of Foreign Affairs, 2018). However, this issue is fiercely debated, as the palm oil industry is known to be a major cause of deforestation and the destruction of local crops in Indonesia, as well as other countries with dwindling expanses of tropical rainforest. While the Indonesian government believes the opposite, in fact, palm oil plantations in Indonesia still not in accordance with the environmental protection standards. Often, though not always, palm oil plantations are conducted through the clearing of protected forests or conservation lands.

TABLE 2.1 Areas of convergence: RPJMN 2015–2019 and SDGs

RPJMN 2015–2019	Sustainable development goals
Chapter 6.1, sub-chapter 10	Goals 3, 10, 16 and 17
Chapter 6.2, sub-chapter 5	Goal 16
Chapter 6.3, sub-chapter 3	Goals 1–11
Chapter 6.4, sub-chapter 6	Goals 14–16
Chapter 6.5, sub-chapter 7	Goals 1–6
Chapter 6.6, sub-chapter 11	Goal2 1–10
Chapter 6.7, sub-chapter 9	Goals 1–5, 8, 9, and 12–15
Chapter 6.8, sub-chapter 1	Goals 3–4 and 11
Chapter 6.9, sub-chapter 1	Goals 5, 10, 16, and 17

SOURCE: UNDP (2015)

In addition, the land management process for this business was also found to violate the rights of local communities. According to a European Parliament report on palm oil and deforestation, the largest land fires in Indonesia are the result of clearing land for palm oil plantations (EU, 2017). On record as the worst for nearly two decades, forest fires in Indonesia and Borneo in 2015 turned the country into the single largest contributor to global warming, exceeding the average daily emissions of the USA, as well as causing air pollution affecting millions of people (EU, 2017; Harris, Minnemeyer, Stolle & Payne, 2015). It is therefore clear that in an effort to eradicate poverty, Indonesia is moving even further away from realizing SDG 13 on climate action, and from protecting endangered animals such as orangutans that are threatened by deforestation.

The fundamental problem with palm oil investment in Indonesia is the fact that the plantation process does not meet environmental protection standards, but is conducted largely by clearing protected forest or conservation land. This type of business is clearly inconsistent with the ideals and mission stipulated in RPJMN 2015–2019, Chapter 6.7.4, on forest conservation through fire prevention and control:

To date, and regardless of the various management strategies the government has attempted to implement, the case of forest fires and deforestation represents an acute and protracted problem that is difficult to resolve. There are many aspects to this complex issue, one of which is the high economic reward for the burning of land, prompting many people to carry out forest burning for a variety of different interests (Artharini, 2015). The clearing of land, either mechanically or by burning, brings substantial revenue to the clearers. Land cleared through burning is worth US$10–20 per ha, and is therefore significantly cheaper than land cleared by mechanical means, at US$200 per ha (ibid.), making it attractive for companies to purchase. Apart from this financial incentive for clearing, there is also a lack of coordination between relevant government agencies at the central and regional levels in following legal processes and eradicating the practice of clearing new land by burning. According to Purnomo et al. (2015), many of those who seek profits through this activity have close relationships with important stakeholders at the local, national, and even ASEAN level, so that finding a solution to the complex problem of forest fires becomes further complicated by the issue of political networks.

It is therefore clear that realizing the goals of sustainable development is not easy. Not all of the SDGs can be incorporated into national development plans, and even for those that are, the path to meeting those goals may not run smoothly because, in practice, Indonesia's capabilities are limited. If Indonesia is serious about eradicating poverty while meeting its climate commitments, investing in ways of achieving economic growth without sacrificing environmental and forest conservation efforts would be a major step forward.

4 The Sustainable Development Goals, New Developmentalism, and Local Adaptive Capacity

Based on the review of the two planning documents, it can be said that Indonesia has so far established a strong commitment to sustainable development and incorporated this commitment into relevant development policies, as reported in the Voluntary National Review. However, when the VNR is assessed against the broader development framework, as exemplified in the RPJP and RPJMN, it appears that the SDGs have *not yet* been placed at the heart of Indonesia's development policies due to some inconsistencies: there is a lack of coherence between the SDG regime and the already established development framework in the national context. In the case of RPJP 2005–2025, such inconsistency is understandable since this platform was designed prior to the establishment of the SDGs in global politics. However, the disconnects between the SDGs and the RPJMN give a clearer indication of the complexity of development at the national level, resulting in the incomplete adoption of the SDGs within the national development strategy.

It can be argued that this incoherence occurs due to the different development discourses that dominate the planning and implementation of development policies, particularly since the election of Joko Widodo as the Indonesian President in 2014. There are at least two such narratives: the first is the sustainable development narrative put forward by BAPPENAS. The second is the growth–productivity narrative, which has now shifted in focus to post-2015, infrastructure-led development. Since these contending narratives underpin Indonesia's approach to development, it is useful to analyse them in more detail and depth. Warburton (2016) argues that Jokowi's development strategy reveals what she calls a "New Developmentalism", which to some extent follows the state-centric development discourse of the Soeharto era but is framed under a new strategy and political narratives. With this new developmentalist discourse, Jokowi has put forward a state-centric narrative that is based upon the maximization of public service delivery and infrastructure development. Although he does not abandon other discourses of development, such as the sustainable development agenda (which he puts into BAPPENAS's technocratic planning system), it appears that Jokowi's new developmentalism gives sustainable development a less central role, secondary to physical-infrastructural progress and political stability (ibid.; see also Mietzner, 2015).

Jokowi's new developmentalist agenda raises questions with regard to how the SDGs are articulated in Indonesia's development policies. The analysis above has produced two important insights. First, it appears that Jokowi has not set aside the SDGs. The VNR shows that the Indonesian government is making efforts to implement the SDGs as a part of its development planning, as well as integrating the goals into its ministerial framework (namely, the Ministry of Forestry and Environment). Second, whilst Jokowi is still managing to implement some of the SDGs, it seems clear that this is only part of the bigger picture of the Indonesian development agenda. When the SDGs are put into the wider story of national development policies, it could be argued that Jokowi does not put the SDGs at the core of his national development framework; they remain at the margins, pushed there by the rise of infrastructural development and public service delivery.

Given this political context, the next question is how the SDGs could become embedded in Indonesia's development policies at a local level. Localizing the SDGs will involve at least six important variables, namely economic stability; access to technology; a delineated method of coordination between each governance level; access to scientific information; distributive and participatory allocation of power; and the maintenance of adaptive systemic capacity (IPCC, 2007). Indonesia's VNR shows that access to technology and economic stability have been pursued or are at least in progress. However, it also reveals the lack of a multi-level method of coordination between the national and subnational levels of governance, as well as a lack of participation in the implementation process, given the technocratic nature of SDG implementation. Improving the implementation process and fulfilment of the SDGs, therefore, requires the development of local adaptive capacity, supported by a coherent and participatory strategy of governance at both national and subnational levels.

Drawing upon the understanding of the SDGs as goal-setting, discussed earlier, this chapter recommends four strategic steps to further localize the SDGs within a national and subnational development framework. The first step is to improve the current technocratic framework of SDG implementation. This means that the technocratic fulfilment of the goals, which is now led by BAPPENAS and handled to some extent by relevant ministries, should be internally optimized, despite the new developmentalist turn in national politics. The best way to improve the government's technocratic capability is by developing a capacity for "policy learning" (see Rietig, 2018). For example, besides the usual check-list measurements of implementation, the government could establish centres for institutional learning, such as policy-relevant research or scientific inputs, to improve its capacity to implement the SDGs. The role of researchers working on sustainable development would be instrumental at this stage.

The second step is to establish guidelines for a sustainable development framework in provincial and district/city-level governance. With the process of decentralization and governance reform underway since 1999, Indonesia now has a multi-level system of governance with the devolution of power from a centralized authority at the national level to the provincial, district/city, and village levels. This decentralized form of governance has opened some possibilities, and also challenges, for the implementation of the SDGs. While the central government in Jakarta remains pivotal in leading the implementation process, some authority (for example, regarding land acquisition for extractive industries or conservation) belongs to the local governance institutions. In optimizing the implementation process, it is important to acknowledge this multi-level structure of governance and to start considering local government not only as a beneficiary of the SDGs but also as an active player in their implementation. This implies the need for a comprehensive set of implementation guidelines for local governance that take account of the diversity of local governance structures, to accompany national-level implementation.

The third step is to incorporate insights from civil society, small and medium enterprises (SMEs), industry, and broader non-state actors who have expertise in sustainability, and to encourage their participation in the implementation of the SDGs. Participation plays an important role in ensuring that actions undertaken by policy makers match the needs and resources available in the local context (IPCC, 2007; see also Ramakrishnan, 1999; Rayner & Malone, 1998). Embracing participation means providing spaces for academics, civil society organizations, SMEs, and industry to engage in the implementation process. Goal 17 of the SDGs explicitly requires the government to engage with non-state actors; however, in the VNR, the space for business actors or civil society organizations is relatively limited. Greater engagement, particularly at provincial, district, and city level could improve the implementation process.

The fourth step is to establish a policy network to measure the implementation of the SDGs at both national and subnational level by incorporating the insights of non-state actors. SDG 17 also explicitly mentions the need for "partnership" in implementing the goals. In this context, the establishment of a policy network is important, in which the government does not take a dominant role in the implementation process, but instead acknowledges other actors' roles (Rhodes, 1997; Rhodes & Marsh, 1992). The government could use nonformal settings, alongside the bureaucratic and technocratic governance process, to communicate with other actors and gain their insights. Alternatively, the government could develop a communications strategy with non-state actors in order to socialize the government programmes and receive feedback. This could be done by engaging local leaders in the implementation of the SDGs, in parallel with the standard bureaucratic process in BAPPENAS.

To summarize, it could be argued that Indonesia's implementation of the SDGs is 'on track', in the sense that the process has been implemented in accordance with the global goals, as laid out by the United Nations. However, in order to localize the SDGs and embed them into the broader development context at the national and subnational levels, a better developmental strategy is required. Thus far, in spite of the positive message of the government's Voluntary National Review, recent developments in the politics and economy of Indonesia indicate that the SDGs, both as idea and as goals, have become relatively marginalized. In order to overcome this, an alternative strategy seems necessary, which might include: optimizing the technocratic process at the ministerial level through policy-learning; setting out guidelines for implementation at subnational levels; encouraging participation; and establishing policy networks to capture insights from diverse actors and institutions.

5 Conclusion

This chapter has sought to understand the relationship between Indonesia's development policies and its implementation of the SDGs. While Indonesia has produced a compelling VNR that discusses its technocratic effort to implement the SDGs, it is not clear whether Indonesia's broader development framework at either the national or subnational level supports this effort. Through a review of key documents — the Voluntary National Review, and the Medium-term and Long-term Development Plans — as well as a deeper analysis of the broader trends in Jokowi's developmental policies, it can be concluded that the goal-setting aspect of the SDGs has not yet been coherently incorporated into the national development framework. This requires a new strategy to 'localize' the global goals. The analysis suggests four interrelated steps: (1) optimizing the current technocratic framework of SDG implementation; (2) establishing guidelines for a sustainable development framework at provincial and district/city level; (3) incorporating insights from civil society, SMEs, industry, and a range of non-state actors who work on sustainability, establishing a forum for their participation in the SDG implementation processes; and (4) establishing a policy network to measure the implementation of the SDGs at the national and subnational levels by incorporating the insights of non-state actors.

These strategies require further research as to how the SDGs might be incorporated into Indonesia's future development policies, at the technocratic level in BAPPENAS, the inter-institutional level in Indonesia's broader bureaucratic polity, or the political level of the core decision makers — in this case, Jokowi's elite circles. Indonesia's ability to substantively implement the SDGs will depend on both technocratic and political processes. President Jokowi has so far prioritized development in the rural areas — especially in the eastern part of Indonesia — as his main policy agenda (see Negara, 2016). While this might create opportunities for further embedding the SDGs at the local level, institutional changes in the context of the new developmentalist turn (Warburton, 2016) might constrain such opportunities. It is therefore important for policy makers working in SDG-related sectors to maintain the institutional compatibility of the SDGs with the broader development policy framework as a prerequisite to the substantive implementation of the global goals in Indonesia.

Acknowledgement

An earlier draft of this article was presented at the International Working Conference 'Sustainable Development Goals in Southeast Asia and ASEAN', Universitas Gadjah Mada, Yogyakarta, 3–5 October 2017. We would like to thank Bakti Setiawan, Ronald Holzhacker, and all workshop participants for their constructive feedback.

References

Artharini, I. (2015). Siapa 'aktor' di balik pembakaran hutan dan lahan?. Retrieved from http://www.bbc.com/indonesia/berita_indonesia/2015/09/150923_indonesia_pembakaranlahan.

BAPPENAS (2005). *Visi dan Arah Pembangunan Jangka Panjang (PJP) Tahun 2005–2025*. Jakarta: BAPPENAS.

BAPPENAS (2014). *Rencana Pembangunan Jangka Menengah Nasional 2015–2019: Buku I Agenda Pembangunan*. Jakarta: BAPPENAS.

BAPPENAS (2017). *Voluntary National Review (VNR): Eradicating poverty and promoting prosperity in a changing world*. Jakarta: BAPPENAS.

Biermann, F. (2012). Planetary boundaries and earth system governance: Exploring the links. *Ecological Economics*, 81, 4–9.

Biermann, F., Kanie, N., & Kim, R. (2017). Global governance by goal-setting: The novel approach of the UN Sustainable Development Goals. *Current Opinion in Environmental Sustainability*, 26–27, 26–31. http://dx.doi.org/10.1016/j.cosust.2017.01.010.

BRPAMDKI (2016). *Tarif Air Minum DKI Jakarta.* Jakarta: BRPAMDKI.

EU (2017). Report on palm oil and deforestation of rainforests. Brussels: European Parliament. Retrieved from http://www.europarl.europa.eu/sides/getDoc.do?pubRef=-//EP//TEXT+REPORT+A8-2017-0066+0+DOC+XML+V0//EN.

Geels, F. W. (2010). Ontologies, socio-technical transitions (to sustainability), and the multi-level perspective. *Research Policy*, 39(4), 495–510.

Geels, F. W. (2012). A socio-technical analysis of low-carbon transitions: Introducing the multi-level perspective into transport studies. *Journal of Transport Geography*, 24, 471–482.

Griggs, D. (2015). Sustainability science: Exploiting the synergies. *Nature*, 519 (7542), 156.

Griggs, D., Satfford-Smith, M., Gaffney, O., Rockström, J., Öhman, M., Shyamsundar, P., … Noble, I. (2013). Policy: Sustainable Development Goals for people and planet. *Nature*, 495 (7441), 305–307.

Harris, N., Minnemeyer, S., Stolle, F., & Payne, O. (2015). Indonesia's fire outbreaks producing more daily emissions than entire US economy. Washington, DC: World Resources Institute. Retrieved from http://www.wri.org/blog/2015/10/indonesia%E2%80%99s-fire-outbreaks-producing-more-daily-emissions-entire-us-economy.

Hooghe, L., & Marks, G. (2001). *Multi-level governance and European integration.* Lanham, MD: Rowman & Littlefield.

Indonesian Ministry of Foreign Affairs (2018). Kementerian Luar Negeri Indonesia: Kelapa Sawit Memperkuat Upaya Pencapaian SDGs. Retrieved from https://www.kemlu.go.id/id/berita/Pages/Kelapa-Sawit-Memperkuat-Upaya-Pencapaian-SDGs.aspx.

IPCC (Intergovernmental Panel on Climate Change) (2007). *Climate change 2007 — Impacts, adaptation and vulnerability. Contribution of Working Group II to the Fourth Assessment Report of the IPCC.* Cambridge: Cambridge University Press.

Kanie, N., & Biermann, F. (Eds) (2017). *Governing through goals: Sustainable Development Goals as governance innovation.* Cambridge, MA: MIT Press.

Lafferty, W. M. (1996). The politics of sustainable development: Global norms for national implementation. *Environmental Politics*, 5(2), 185–208.

Le Blanc, D. (2015). Towards integration at last? The Sustainable Development Goals as a network of targets. *Sustainable Development*, 23(3), 176–187.

Mietzner, M. (2015). *Reinventing Asian populism: Jokowi's rise, democracy, and political contestation in Indonesia.* Policy Studies 72. Honolulu, HI: East-West Center; Singapore: Institute of Southeast Asian Studies.

Negara, S. D. (2016). Indonesia's infrastructure development under the Jokowi administration. *Southeast Asian Affairs*, 2016(1), 145–165.

Newig, J., & Fritsch, O. (2009). Environmental governance: Participatory, multi-level — and effective? *Environmental Policy and Governance*, 19(3), 197–214.

Norström, A. V., Dannenberg, A., McCarney, G., Milkoreit, M., Diekert, F., Engström, G.... Sjöstedt, M. (2014). Three necessary conditions for establishing effective Sustainable Development Goals in the Anthropocene. *Ecology and Society*, 19(3), Article 8. https://www.ecologyandsociety.org/vol19/iss3/art8/.

Pahl-Wostl, C. (2009). A conceptual framework for analysing adaptive capacity and multi-level learning processes in resource governance regimes. *Global Environmental Change*, 19(3), 354–365.

Purnono, H., Shantiko, B., Sitorus, S., Salim, M., Achdiawan, R., & Gunawan, H. (2015). Ekonomi Politik Kebakaran Hutan dan Lahan: Sebuah Pendekatan Analisis. Presented at the conference: Limited discussion "Preventing Land and Forest Fires", Gedung Manggala Wana Bakti, Jakarta.

Ramakrishnan, P. S. (1999). The impact of globalization on agricultural systems of traditional societies. In A. K. Dragun & C. Tisdell (Eds), *Sustainable agriculture and environment: Globalization and the impact of trade liberalisation* (pp. 185–200). Geneva: FAO.

Rayner, S., & Malone, E. L. (1998). *Human choice and climate change: An international assessment*. Columbus, OH: Battelle Press.

Reid, W. V., Chen, D., Goldfarb, L., Hackmann, H., Lee, Y. T., Mokhele, K., ... Whyte, A. (2010). Earth system science for global sustainability: Grand challenges. *Science*, 330 (6006), 916–917.

Rhodes, R. A. (1997). *Understanding governance: Policy networks, governance, reflexivity and accountability*. Buckingham and Philadelphia, PA: Open University Press.

Rhodes, R. A., & Marsh, D. (1992). New directions in the study of policy networks. *European Journal of Political Research*, 21(1–2), 181–205.

Rietig, K. (2018). The links among contested knowledge, beliefs, and learning in European climate governance: From consensus to conflict in reforming biofuels policy. *Policy Studies Journal*, 46(1), 137–159.

Sachs, J. D. (2012). From Millennium Development Goals to Sustainable Development Goals. *The Lancet*, 379(9832), 2206–2211.

Sachs, J. D. (2015). *The age of sustainable development*. New York, NY: Columbia University Press.

SDSN (2016). Getting started with the SDGs in cities: A guide for stakeholders. UN Sustainable Development Solutions Network. Retrieved from https://sdgcities.guide/.

SDSN (2017). SDG index and dashboards report 2017: International spillovers in achieving the goals. Bertelsmann Stiftung and UN Sustainable Development Solutions

Network. Retrieved from http://www.sdgindex.org/assets/files/2017/2017-SDG-Index-and-Dashboards-Report--full.pdf.

Smit, B., & Wandel, J. (2006). Adaptation, adaptive capacity and vulnerability. *Global Environmental Change*, 16(3), 282–292.

Smith, A., Voß, J. P., & Grin, J. (2010). Innovation studies and sustainability transitions: The allure of the multi-level perspective and its challenges. *Research Policy*, 39(4), 435–448.

Steffen, W., Richardson, K., Rockström, J., Cornell, S. E., Fetzer, I., Bennett, E. M., ... Sörlin, S. (2015). Planetary boundaries: Guiding human development on a changing planet. *Science* 347(6223), 1259855.

Tempo (2015, January 11). Expert: Water tariff in Jakarta highest in South East Asia. Tempo.co. Retrieved from https://en.tempo.co/read/news/2015/01/11/057634142/Expert-Water-Tariff-in-Jakarta-Highest-in-South-East-Asia.

UN (2015). Transforming our world: The 2030 Agenda for Sustainable Development. Sustainable Development Knowledge Platform. Retrieved from https://sustainabledevelopment.un.org/post2015/transformingourworld.

UN Development Group (2016). Indonesia: Promoting inclusive approaches to localize the SDGs. Retrieved from https://undg.org/indonesia-promoting-inclusive-approaches-to-localize-the-sdgs/.

UN News Centre (2015). UN decoded: UN Millennium Development Goals (MDGs). Retrieved from http://www.un.org/apps/news/infocus/UNdecoded/UNdecoded.asp?NewsID=1330&sID=48.

UNCED (1992). Rio Declaration on Environment and Development. New York: United Nations Conference on Environment and Development. Retrieved from http://www.unesco.org/education/pdf/RIO_E.PDF.

UNDP (2015). Converging development agendas: "Nawa Cita", "RPJMN" and SDGs. Jakarta: United Nations Development Programme Indonesia Country Office.

UNICEF (2015). Sustainable Development Goals: All you need to know. UNICEF Zimbabwe. Retrieved from https://www.unicef.org/zimbabwe/media_16207.html.

Warburton, E. (2016). Jokowi and the new developmentalism. *Bulletin of Indonesian Economic Studies*, 52(3), 297–320.

World Bank (2018). Poverty overview. Washington, DC: World Bank. Retrieved from http://www.worldbank.org/en/topic/poverty/overview.

WCED (1987). *Our common future*. Oxford: Oxford University Press for the World Commission on Environment and Development.

Young, O. R. (2017). Conceptualization: Goal setting as a strategy for earth system governance. In N. Kanie & F. Biermann (Eds), *Governing through goals: Sustainable Development Goals as governance innovation* (pp. 31–52). Cambridge, MA: MIT Press.

Zamzami, I., & Ardhianie, N. (2015). An end to the struggle? Jakarta residents reclaim their water system. In S. Kishimoto, E. Lobina & O. Petitjean (Eds), *Our public water future: The global experience with remunicipalisation* (pp. 40–49). Amsterdam: Transnational Institute.

van Zeijl-Rozema, A., Cörvers, R., Kemp, R., & Martens, P. (2008). Governance for sustainable development: A framework. *Sustainable Development*, 16(6), 410–421.

CHAPTER 3

Partnerships for Implementing the 2030 Agenda for Sustainable Development in Lao PDR

Phanthanousone Khennavong

1 Introduction

Like many countries, Lao PDR (hereafter Laos) has put in place a number of tools and institutional arrangements to ensure the implementation and localization of the 2030 Agenda for Sustainable Development (UN, 2015), including the SDGs, through multi-stakeholder partnerships (MSPs). An important task now is determining how these MSPs can best be mobilized, based on the specific country context, so that all stakeholders will buy into the 2030 Agenda and move the SDGs forward. With that task in mind, this chapter asks what lessons can be learned from the experience of Laos — one of the few one-party states in the world — highlighting factors necessary to make MSPs transformative, inclusive, and accountable.

This chapter describes how the 2030 Agenda and the SDGs have helped Laos reshape and enhance the country's interactions with its various stakeholders, moving from a closed door mentality to a more open, transformative, inclusive, and accountable approach. How this has happened lies in the realm of politics. The findings reported in this chapter reinforce the view in the literature that local context and factors are of great importance in realizing the potential of MSPs in the achievement of the SDGs.

The following section outlines the current MSP framework as set out by the United Nations, presents the study questions and documents the methodology. The subsequent section describes the history of MSPs in the context of the national planning process in Laos and discusses the evolution of MSPs in the country. This is followed by a discussion of the factors that have helped promote transformative, inclusive, and accountable MSPs in the Lao context, before some brief concluding remarks are offered.

2 Multi-stakeholder Partnerships for Sustainable Development

At the broadest level, this chapter addresses two questions related to multi-stakeholder partnerships. First: how does the government of a developing

 | DOI:10.1163/9789004391949_004

country manage MSPs for sustainable development? Second: what are the factors that underpin successful partnerships for sustainable development in a developing country? Broadly speaking, there are two strands within the MSP literature. One looks at what needs to be done, and usually takes the region as its unit of analysis; the other looks how things are managed and factors that enhance greater partnerships, and typically takes the country as its unit of analysis. This chapter belongs to the second strand.

The most recent framework guiding MSPs for sustainable development was issued by the United Nations Economic and Social Commission for Asia and the Pacific in March 2018 (UNESCAP, 2018). It offers guidelines that can be applied by every country around the world to promote transformative, inclusive, and accountable MSPs. However, how individual countries have been managing the business of MSPs has, so far, been little documented. Even less is known about how a developing country with a one-party state system, like Laos, manages its MSPs. This chapter therefore adds to the very limited literature on how multi-stakeholder partnerships are managed at the country level.

2.1 *Methodology*

In order to address the questions that this chapter poses, a case study approach was selected. To discover how the Lao government manages MSPs in the country, a descriptive analysis is the most suitable technique. Information included in this chapter is drawn from informal discussions, observation, and analysis of reports, documents, and other related materials. The author is familiar with the issues covered here, firstly through his work as a programme analyst for the United Nations Development Programme (UNDP), managing aid effectiveness and partnership development cooperation in Laos between 2003 and 2004 and between 2006 and 2010; and secondly through his position as a technical advisor to the Lao government since 2014. In that capacity, he has been deeply involved in SDG implementation and localization and in the Lao national partnership for effective development cooperation.

3 Multi-stakeholder Partnerships in Laos

Since the adoption of the 2030 Agenda for Sustainable Development, in 2015, Laos has restructured and revitalized its partnership framework, mechanisms, and relevant arrangements to help ensure SDG implementation and localization, in line with the UN partnership framework. This section describes how Laos has been working with its partners in promoting transformative, inclusive, and accountable MSPs.

In November 2015, Laos and its 32 development partners adopted the Vientiane Declaration on Partnership for Effective Development Cooperation (Government of Lao PDR, 2015). The adoption of this Declaration places a joint responsibility on the government and its partners to transition from the previous aid effectiveness agenda to the new agenda for transformative, inclusive, and accountable MSPs. The government and partners undertake to apply eight principles for effective partnerships in Laos to support the realization of the 2030 Agenda and SDGs: these eight principles are summarized in Table 3.1 and are the focus of this section. Since the signing of the Declaration, MSPs in Laos have been managed according to these principles.

TABLE 3.1 Description of Eight Principles of the Vientiane Declaration in support of the 2030 agenda and SDGs

Principles and action areas	
I. Ownership & II. Alignment	
1.	Greater aligning and tailoring of development finance to the national development agenda and context through results-based planning practices
2.	Strengthening of linkages between national budget and national planning processes
3.	Ensuring Official Development Assistance (ODA) and other official flows are on budgets which are endorsed by the National Assembly (NA)
4.	Within the national regulatory framework, strengthening country systems which are used by providers of development cooperation to the extent possible: budget execution, financial reporting, audit, and procurement
III. Harmonization & Simplification	
5.	Adopting Programme Based Approaches (PBAs) and Sector Wide Approaches (SWAp) to support the implementation of the National Socio-Economic Development Plans (NSEDPs) including SDGs
6.	Under government leadership, providers of development cooperation to coordinate their strategies and actions, simplify procedures vis-à-vis the government, and enhance collaboration to avoid duplication
IV. Inclusive Partnership for Development Results	
7.	Enhancing capacities of the NA, including the newly established Provincial People's Assembly (PPA) to approve and monitor national development plans and budget for greater development results
8.	Advancement of the Round Table Process, moving towards greater partnership for effective development cooperation

TABLE 3.1 Description of Eight Principles of the Vientiane Declaration (*cont.*)

Principles and action areas
9. Reviewing legal and regulatory frameworks for non-profit associations (NPAs) and international non-governmental organizations (INGOs) as necessary and improving engagement of NPAs and INGOs in national and provincial development planning and coordination, including the implementation and reporting of development results in accordance with the national legal and regulatory framework
V. Transparency, Predictability, and Mutual Accountability
10. Ensuring a more predictable and accountable national budget that becomes the principal tool of effective development cooperation
11. Ensuring availability and public accessibility of information on development cooperation and other development resources (commitments, actual disbursements and results)
VI. Domestic Resource Mobilization
12. To the extent possible, ensuring adequate mobilization of government revenues which is required for leveraging development cooperation funds for investments in various development activities
VII. South–South Cooperation, Triangular Cooperation, and Knowledge Sharing
13. Improving understanding of the nature and modalities of South–South cooperation partners and their significant participation in development cooperation management through a clear coordination and reporting mechanism
VIII. Business as a Partner in Development
14. Improving engagement of private sector in national and provincial development planning and coordination, including the implementation, reporting and oversight of development results and outcomes through a proper analytical framework/mechanism

SOURCE: VIENTIANE DECLARATION ON EFFECTIVE DEVELOPMENT COOPERATION (GOVERNMENT OF LAO PDR, 2015)

3.1.1 Creating Ownership of the SDGs

The National Committee on SDG implementation and localization is currently led by the Prime Minister of Laos and includes all ministers in the government, to ensure that effective leadership is exercised over the country's development policies and strategies, in line with the 2030 Agenda and SDGs, and to coordinate nation-wide efforts to build stronger partnerships with various development actors working in the country. To help deliver progress on the SDGs, Laos has integrated the 2030 Agenda on Sustainable Development into its own long-term vision for 2030, as expressed in the National Socio-Economic Development Plans (NSEDPs), published every five years. The period from 2015 to 2030 is covered by three NSEDPs: the 8th NSEDP (2016–2020), the 9th NSEDP (2021–2025), and the 10th NSEDP (2026–2030). Table 3.2 provides some background to the history and evolution of the national planning process in Laos.

TABLE 3.2 The evolution of the National Planning Process in Laos

National Plans	Theme of each National Plan	Comments
1976–1985: Central Planning		
– One-year plan (1976)	Rehabilitation	The plan was prepared to guide the country after Lao People's Republic was proclaimed in 1975
– Three-year plan (1978–1980)	Reconstruction of economy	The plan was a framework for the country to build major infrastructure after the Viet Nam war
– 1st five-year plan (1981–1985)	Strengthening socio-economic development base	This plan was prepared as a preparatory document for opening up the country
1986–Present: Market-Oriented Economic Planning		
– 2nd five-year plan (1986–1990)	Open cooperation with other countries	This was the plan to introduce economic reform (the New Economic Mechanism)
– 3rd five-year plan (1991–1995)	Broader cooperation and investment promotion	This plan was prepared to guide the country in joining ASEAN
– 4th five-year plan (1996–2000)	Implementation of the eight national priority programmes and cooperation with international and regional organizations	The eight priority programmes were: 1) food production, 2) commodity production, 3) stabilization of shifting cultivation, 4) rural development, 5) infrastructure development, 6) expansion of external economic relations and cooperation, 7) human resource development, and 8) services development
2000–2015: The Poverty Reduction Strategy (PRS) Period		
– 5th five-year plan (2001–2005)	Merging national goals with global frameworks within the context of globalization	A ten-year strategy was prepared; this was the first five-year plan to operationalize the ten-year strategy

TABLE 3.2 The evolution of the National Planning Process in Laos (*cont.*)

National Plans	Theme of each National Plan	Comments
– 6th five-year plan (2006–2010)	The introduction of poverty-reduction focus areas	The PRS approach was merged with the five-year planning process in 2005. The World Bank classified the 6th five-year plan (2006–2010) as the second PRS.
– 7th five-year plan (2011–2015)	Accelerating the achievement of national goals in line with the MDGs	A Joint Staff Assessment Note on this five-year plan (and third PRS) was discussed by the World Bank Board on 12 January 2012
2016–2030: The 2030 Agenda and SDG Period		
– 8th five-year plan (2016–2020)	Integration of the Sustainable Development Goals and foundation for graduating from the status of Least Developed Country (LDC)	For the first time in Lao history, the five-year plan was prepared and implemented through the application of a results-based approach and the introduction of multi-stakeholder partnerships for sustainable development through adoption of the Vientiane Declaration on Partnerships for Effective Development Cooperation
– 9th five-year plan (2021–2025)	Under formulation	Initial discussion within the government is to ensure effective implementation of the 2030 Agenda for Sustainable Development, greater effort for SDG achievement, and graduation from LDC status through more effective partnerships
– 10th five-year plan (2026–2030)	Only in discussion at this stage	This will be the last five-year plan for the realization of the national development agenda including SDGs and LDC graduation

Notes: The themes of the Plans are a translation of the way each Plan is described in later Plans, which typically include a note summarizing earlier Plans. The comments are the author's own. The classification of the National Plans into four periods is based on the author's discussions with the Lao government, and reflect the way the government thinks about the history of Laos's planning process.

SOURCES: GOVERNMENT OF LAO PDR (2004, 2006, 2011, 2016)

The integration of the 2030 Agenda and NSEDP has also meant changes for the latter, since the whole planning process now has to be based on an MSP framework. First, this has made the planning process more transparent: the government now shares with donors, provinces, and the public, information about how the NSEDP is formulated, implemented, and monitored. Annual and mid-term progress reports are also publicly available and there is media coverage of the government's annual reports to the National Assembly. Selected donors are also invited to be observers during these National Assembly sessions.

Second, the planning process is now more participatory. During the preparation of the most recent plan — the 8th NSEDP 2016–2020 (Government of Lao PDR, 2016) — three regional consultations were organized for the north, the south, and the central provinces as input into the first draft. Most participants in these regional consultations were representatives from national line ministries and provincial offices. The draft was then shared with the Sector Working Groups (see below) for comments and feedback. This gave not only donors, but also line ministries, provinces, and selected international and national non-governmental organizations (NGOs) an opportunity to comment. A mid-term review of the 8th NSEDP is currently underway, and has applied the same approach as that adopted during the formulation phase.

Third, the planning process is now managed by a national taskforce, rather than by the Department of Planning within the Ministry of Planning and Investment (MPI) as used to be the case. Although still led by MPI, the taskforce for the preparation of the national five-year plan consists of all line ministries, the Ministry of Finance, and the Ministry of Foreign Affairs. Fourth, the content of the plan also changed. Grawe (2010, p. 7) notes that the 2006–2010 NSEDP "took on the poverty reduction metric of the growth strategy, shifting from a traditional input-output framework to an outcome-oriented, policy-based approach". As a result of the 2030 Agenda, the focus of the NSEDPs has shifted again, towards the achievement of the SDGs through the effective use of MSPs.

To further facilitate MSPs at the sectoral level, Laos has created partnership and coordination forums called Sector Working Groups (SWGs). There are currently 10 SWGs covering various key sectors for implementation of the SDGs. The creation of the SWG mechanism has also helped improve collaboration between the government and its partners. Chaired by the relevant government ministries, each SWG now plays a meaningful role in supporting the preparation of the sector development plans which feed into NSEDP formulation, implementation, and review, ensuring that sector priorities are reflected in national plans.

While there have thus been numerous positive developments, challenges remain in instituting broad-based and robust consultation mechanisms. Coordination and partnership structures and processes will need to be made more effective, including through further efforts to widen stakeholder participation to include civil society, South–South partners, and the private sector. While these partners are invited to participate in and contribute to cooperation and planning initiatives and many report good cooperation with the government, especially at the local level, more space could be given to enable these groups to make a meaningful contribution to Lao's SDG-based national development planning.

3.1.2 Alignment of Partner Supports with the Government's 2030 Agenda-Based Priorities

To further enhance the effectiveness of MSPs in the country, the Lao government has called for greater efforts from partner countries to align their support with the government's aims and objectives, as highlighted in it national development plan (NSEDP). In response, 95% of development cooperation reported in the OECD's 2016 survey was found to be aligned to the objectives outlined in the SDG-based NSEDP and sector plans (OECD, 2016). In another effort to enhance MSPs, the Lao government is working toward the improvement of public sector management and initiatives that can increase partner confidence in using national systems, which include a monitoring and evaluation framework, budget execution, financial reporting, auditing and procurement. Some progress has been made in improving the country's policy and institutional capacity: according to the Country Policy and Institutional Assessment (CPIA) by the World Bank,[1] the CPIA score for Laos rose from 2.5 in 2005 to 3.5 in 2015 and around 3.6 in 2017. Comparing to its neighbour Cambodia, which is on a similar development path, Laos has been doing much better: Cambodia's CPIA score rose from 2.6 in 2005 to 2.7 in 2015 and stayed unchanged in 2017.

1 The Country Policy and Institutional Assessment (CPIA) assesses the quality of a country's policy and institutional framework, in terms of how conducive that framework is to fostering poverty reduction, sustainable growth, and the effective use of development assistance. For each criterion, countries are rated on a scale of 1 (low) to 6 (high). An overall rating of 1 corresponds to a very weak performance, and a rating of 6 to a very strong performance. The 16 CPIA criteria are grouped into four clusters: economic management; structural policies; policies for social inclusion and equity; and public sector management and institutions. Country scores reflect a variety of indicators, observations, and judgements that are based on country knowledge within the World Bank, analytic work or policy dialogue, or work done by partners, as well as relevant publicly available indicators (World Bank, 2017).

Nevertheless, the use of Lao national systems by its development partners remains low due to concerns about reliability and about the country's limited capacity in budget execution, financial reporting, and transparency (OECD, 2016). Within the MSP context, the government and its partners have recently agreed to undertake some measures to address this problem: for instance, the use and strengthening of national systems will be placed within the overall context of national capacity development for sustainable outcomes. The Lao government and its partners will jointly assess key country systems, using mutually agreed diagnostic tools. Based on the results of these assessments, a decision could then be taken on the extent to which the partners can use country systems.

3.1.3 Harmonization and Simplification

The Vientiane Declaration of 2015 also includes a commitment among national and international stakeholders to harmonize aid delivery procedures and to adopt common arrangements in order to help reduce duplication of procedures and lower the transaction costs associated with aid management. Currently, efforts to coordinate and facilitate joint missions have mainly taken place within 'like-minded' groups of donors, for example within the European donors' group, the international financial institutions, or UN agencies. Coordination has been based around joint programming and joint portfolio review. In this regard, the integration and localization of the SDGs seems to have helped the donor community to harmonize its approaches to supporting the government's 2030 Agenda-based priorities. Since the adoption of the SDGs in 2015, an informal donor meeting has been organized quarterly, convened by the United Nations Resident Coordinator, to ensure that the donor community gives common and united messages during individual government–donor SWGs and Round Table Meetings. For the current 8th NSEDP (2016–2020), donors prepared joint analytical inputs and policy advice during the formulation and review process.

Furthermore, the relationship between the government and the donor community is not confined to the technical level, but now operates at the political level as well. Since the adoption of the 2030 agenda and SDGs in 2015, government–donor face-to-face interaction to discuss issues relating to the SDG-based national planning process has been taking place at three levels: political, policy, and technical. Considering Laos's closed political system, it is a major shift that senior government officials such as Directors General, Vice-Ministers and Ministers now hold direct dialogue with the donor community and other stakeholders in open forums, and it has improved trust between the two parties.

3.1.4 Mutual Accountability

Mutual accountability is the core principle of the MSPs in Laos. To help ensure that the country is making progress with MSPs, the Lao government — with support from its partners — restructured its national aid coordination and partnership mechanism. From 1983 until 1997, the aid coordination process focused on Round Table Meetings (RTMs) held every three years in Geneva. These meetings were coordinated and managed by UNDP. However, in 1999, at the suggestion of UNDP, the Lao government made aid coordination an in-country process, now called the Round Table Process or RTP. The RTMs continued on their three-yearly cycle (until 2015), but they now met in Laos, chaired by the Lao government with the UNDP acting as co-chair on behalf of donors. With the integration and localization of the SDGs in Laos, the time-frame for RTMs was changed in 2016, moving from every three years to every five years. The purpose of this change is to create a synergy between the coordination/partnership forum and the national planning process. Round Table Implementation Meetings (RTIMs) are now held annually, mainly to report on progress in relation to the Poverty Reduction Strategy (PRS). The RTP also includes various SWGs, established in 2005 to promote coordination in particular sectors. Finally, since 2008, provincial consultations have been organized on an annual basis to extend aid coordination and consultation to provincial officials.

The Round Table Process is the national partnership platform to promote and increase collaboration around the implementation and assessment of progress of national development plans and the timely sharing of information on aid flows. The Lao government commissioned an independent review in 2016, which showed that the process has become more inclusive and confirmed the relevance and effectiveness of the process for further enhancing MSPs in the context of achieving the SDGs in Laos (UNDP and the Government of Lao PDR, 2016).

As part of its efforts to make MSPs more transformative, inclusive, and accountable, the government and its partners have agreed to work closely together to increase the predictability and transparency of development financing. This involves a rethink by both sides as to how development finance should be spent, in ways that are consistent with agreed international rights, norms, and standards, so that different sources of finance catalyse development. At the time of writing, the Department of International Cooperation/Ministry of Planning and Investment — the national aid coordinating agency — has received support from Australia in creating a national aid database to track, monitor, and assess not only aid but various sources of development finance.

There is also a sense that the MSP approach promoted by the 2030 Agenda has helped strengthen the accountability framework in Laos, even though progress is still limited and tentative. The government has not only embraced this as part of NSEDP preparation, but as part of NSEDP monitoring as well. The government has now presents an NSEDP progress report during annual government–donor meetings. Some might criticize this as promoting external rather than internal accountability: however, given the culture of closed policy making, an opening up of the process represents a major change which helps sensitize government officials to the concept of accountability.

3.1.5 Managing for Results

Laos adopted a results-based management framework during the formulation of its current five-year plan (8th NSEDP 20160–2020). With the integration of the SDGs into the national planning process, the country has been making some changes to its policy making. First, the Lao government and its partners have started working closely together towards common development results. At the time of writing, around 60% of the SDG indicators have been integrated into the Lao national Monitoring & Evaluation (M&E) framework. Second, various partnership forums in the country (from policy to technical levels) have been (re)structured around outcomes and outputs of the SDG-based 8th NSEDP. At a policy level, Round Table Meetings and Round Table Implementation Meetings address matters relating to outcomes; at a technical level, the 10 SWGs, informal groups and provincial meetings focus on actions and activities for the achievement of the national development plan. Third, guidelines for a systematic way of reporting have been put in place. The central ministry responsible for MSPs, the Ministry of Planning and Investment, collects periodic reports (monthly, quarterly, semi-annual, and annual) from line ministries, central and subnational agencies (provinces, districts, and villages) to gather information on progress. Summary reports are then produced for submission to the government and the National Assembly, as well as for discussion at MSP forums such the SWGs and presentation at the Round Table Meetings. So far, this practice is proving effective in preventing the use of additional frameworks and the introduction of performance indicators that are not consistent with SDG-based national development priorities.

3.1.6 Domestic Resource Mobilization

Making effective use of MSPs in Laos also means gaining support and mobilizing all forms of resources from domestic partners and citizens. Rules, regulations, and initiatives govern the mobilization of domestic sources of development finance, such as tax reform, budget law and budget norms, public administration, and so on. Various platforms and mechanisms at national and sectoral level seem to be moving in the right direction as indicated above. However, creating transformative, inclusive, and accountable MSPs requires coalition building and local community involvement. A critical task for MSPs in Laos is thus the formulation of simple messages, rational and educational campaigns about various aspects of the 2030 Agenda and the SDGs, that are easy to understand and to visualize by local communities. A bottom-up approach to domestic resource mobilization will be needed in order to help build MSPs that work for sustainable development from local perspectives rather than translating what has been identified as important at the national level.

3.1.7 Participation of Other Stakeholders through Collective Platforms

Multi-stakeholder partnerships for sustainable development within the Lao context also means representation of partners that have not previously been involved in the Round Table Process. One such newly active group, which has a shared development path or development status, but was not part of the traditional donor club, are the so-called South–South partners. A second group is the private sector; a third group of partners is civil society. The Lao government and its existing partners agreed on a framework for involvement of these three groups as part of the Vientiane Declaration in 2015; since then, South–South partners, the private sector, and civil society have been invited to take part in the Round Table Process. However, partnerships with these groups need to go beyond invitation and representation, and to include a substantive role at national and sectoral coordination forums. There needs to be a consistent policy regarding the involvement and contribution of these partners.

4 Factors Underpinning Multi-stakeholder Partnerships in Lao PDR

Why was the MSP approach for sustainable development mainstreamed rather than isolated and, eventually, discarded? The answer lies in the realm of politics. Three things conspired to make the government of Laos embrace the MSP approach for sustainable development.

The first was the adoption by the Party Congress in 2001 of a commitment to graduate from the status of Least Developed Country (LDC) by the 2020s, and the integration of the sustainable development and SDG Agenda into the government's long-term vision to 2030. Through the preparation of the 8th NSEDP (2016–2020) — which is the first results-based five-year development plan for the country and shares various elements of the 2030 Agenda and SDGs — the

Party came to realize that the MSP approach would be a useful instrument to help it to meet its own goals and long-term vision, as the involvement of various partners would open up different sources of development finance for the country.

The second factor leading to the embrace of the MSP approach by the government was the rise to power of a new prime minister, His Excellency Thongloun Sisoulith, in 2016. This Prime Minister was seen as a reformer who represented the new guard. Prior to his appointment as Prime Minister, he had been involved with revitalizing the national planning process and had a general oversight role in relation to donor coordination and partnerships. It is plausible that as Prime Minister, looking for ways to introduce reform of the planning process in a highly change-averse system, he supported the mainstreaming of the MSP approach because reform could then be seen as a response to pressure from the international community. In other words, he could pursue reform without appearing to his government colleagues to be a maverick. A third reason is a strategy of the Lao government to balance and counter the power and influence of some regional partners such as China, Thailand, and Vietnam.

5 Concluding Remarks

The story of multi-stakeholder partnerships in Laos is an integral part of the country's national story. What is the future of MSPs in Laos? At the time of writing, the next five-year plan (9th NSEDP, 2021–2025) is under preparation. As Table 3.2 above indicates, ongoing discussions within the government suggest that the plan will encapsulate elements of the 2030 Agenda for Sustainable Development including full integration of the SDGs as well as Laos's transition out of LDC status. It is thus likely that the picture of MSPs in Laos will come fully into focus in the course of the next few years. The effectiveness of MSPs and their impact in terms of development in Laos remain to be seen. Whatever the future, however, it seems certain that the MSP approach will help to bring new ideas and resources to support the development of the country.

References

Government of Lao PDR (2004). National Growth and Poverty Eradication Strategy (NGPES). Vientiane: Government of Lao PDR.

Government of Lao PDR (2006). National Socio-economic Development Plan (NSEDP 2006–2010). Vientiane: Government of Lao PDR.

Government of Lao PDR (2011). National Socio-economic Development Plan (NSEDP 2011–2015). Vientiane: Government of Lao PDR.

Government of Lao PDR (2015). Vientiane Declaration on partnerships for effective development cooperation (2016–2025). Vientiane: Government of Lao PDR. Retrieved from http://www.la.one.un.org/images/publications/Vientiane-Declaration-II-Final_Signatories-and-Endorsement.pdf.

Government of Lao PDR (2016). 8th National Socio-economic Development Plan (2016–2020). Vientiane: Government of Lao PDR. Retrieved from http://www.la.one.un.org/images/publications/8th_NSEDP_2016-2020.pdf.

Grawe, R. (2010). Poverty reduction support credit: Lao PDR country study. Working Paper 2010/2011. Washington, DC: Independent Evaluation Group, World Bank.

OECD (2016). Monitoring profile: Lao PDR. Paris: OECD DAC.

UN (2015). Transforming our world. The 2030 Agenda on Sustainable Development. New York: United Nations. Retrieved from https://sustainabledevelopment.un.org/content/documents/21252030%20Agenda%20for%20Sustainable%20Development%20web.pdf.

UNDP and Government of Lao PDR (2016). Final report of the review and assessment of the programme of support to the Ministry of Planning and Investment and the Ministry of Foreign Affairs. Vientiane: United Nations Development Programme and the Government of Lao PDR.

UNESCAP (2018). Partnering for sustainable development: Guidelines for multi-stakeholder partnerships to implement the 2030 Agenda in Asia and the Pacific. Bangkok: United Nations Economic and Social Commission for Asia and Pacific.

World Bank (2017). *CPIA Criteria 2017*. Washington, DC: World Bank.

PART 2

Accountability to Citizens and Human Rights to Ensure Progress toward the SDGs

CHAPTER 4

Accountability Challenges to Sustainable Development Goals in Southeast Asia

Julio C. Teehankee

1 Introduction

The Sustainable Development Goals (SDGs), mandated by the United Nations (UN), is an action plan for people, planet, and prosperity that "seeks to strengthen universal peace in larger freedom" (UN, 2015). The 17 sustainable development goals and 169 targets offer a comprehensive template to assess and monitor the multifaceted task of development work in UN member countries. Goal 16 aims at the promotion of "just, peaceful and inclusive societies". Specifically, Target 16.6 seeks to develop effective, accountable, and transparent institutions at all levels. This is the first time that the international community has endorsed a shared vision on the critical role of effective and inclusive institutions in the pursuit of sustainable development (Whaites, 2016).

Southeast Asia is a region characterized by institutional variations of governance. Most governments in the region are held accountable by means of vertical accountability through formal elections, horizontal accountability through institutional restraints, and diagonal accountability through the media and civil society. However, the existence of 'accountability deficits' in one, two or all of these dimensions in some countries continues to be a major challenge to realizing SDG Target 16.6. This chapter delineates the accountability challenges facing the Philippines, Indonesia, Thailand, and Singapore. It argues that the presence of accountability deficits serves as a potential hurdle in fully realizing the SDGs in the region.

2 Effective, Accountable, and Transparent Institutions

The SDGs emerged out of the momentum generated by the Millennium Development Goals (MDGs) in 2015.[1] In order to further operationalize the

1 The United Nations Millennium Declaration, signed by world leaders in 2000, was the historic commitment to pursue eight measurable development goals. After assessing the

 | DOI:10.1163/9789004391949_005

systematic changes that are essential in the attainment of the SDGs, effective governance that will recognize the significance of access rights at all levels was emphasized. Governance plays an integral role in the realization of all the SDGs, particularly in transforming "political commitments into actual realities for people on the ground" (Orellana, 2016, p. 50). Thus, the principles of effective, accountable, and transparent governance must be integrated in the three traditional dimensions of sustainable development (i.e., economic development, social inclusion, and environmental sustainability). Governance, which is about the rules of behaviour in organizations, will play a central role in the success or failure of the SDGs (Sachs, 2015). Hence, SDG 16 "is more than simply a normative agenda, and equally it is more than just technical capacity building. It lays out the central role for institutions, encompassing both their function and their ethos and logically puts politics at the heart of institutions" (Whaites, 2016, p. 2).

2.1 *Mapping the Concept of Accountability*

Although accountability has had a long tradition of academic scholarship in both political science and financial accounting, it only recently came to prominence as a research focus in the fields of public administration and international development (Lindberg, 2013). It has also become a central element to most theories of democracy. With this proliferation of interest, innumerable meanings and conceptualizations have been ascribed to the notion of accountability, including the prescriptions to the global South made by donor agencies and partner governments from the global North in relation to 'good governance' and other factors found in democratic regimes (Bovens, 2007; Dubnick, 2007; Lindberg, 2013).

Accountability is often held to be more or less synonymous with other loosely defined concepts and notions of good governance, such as democracy, transparency, equity, efficiency, integrity, and responsiveness (Bovens, 2007; Dubnick, 2007; Mulgan, 2000).[2] The study of accountability has spurred

---

substantial progress achieved under the MDGs, the United Nations Conference on Sustainable Development spearheaded the process of drafting a new set of sustainable development goals at the Rio+20 Conference held in Rio de Janeiro in 2012. The UN General Assembly Open Working Group (OWG) completed a draft in 2014 that was a product of global consultation online and offline. The draft was approved by the UN General Assembly in September 2015 (UN, 2015).

2 Historically and semantically, however, the concept of 'accountability' was closely related to accounting or bookkeeping (Bovens, 2007; Dubnick, 2007). The shift in the semantic use of the term can be attributed to the introduction of New Public Management in the United Kingdom under Margaret Thatcher and the 'reinventing government' reforms implemented under Bill Clinton in the United States (Bovens, 2007).

a myriad of discourses, definitions, uses, and measurements (Bovens, Goodin, & Schillemans, 2014; Lindberg, 2013; Mulgan, 2014). As a multi-faceted phenomenon, it can be studied from a wide range of theoretical and methodological perspectives (Mulgan, 2014; Patil, Vieider, & Tetlock, 2014). For instance, accountability can be conceptualized as a virtue which represents "a set of standards for the behavior of actors" (Bovens, 2010, p. 949) or as a mechanism which "involves an obligation to explain and justify conduct" (ibid., p. 951). Schedler defines it as "the ability of institutions or mechanisms to ensure that officials are answerable for their behavior and are sanctioned for misbehavior" (Schedler, 1997, as cited by Arugay, 2005, p. 64). Some have also studied the interconnectedness of democracy and accountability (Mulgan, 2014; Schmitter, 2004). Lindberg (2013) proposes the conceptual organization of accountability into three dimensions: the *source* of the accountability relationship; the *degree of control* that the principal exercises over the power holder; and the *spatial direction*, which could be vertical upwards, vertical downwards, or horizontal.

One of the main issues in accountability studies is the notion of 'accountability deficits' which are particularly associated with the absence of political control over democratically elected representatives. Mulgan (2014, p. 545) identifies two governmental areas where the deficit arguably exists: "the international sphere which lacks any clear overriding political authority that could undertake the obligations of accountability and ... certain aspects of the domestic government where executive leaders have decoupled themselves from responsibility and accountability through mechanisms such as outsourcing and the establishment of executive agencies". He further argues that the predicament of accountability deficits may stem from a simplistic assumption about the connection between democracy and public accountability. In reality, public accountability is much more complex than the principal–agent model suggests: a whole panoply of accountability mechanisms does not automatically guarantee citizen approval of government actions. From this perspective, the public sector's efforts to identify accountability mechanisms and governance measures are seen as a response to popular demands rather than as an institutionalized and complex set of rules that should be translated into governmental action. This study looks at the issue of accountability deficits in relation to the compliance or observation of accountability mechanisms among selected Southeast Asian states in support of the SDG Target 16.6.

2.2 *Measures of Accountability*

Schmitter (2004, p. 47) identifies accountability as the "relationship between two sets of persons or (more often) organizations ... [an] exchange of responsibilities and potential sanctions between rulers and citizens, made all the more complicated by the fact that a varied and competitive set of representatives typically interposes between the two". While this definition renders any type of regime subject to the concept of accountability, it highlights that, in a democracy, the relationship between ruler and citizen has been complicated by the obtrusive existence of "representatives". Schmitter further asserts that measuring the effectiveness of political accountability is a complex process — more complex than, for example, a presidential impeachment or resignation. In fact, the most accountable leaders are those who have never experienced or faced the threat of such measures.

In order to address the shortcomings of existing approaches, Lührmann, Marquardt, & Mechkova (2017) suggest the use of a "novel way" to effectively conceptualize and systematically measure public accountability. Following Schedler (1999) and Lindberg (2013), accountability is defined "as constraints on the government's use of political power through requirements for justification of its actions and potential sanctions" (Lührmann et al., 2017, p. 4). Lührmann et al. organize sub-types of accountability according to the positions of groups of actors in relation to the common distinction between vertical, horizontal, and diagonal components of accountability. They use the Varieties of Democracy Institute (V-Dem) data set to create accountability indices based on an unprecedented cross-national and temporal coverage that represents a precise model of the conceptual notions of their research.[3]

Thus far, there are only two official indicators recognized by the UN Statistical Commission for SDG Target 16.6, namely: (1) primary government expenditures as a percentage of the originally approved budget; and (2) percentage of recommendations to strengthen national anti-corruption frameworks implemented. While these indicators are important, they do not capture how accountable and transparent state institutions actually are. In order to provide a more comprehensive picture, it is necessary to demonstrate how accountability can be realized through stable mechanisms that hold decision makers responsible for their actions (Andersson, Mechkova, & Lührmann, 2015, p. 2).

The V-Dem data set offers complementary measures that look beyond formal processes and investigate how these processes are actually put into practice by incorporating the following dimensions of accountability: (1) *vertical accountability* captures the mechanisms of formal political participation in

3 The Varieties of Democracy Institute (V-Dem) data set covers 174 countries and dependent territories from 1900 to 2012 and provides an estimate of measurement reliability for each rating. In relation to the Sustainable Development Goals, particularly SDG Target 16.6, the V-Dem data set "provides precise supplementary indicators for all democracy-related SDG targets" (Andersson, Mechkova, & Lührmann, 2015, p. 2).

the exercise of accountability (i.e., clean elections, percentage suffrage, chief executive, parties); (2) *horizontal accountability* represents the extent to which state institutions (judiciary, legislature, and other bodies) are able to hold to account the executive branch of the government (head of state, head of government and cabinet ministers); (3) *diagonal accountability* represents the extent to which citizens are able to hold a government accountable outside of formal political participation (i.e., media freedom, civil society organizations, freedom of expression, engaged society). The following sections delineate the accountability challenges facing Southeast Asia.

3 Accountability Challenges in Southeast Asia

A "new accountability agenda" is currently emerging in Southeast Asia (Goetz & Jenkins, 2005, as cited by Rodan & Hughes, 2014, p. 1). The establishment of accountability mechanisms within the state includes setting up human rights commissions, special auditing bodies, competitive commissions, and environmental monitoring agencies, as well as expanded scope for administrative and judicial powers. Outside the state, accountability initiatives include active civil society participation in corruption watchdogs, establishing transparency networks, formulating labour standards, and promoting corporate ethics, election monitoring, and consumer protection (Rodan & Hughes, 2014).

However, the outcomes of these accountability initiatives have varied for different countries in the region. The weakness of accountability in Thailand, Indonesia, and the Philippines, for example, can be traced to "the mutual reinforcement between flawed elections, weak institutional oversight agencies, and an unclarified division of power between the national and subnational governments" (Zeigenhain, 2015, p. 2). These countries have gone through their respective democratization processes but have moved forward only slowly, stagnated, or even deteriorated as a result of the failure of accountability mechanisms to work in actual practice (Zeigenhain, 2015). The collapse of democracy in Thailand, the rise of populist illiberalism in the Philippines, and the continuing threat of authoritarian revival in Indonesia clearly demonstrate the ill effects of accountability deficits in the region.

While the concept of accountability has been closely linked to liberal democracy, this relationship is not automatic. The current trend towards pursuing accountability in the region is not only strong among nascent liberal democracies, but also within authoritarian and post-authoritarian regimes (Rodan & Hughes, 2014). Thus, "Southeast Asia provides an excellent opportunity for examining the utility and contestability of accountability ideas across different regime types, given it contains a complex mix of authoritarian and democratic systems where the concept of accountability is assuming growing importance in politics" (ibid., p. 2). Singapore, for example, has privileged 'good governance' over democracy. The city-state emphasizes economic development and the establishment of a strong state anchored in the rule of law at the expense of democratization (Verweij & Pelizzo, 2009). In Southeast Asia, the underlying ideology in accountability politics is moral. As Rodan and Hughes (2014, p. 3) argue "struggles over accountability reform are more influenced by moral conservatism than democratic assertiveness. Morally conservative approaches to accountability may be promoted, often in alliance with liberal approaches, to challenge or co-opt incipient democratic movements". The following sections assess the vertical, horizontal, and diagonal accountability dimensions in the Philippines, Indonesia, Thailand, and Singapore.

4 Philippines

The Philippines takes pride in being one of the oldest democracies in the Asia-Pacific region. However, the quality of its democracy has been found wanting. Despite its rich democratic tradition, the country has not been able to reduce centuries-old socio-economic inequalities that have sustained clientelist politics and morphed into an "elite democracy" (Timberman, 2016). The USA transplanted its own constitutional model and design to its former colony through the promulgation of the 1935 Philippine Constitution. In 1946, the Philippines gained independence after 48 years of American colonial rule. The collapse of the 1935 Philippine Constitution in 1972 ushered in 14 years of authoritarian rule under Ferdinand Marcos. The current (1987) Constitution — the country's fourth within 100 years — restored institutional continuity with the 1935 Constitution. The continuing failure to consolidate and deepen democracy has impacted on the quality of governance (De Dios & Hutchcroft, 2003).

The Philippine's democratic transition has been marked by accountability deficits which have hobbled its ability to capitalize on its democratic dividends. As Arugay (2005, p. 81) asserts, the "challenge for Philippine democracy is how to address the accountability deficit of the country's institutions and at the same time balancing it with the emergence of societal accountability carried by a relatively powerful civil society". These deficits are manifested vertically through a flawed electoral system, horizontally by a strong presidency, and diagonally by an empowered civil society (see Figure 4.1).

Several accountability institutions established in pursuit of democratic consolidation have been created and challenged in the post-authoritarian

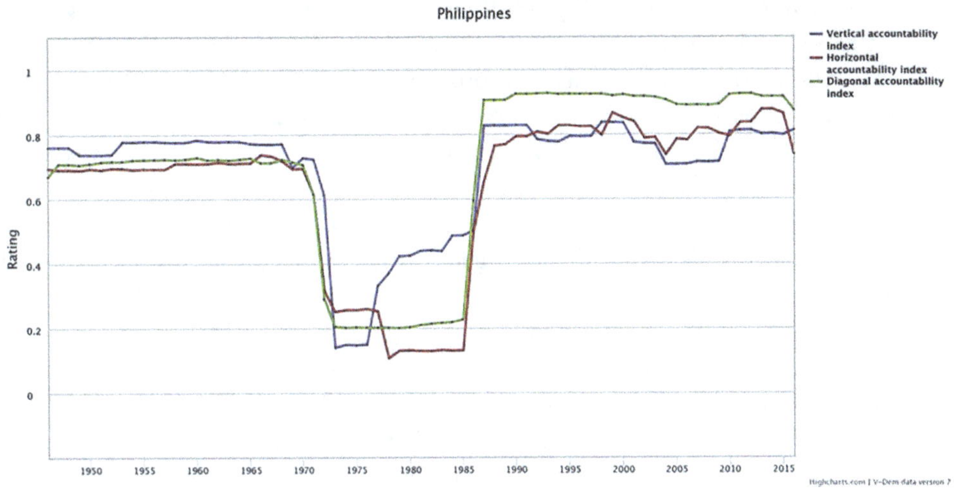

FIGURE 4.1 Accountability measures for the Philippines (1946–2016)
SOURCE: VARIETIES OF DEMOCRACY DATA SET (2018)

period. The re-introduction of democratic elections, for example, was supposed to serve as a mechanism for vertical accountability. However, despite a long history of electoral politics dating back to the American colonial period, elections have failed to ensure public accountability. As Figure 4.1 illustrates, vertical accountability through elections and political parties scored higher in the post-war period (1946–1972), dropped dramatically during the authoritarian period (1972–1986) due to electoral manipulation and the rise of a dominant state party, and climbed again with the restoration of democratic, albeit flawed, elections since 1986. The Commission on Elections — mandated to be "the prime institution for managing elections" in the Philippines — has been "blamed for the inability of elections to reflect the will of the people" (Arugay, 2005, pp. 70). Allegations of electoral fraud and incidents of election-related violence have hounded Philippine elections since the return to democracy in 1986.

In terms of horizontal accountability, the 1987 constitution restored the formal separation of powers among the three branches of government after Marcos had exercised both executive and legislative powers under the 1973 constitution. Yet, although the current constitution clearly delineates the executive from the legislature and judiciary, the Philippine chief executive — as a political institution — has been rendered enough constitutional power to be characterized as a "strong presidency", in an echo of the 1935 constitution, but apparently not enough to totally control strategic interests in Philippine society. The lack of horizontal accountability in state institutions and limited electoral vertical accountability strengthen the president through patronage control of legislators and other state agencies (ibid., p. 66). Figure 4.1 shows that horizontal accountability, measured through the capacity of legislatures and judiciaries to hold the government to account, scored low during the post-war period (1946–1972) and erratically in the post-authoritarian period after 1986. While practically all the post-Marcos presidents have been embroiled in political scandals and controversies, not one has been held accountable through the courts. One president, the movie-actor-turned-populist politician Joseph Estrada, was impeached but his trial was aborted and he was instead ousted extra-constitutionally through a military-backed 'people power' uprising.

In terms of diagonal accountability, the Philippine state continues to face a strong and vibrant civil society that includes: politically active religious groups; business groups organized as various chambers of commerce, cosmopolitan business clubs and other social associations; several labour organizations; and thousands of non-governmental organizations (NGOs) that have emerged over the years (Miranda, 1992). Historically, the media has played a big role in the Philippines. Traditionally, the country's print and broadcast media have challenged censorship; investigative journalists have used their skill to expose corruption and demand accountability. And while there is a tradition of media freedom, there is also a dark side: the Philippines ranks high in terms of violence against those working in the media (Coronel, 2003), while several presidential administrations have attempted to curtail media freedom. More recently, the freedom of the press enjoyed by the traditional media companies has been challenged by the widespread use of social media. In investigating the rise of troll accounts and fake news in the Philippines, Ong and Cabañes (2018, p. 1) report: "[n]o technology has been weaponized at such an unprecedented global scale as social media. Diverse research approaches now attempt to decipher how laptop screens and smartphones around the world are used to manipulate public debate, hijack mainstream media agenda, and influence political outcomes".

5 Thailand

Since it became a constitutional monarchy in 1932, Thailand has undergone a tumultuous period of coups and constitutions. The turbulent trajectory of Thai politics can be traced to "the preponderance of military and civil bureaucrats in political office [with] roots in the [modernizing] military and civilian bureaucratic elite which overthrew the absolute monarchy in 1932" (Wilson, 1932, as cited in Leftwich, 1995, p. 406). Political elites use a continuous cycle

FIGURE 4.2 Accountability measures for Thailand (1932–2016)
SOURCE: VARIETIES OF DEMOCRACY DATA SET (2018)

of elections and constitutional reforms to legitimize whatever regime they put in place. Thus, constant constitution making is simply part of consolidating the elite's political power and dissipating opposition voices (Dressel, 2009, p. 297). Surprisingly, perhaps, significant accountability mechanisms have been institutionalized within the Thai state through its numerous constitutions. Figure 4.2 illustrates the cycle of development and decay of vertical accountability mechanisms in Thailand. This trend oscillates with the trajectory of horizontal accountability mechanisms. A steady growth of horizontal accountability mechanisms can also be observed in the figure.

From 1932 to 1992, Thailand was ruled exclusively by bureaucratic elites, whose political struggles usually occurred within the centralist state and whose dominance was fortified by frequent military interventions and extended authoritarian rule (Zeigenhain, 2015). In terms of vertical accountability, this timespan can be subdivided into a period of military and bureaucratic control of political parties (1946–1988); an era of rural capitalists and network politicians (1988–2000); business conglomerate control over political parties (2001–2014); and the return of military intervention (2014 to present) (Sawasdee, 2006).

A pro-democracy uprising in 1992 brought a temporary end to 60 years of military interventionism in Thai politics. This period marked the steady growth of diagonal accountability in Thailand, as reflected in Figure 4.2. The 1997 Constitution was a product of the opening of democratic space. Drafted by a political reform coalition (composed of civil society reformers, the Left, and the Bangkok middle class) that challenged the clout of the military and the bureaucracy, it reflected the aspiration of Thailand's urban middle class and its liberal elite for greater democracy and good governance (Haggard, 2000). Then came the 1997 Asian financial crisis, and thereafter the meteoric rise of populist billionaire Thaksin Shinawatra to power. As prime minister, Thaksin enjoyed resounding electoral victories in 2001 and 2005. Moreover, he governed with an unprecedented single party dominance and executive stability. Ironically, Thaksin was the beneficiary of institutional reforms under the 1997 Constitution designed to strengthen horizontal accountability. According to Zeigenhain (2015 p. 177), the "drafters of the 1997 Constitution attempted to strengthen the executive power for stable governments and at the same time provide checks and balances to control government action. The position of the prime minister was strengthened in comparison to previous constitutions in order to provide more stability". However, Thaksin's strong leadership style, populism, and rumoured corruption threatened the royal networks and the Thai elite. In the end, he was ousted by the urban middle class, supported by the traditional elites and the military, in the name of "good governance" (Thompson, 2016).

Although provisions inserted in the military-sponsored 2007 Constitution thwarted the return of Thaksin himself to power, the former prime minister continued to win elections by renaming his party and fielding his proxies — including his sister, Yingluck Shinawatra, who was elected prime minister in 2011. In 2014, the Thai Constitutional Court dismissed Yingluck and several senior ministers based on the illegal transfer of the Chief of National Security (Bjarnegård & Melander, 2014). To prevent renewed confrontations between the pro-Thaksin supporters ("Red Shirts") and the royalist middle class ("Yellow Shirts"), the Thai military, headed by General Prayuth Chan-ocha, declared a coup and assumed political power under the junta named National Council for Peace and Order (McCargo, 2014). The 2006 and 2014 coups effectively rolled back the democratic gains made in Thailand since 1992.

6 Indonesia

In the aftermath of the 1997 Asian financial crisis, Indonesia became the last country to ride the crest of the democratization wave that swept the region. In 1998, pro-democracy demonstrators successfully forced the resignation of Indonesian strongman Suharto after 32 years in power. Since the inception of democracy in post-Suharto Indonesia, enforcement of public accountability has remained a pressing issue in the world's largest Islamic nation.

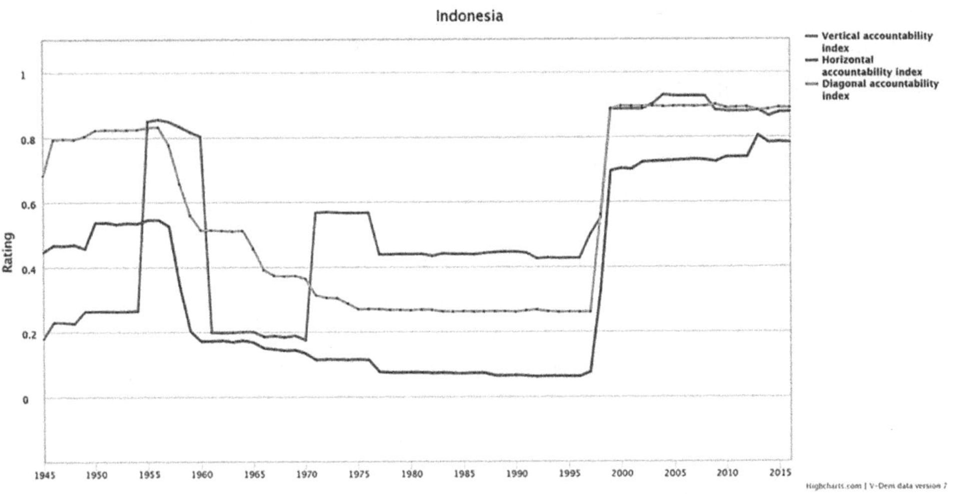

FIGURE 4.3 Accountability measures for Indonesia (1945–2016)
SOURCE: VARIETIES OF DEMOCRACY DATA SET (2018)

Figure 4.3 illustrates the trajectory of accountability mechanisms in Indonesia, roughly corresponding with the country's varying political regimes: Sukarno's 'Guided Democracy' (1957–1965); Suharto's 'New Order' (1965–1999); and the 'Reformasi' period (1999 to present). Despite the emergence of a discourse on good governance, accountability, and transparency of public institutions in response to the Suharto regime, the political configuration in Indonesia has not succeeded in making a complete break with the New Order (Antlov, Ibrahim, & van Tuijl, 2006).

It has been noted that the weakest link in Indonesia's democratization is its ineffective accountability (Tan, 2006). Indeed, "Indonesia's most important democratic deficit is its accountability deficit, and the responsibility for this deficit lies with the party elites who have forged it through their own deficient political strategies" (Slater, 2014, p. 293). Low levels of party institutionalization have contributed to accountability deficits and unstable governance. Party cartelization co-opts all major parties into a national alliance, stunting the development of a viable opposition party and constricting vertical accountability.

Institutions established within the Indonesian state, such as Komisi Pemberantasan Korupsi (KPK), and the Judicial Mafia Eradication Task Force, were intended to serve as the state's apparatus to ensure public accountability. However, these institutions appear to be hotbeds of elite contestation; Rodan and Hughes (2014, p. 154) describe them as "tolerable compromises for elite actors". This has been manifested in the numerous corruption scandals that have rocked Indonesia and that led to "the government and the major political parties act[ing] in concert to initiate regressive measures, with both sharing a common interest either in lessening scrutiny of their actions, particularly the risk of malfeasance being exposed, or in expanding their influence over political processes" (Fealy, 2011, p. 338).

Such conditions can severely restrict the development of accountability within a state. In terms of vertical accountability, this can be manifested through the limitation and cooptation of oppositional groups within smaller and/or non-compliant political parties. Power-sharing mechanisms between elites, in this context, imply a loss of choice for the electorate in challenging party cartels within formal political processes such as elections, and other related processes. In this way, "the public willingness of all parties to consider power-sharing alliances with all other parties means that Indonesia's voters can never be confident that a vote for one party means a vote against any other" (Slater, 2018, p. 32).

This situation — characterized by Slater and Simmons (2013) as "promiscuous power sharing" — creates "arrangements [that] prove so comprehensive as to make a mockery of the putative partisan differences that shaped voter choice" (ibid., p. 1367). The small pool of possible candidates and/or parties for the electorate to choose or support limits the voters' agency within the electoral process. This also affects diagonal accountability, for smaller parties will have a reduced chance of surviving in the political arena, given the muddled and borderless dimension of electoral contestation. The hazy boundaries between parties and policies limit the opportunities for mobilizing opposition. The only accountability that does exist in this context is horizontal accountability between political elites, which are able to check and respond to mechanisms among contending elite factions through both informal and formal means (O'Donnell, 1994, in Slater, 2018, p. 29).

Ellis (2005, as cited by Croissant and Laurenz, 2018, p. 81) posits that "constitutional amendments after the downfall of Suharto strengthened the system of horizontal accountability … but hardly reduced the constitutional powers of the presidency". This is highly problematic, considering that power-sharing measures combined with pliable institutions further entrench elites through their use of these institutions for their own benefit. Examples of this can be found in the use of accountability-checking institutions as an avenue for elite bargaining, as well as in the dilution of the institutions' capacity to pursue officials beset with corruption scandals. Institutions such as the KPK, among many others, are constantly challenged by elites through agency-removing measures (Rodan & Hughes, 2014, p. 155). Thus, vertical accountability is weakened by the political elite's ability to bend institutions to serve its own ends. Finally, in addition to pliable cartels of coalition parties, the massive commercialization of electoral politics has had negative implications for accountability and representation (see Fionna & Tomsa, 2017).

6 Singapore

Singapore has been heralded as the paragon of the Weberian (if not Confucian) administrative state in the region. Quah (2001, p. 316) describes "governance Singapore-style" as follows: "[a]ccountability, transparency, predictability, and participation are the basic elements of good governance in Singapore, but the level of political participation is low". Singapore's success is attributed to meritocracy and the integrity of its elite. Hence, "a meritocratic elite largely obviates the need for political accountability beyond general elections — including through meaningful horizontal accountability" (Rodan & Hughes, 2014, pp. 75–76). According to Case (2011, p. 57), "[t]he government curbs freedom of speech and the press through its vigilant deployment of the judiciary and its tireless lodging of defamation suits. It also limits freedom of assembly". Thus, the country's seemingly authoritarian set-up has been variously described as "soft", "electoral", or "competitive".

Singapore has been ruled by the People's Action Party (PAP) since 1959. Rodan (2009, p. 191) suggests that, "the merger of state and party has seen over time the emergence and consolidation of a virtual class of politico-bureaucrat state capitalists with tightly shared world views and interests". The system of single-party dominance in Singapore has proven to be both boon and bane for the institutionalization of accountability mechanisms for the city-state (see Figure 4.4). Quah (2008, p. 18) asserts, "Singapore has been able to ensure good governance by emphasizing accountability and administering reforms successfully". The first step that the PAP government took on assuming power in 1959 was to systematically improve the Singapore Civil Service. A British practice that the PAP continues to use is the reliance on merit "ensuring that recruitment and promotion are based on merit and not patronage" (ibid., p. 24). Aiming to attract the "best and the brightest" to work in the public sector, "the PAP government has raised the salaries of civil servants and political leaders from 1972 onwards" (ibid., p. 27). To safeguard good governance, PAP leaders began an extensive anti-corruption scheme in 1960 by ratifying the Prevention of Corruption Act (POCA) and strengthening the Corrupt Practices Investigation Bureau (CPIB). The PAP government has sought to maintain the effectiveness of the POCA by introducing amendments whenever necessary (1963, 1966, 1981, and 1989), and by enacting a revised version of the law in 1989 to eliminate loopholes and manage unforeseen issues.

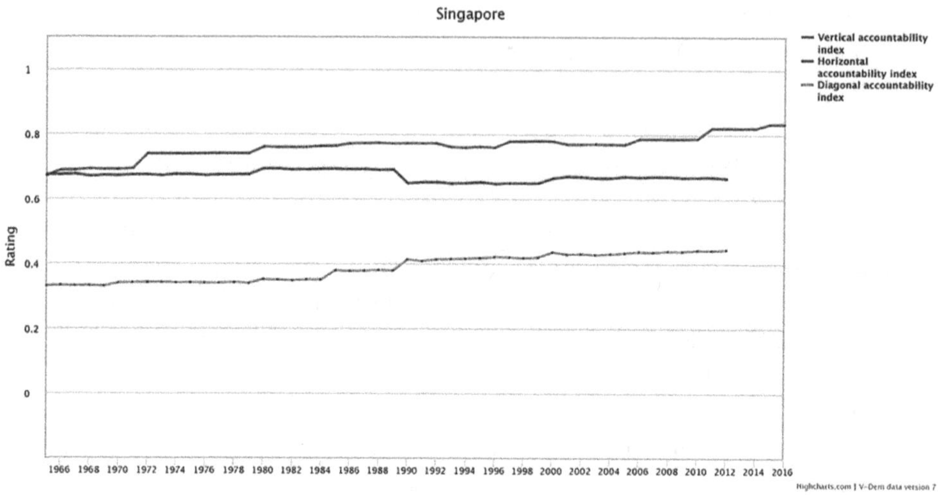

FIGURE 4.4 Accountability measures for Singapore (1965–2016)
SOURCE: VARIETIES OF DEMOCRACY DATA SET (2018)

Singapore has emphasized good governance and meritocracy over democracy and human rights. As Rodan (2009, p. 182) argues, the foundational myth for Singapore's success "is grounded in the meritocratic nature of its elite and the integrity with which the elite is imbued … Rather than concessions to the idea of institutionalising checks and balances on state power, the PAP insists that the recruitment of similarly endowed elites is what secures good governance in Singapore. It makes no exception for human rights". Hence, the legitimation of the ruling party is antithetical to the premise of horizontal political accountability. The PAP was able to short-circuit vertical accountability by its "marriage with the state" through a durable coalition between political elites and civilian bureaucrats (Slater, 2012, p. 23). In effect, for more than half a century, the PAP has succeeded in branding itself as the only party capable of running Singapore and has continuously asserted that "Western-style democracy" will lead to chaos, as it supposedly has in other Asian countries (Ortman, 2011, p. 156). Moreover, the PAP "embraces a very centralized concept of horizontal political accountability" (Rodan & Hughes, 2014, p. 77).

Clearly, diagonal accountability in Singapore is seriously underdeveloped compared to its vertical and horizontal accountability mechanisms (see Figure 4.4). Verweij and Pelizzo (2009, p. 19) enumerate the symptoms of this:

> [T]he jailing and bankrupting of opposition leaders; the engineered sacking of critical commentators; the withholding of state funds from opposition wards and the redrawing of their boundaries (as well as other, more

> creative forms of gerrymandering); the manipulation of election schedules to deprive the opposition of time to campaign; the restriction of political debate to officially registered parties; the placement of ambiguous limits on any form of public discourse; the curtailing of media coverage of opposition parties; and so on.

Stephan Ortmann (2011, 2015a, 2015b) argues that Singapore is undergoing gradual political transformation. The impetus for this trend is the internet which has enabled civil society activists to organize around issues such as human rights, the condition of migrant workers, and the implementation of the death penalty. The 2011 and 2015 elections have also seen some minor victories for the opposition which are indicative of a decline in the PAP's hegemonic position. These developments underscore that "[t]he majority of Singaporeans still wants a responsive government with sufficient checks on arbitrary power. The ruling party now has to be much more responsive to popular desires and quirks than in the past and can no longer act according to what it claims to be the long-term interests of the country" (Ortmann, 2015a, p. 197).

7 Addressing the SDG Challenges in Southeast Asia

The primary challenge to SDG governance goals in Southeast Asia is to overcome the limitations of current accountability indicators, namely: (1) primary government expenditures as a percentage of the originally approved budget; and (2) the percentage of recommendations to strengthen national anti-corruption frameworks that are implemented. As this chapter has discussed, there is a need to look beyond the formal processes and to delineate the vertical, horizontal, and diagonal components of accountability mechanisms in the region. This approach, accompanied by a more robust set of measures such as the V-Dem data set, can provide a more nuanced account of the challenges of accountability in Southeast Asia (see Figure 4.5).

A dual trajectory of public accountability can be observed in the four countries of Southeast Asia discussed in this chapter. The first is essentially political (liberal and democratic) that asserts the importance of participation, inclusiveness, and democracy. A second trajectory underscores moral accountability that promotes values, integrity, and meritocracy. The former is evident in countries that underwent democratic transitions from authoritarian rule (Philippines, Thailand, and Indonesia), while Singapore's authoritarian regime tends to emphasize moral accountability. These trajectories, however, do not preclude the shift of emphasis from moral to political accountability and vice

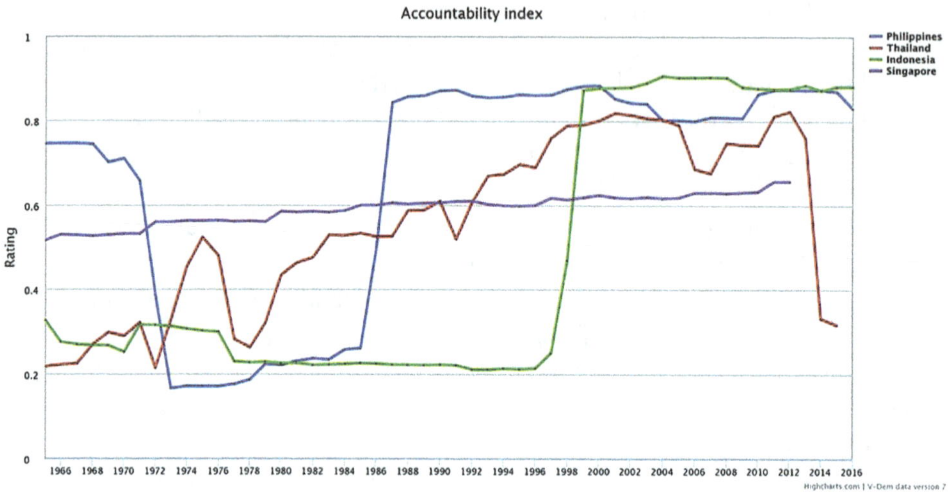

FIGURE 4.5 Comparative accountability measures: Philippines, Thailand, Indonesia, and Singapore
SOURCE: VARIETIES OF DEMOCRACY DATA SET (2018)

versa in these countries. As Rodan & Hughes (2014, p. 179), astutely observe, "regardless of whether the push for accountability employs democratic or non-democratic ideologies that reflect genuine beliefs or tactical considerations, the political effects are in all instances real and consequential".

Acknowledgement

The author would like to acknowledge the assistance of Kevin Agojo, Alain Villegas, Julianna Javellana and Kirk Patrick Pablo in the preparation and writing of this chapter.

References

Andersson, F., Mechkova, V., & Lührmann, A. (2015). Measuring accountability in sustainable development target 16.6 with V-Dem data. V-Dem Institute Policy Brief No. 1. Retrieved from https://www.v-dem.net/media/filer_public/cc/20/cc2025e0-9580-41d1-a2bd-37e399b22fd1/v-dem_policybrief_1_2015_v3.pdf.

Antlöv, H., Ibrahim, R., & van Tuijl, P. (2006). NGO governance and accountability in Indonesia: Challenges in a newly democratizing country. In L. Jordan & P. van Tuijl (Eds.), *NGO accountability: Politics, principles and innovations* (pp. 147–163). London: Earthscan.

Arugay, A. (2005). The accountability deficit in the Philippines: Implications and prospects for democratic consolidation. *Philippine Political Science Journal*, 26(49), 63–88.

Bjarnegård, E., & Melander, E. (2014, May 22). Thailand's missing democrats: Reds, yellows, and the silent minority. *Foreign Policy*. Retrieved from http://www.foreignaffairs.com/articles/141454/elin-bjarnegard-and-erik-melander/thailands-missing-democrats.

Bovens, M. (2007). Analysing and assessing accountability: A conceptual framework. *European Law Journal*, 13(4), 447–468.

Bovens, M. (2010). Two concepts of accountability: Accountability as a virtue and as a mechanism. *West European Politics*, 33(5), 946–967.

Bovens, M., Goodin, R. E., & Schillemans, T. (Eds). (2014). *The Oxford handbook of public accountability*. Oxford: Oxford University Press.

Case, W. (2011). *Executive accountability in Southeast Asia: The role of legislature in new democracies under electoral authoritarianism*. Honolulu, HI: East West Center.

Coronel, S. (2003). The role of the media in deepening democracy. *NGO Media Outreach*. Retrieved from http://unpan1.un.org/intradoc/groups/public/documents/un/unpan010194.pdf.

Croissant, A., & Lorenz, P. (2018). *Comparative politics of Southeast Asia: An introduction to governments and political regimes*. Cham: Springer International Publishing.

De Dios, E. S., & Hutchcroft, P. D. (2003). Political economy. In A. M. Balisacan & H. Hill (Eds), *The Philippine economy: Development, policies, and challenges* (pp. 45–73). Oxford: Oxford University Press.

Dressel, B. (2009). Thailand's elusive quest for a workable constitution, 1997–2007. *Contemporary Southeast Asia*, 31(2): 296–325.

Dubnick, M. J. (2007). Situating accountability: Seeking salvation for the core concept of modern governance. Unpublished manuscript, University of New Hampshire, Durham.

Fionna, U., & Tomsa, D. (2017). Parties and factions in Indonesia: The effects of historical legacies and institutional engineering. ISEAS Working Paper No. 2017. Retrieved from https://www.iseas.edu.sg/images/pdf/WorkingPaper2017_1.pdf.

Fealy, G. (2011). ï . *Bulletin of Indonesian Economic Studies*, 47(3), 333–353.

Haggard, S. (2000). *The political economy of the Asian financial crisis*. Washington, DC: Institute for International Economics. Retrieved from https://piie.com/bookstore/political-economy-asian-financial-crisis.

Leftwich, A. (1995). Bringing politics back in: Towards a model of the developmental State. *The Journal of Development Studies*, 31(3), 400–427.

Lindberg, S. I. (2013). Mapping accountability: Core concept and subtypes. *International Review of Administrative Sciences*, 79(2), 202–226.

Lührmann, A., Marquardt, K. L., & Mechkova, V. (2017). Constraining governments: New indices of vertical, horizontal and diagonal accountability. V-Dem Working

Paper 2017:46. Retrieved from https://ssrn.com/abstract=2956464 or http://dx.doi.org/10.2139/ssrn.2956464.

McCargo, D. (2014, May 29). Thailand's army tears up the script. *New York Times*. Retrieved from https://www.nytimes.com/2014/05/30/opinion/thailands-army-tears-up-the-script.html.

Miranda, F. (1992). Leadership and political stabilization in a post-Aquino Philippines. *Philippine Political Science Journal*, 17, 33–6, 142–222.

Mulgan, R. (2000). Accountability: An ever-expanding concept?. *Public Administration*, 78(3), 555–573.

Mulgan, R. (2014). Accountability deficits. In M. Bovens, R. E. Goodin & T. Schillemans (Eds), *The Oxford handbook of public accountability* (pp. 545–559). Oxford: Oxford University Press.

Ong, J. C., & Cabañes, V. A. C. (2018). Architects of networked disinformation: Behind the scenes of troll accounts and fake news production in the Philippines. The Newton Tech4dev Network. Retrieved from http://newtontechfordev.com/wp-content/uploads/2018/02/ARCHITECTS-OF-NETWORKED-DISINFORMATION-FULL-REPORT.pdf.

Orellana, M. (2016). Governance and the sustainable development goals: The increasing relevance of access rights in principle 10 of the Rio declaration. *Review of European, Comparative & International Environmental Law*, 25(1), 50–58.

Ortmann, S. (2011). Singapore: Authoritarian but newly-competitive. *Journal of Democracy*, 22(4), 153–164.

Ortmann, S. (2015a). Singapore 2011–2015: A tale of two elections. *Asia Maior XXVI*. Retrieved from https://www.asiamaior.org/the-journal/asia-maior-vol-xxvi-2015/singapore-2011-2015-of-two-elections.html.

Ortmann, S. (2015b). Political change and civil society coalition in Singapore. *Government and Opposition*, 50(1), 119–139.

Patil, S., Vieider, F., & Tetlock, P. (2014). Process versus outcome accountability. In M. Bovens, R. E. Goodin & T. Schillemans (Eds), *The Oxford handbook of public accountability* (pp. 69–89). Oxford: Oxford University Press.

Quah, J. T. (2001). Singapore: Meritocratic city-state. In J. Funston (Ed.), *Government and Politics in Southeast Asia* (pp. 291–327). Singapore: Institute of Southeast Asian Studies.

Quah, J. T. (2008). Good governance, accountability and administrative reform in Singapore. *American Journal of Chinese Studies*, 15(1), 17–34.

Rodan, G. (2009). Accountability and human rights in Malaysia and Singapore. *Journal of Contemporary Asia*, 39(2), 180–203.

Rodan, G., & Hughes, C. (2014). *The politics of accountability in Southeast Asia: The dominance of moral ideologies*. Oxford: Oxford University Press.

Sachs, J. D. (2015). *The age of sustainable development.* New York, NY: Columbia University Press.

Sawasdee, S. N. (2006). *Thai political parties in the age of reform.* Bangkok: Institute for Public Policy Studies.

Schedler, A. (1999). Conceptualizing accountability. In A. Schedler, L. J. Diamond & M. F. Plattner (Eds), *The self-restraining state: power and accountability in new democracies* (pp. 13–28), Baltimore, MD: Johns Hopkins University Press.

Schmitter, P. (2004). The ambiguous virtues of accountability. *Journal of Democracy,* 15(4), 47–59.

Slater, D. (2012). Strong-state democratization in Malaysia and Singapore. *Journal of Democracy,* 23(2), 19–33.

Slater, D. (2014). Unbuilding blocs: Indonesia's accountability deficit in historical perspective. *Critical Asian Studies,* 46(2), 287–315.

Slater, D. (2018). Party cartelization, Indonesian-style: Presidential power-sharing and the contingency of democratic opposition. *Journal of East Asian Studies,* 18(1), 23–46.

Slater, D., & Simmons, E. (2013). Coping by colluding: Political uncertainty and promiscuous power-sharing in Indonesia and Bolivia. *Comparative Political Studies,* 46(11), 1366–1393.

Tan, P. J. (2006). Indonesia seven years after Soeharto: Party system institutionalization in a new democracy. *Contemporary Southeast Asia: A Journal of International and Strategic Affairs,* 28(1), 88–114.

Thompson, M. R. (2016). Bloodied democracy: Duterte and the end of liberal reformism in the Philippines. *Journal of Current Southeast Asian Affairs,* 35(3), 39–68.

Timberman, D. G. (2016). Elite democracy disrupted?. *Journal of Democracy,* 27(4), 135–144.

UN (2015). Transforming our world: The 2030 agenda for sustainable development. United Nations. Retrieved from https://sustainabledevelopment.un.org/post2015/transformingourworld.

Verweij, M., & Pelizzo, R. (2009). Singapore: Does authoritarianism pay?. *Journal of Democracy,* 20(2), 18–32.

Whaites, A. (2016). Achieving the impossible: Can we be SDG 16 believers? GovNet Background Paper No. 2. Retrieved from http://www.oecd.org/dac/accountable-effective-institutions/Achieving%20the%20Impossible%20can%20we%20be%20SDG16%20believers.pdf.

Zeigenhain, P. (2015). *Institutional engineering and political accountability in Indonesia, Thailand, and the Philippines.* Singapore: Institute of Southeast Asian Studies.

CHAPTER 5

Human Rights and Sustainable Development Goals: the Philippines in the Postnational

Ulrich Karl Rotthoff

1 Introduction

> [Morality is] a highly contagious object that should be touched only with gloves and sterile instruments, lest one is contaminated with morality and exposes to moral appropriation what had started as scientific analysis.
>
> NIKLAS LUHMANN, 2012a, p. 271, my translation

> Our lives begin to end that day we become silent about things that matter.
>
> Credited to Martin Luther King, Jr.

In September 2015, 193 United Nations member states adopted the 2030 Agenda for Sustainable Development. Sustainable Development Goal (SDG) 16 of that Agenda is intended to "Promote peaceful and inclusive societies for sustainable development, provide access to justice for all and build effective, accountable and inclusive institutions at all levels" (UN, 2015, p. 28); it aims to "reduce all forms of violence and related death rates ... promote the rule of law ... [and] develop effective, accountable and transparent institutions at all levels" (ibid.). SDG compliance is measured by goal-specific sets of indicators. In line with accepted human rights (HR) — and particularly with the right to physical integrity — the relevant indicators for SDG 16 are: "16.1.1 Number of victims of intentional homicide ... 16.3.1 Proportion of victims of violence ... who reported their victimization to competent authorities ... 16.10.1 Number of verified cases of killing, kidnapping, enforced disappearance, arbitrary detention, and torture of journalists, associated media personnel, trade unionists and human rights advocates ..." (UN, n.d.).

In November 2015, Rodrigo Duterte entered the Philippine presidential election campaign with a pledge to "make drug dealers and criminals face the *stiffer penalties*" (Duterte, 2015, emphasis added). After his election to the presidency in May 2016, a campaign of extrajudicial killing (EJK), known as the Drug War

 | DOI:10.1163/9789004391949_006

killings, quickly gained pace in the country. Not for the first time, Philippine state structures have been shown to suffer from vulnerability or a lack of resilience. Checks and balances remain distorted, the rule of law is dysfunctional and the state's monopoly on violence is undermined by family/clan-aligned power centres. State-sponsored or -tolerated killings are observed in the context of numerous sources of conflict, ideological fault lines and moral fanaticisms. The analysis of the dynamics that propel these campaigns requires an understanding of the social landscapes of violence, the characteristics of the victims, the actors within the campaign, and their perceptions of the world.

Reacting tetchily to any reprimands or criticism of the Drug War killings, President Duterte — who has called the UN High Commissioner for Human Rights, Zeid Ra'ad Al Hussein, an "idiot" and a "son of a bitch" (Villamordec, 2016) — made it perfectly clear that he refuses to tolerate such expressions of concern, which he regards as interference in domestic affairs. Duterte is perfectly right: global legal regimes, such as HR treaties and conventions, continuously are setting the norms of *jus gentium*, which structure (i.e., limit) the horizon of options and expectations (Luhmann, 2012a, pp. 25ff). These regimes aim to normatively structure international relations: that is, they narrow legal spaces. In a second, postnational dimension, regimes such as HR treaties or the SDGs define expected behaviour in line with what had formerly been exclusively considered to be 'internal affairs' of a nation state. HR (and, for that matter, the relevant SDG 16 passages) inevitably have a deep impact on domestic affairs, as it is only here — along (domestic) state–citizen interfaces — that violations of HR can occur. Thus, dynamic formation processes of global law regimes gradually narrow down spaces that had previously been guaranteed by the sovereignty doctrine of the Westphalian Order.

Viewed from this perspective, SDG 16 links two discourses that revolve around development issues and HR issues, respectively. This is an important cornerstone for development debates, as the inclusion of SDG 16 appropriates issues that are also the focus of legal discourses. Indeed, the HR caucus has already reached a quasi-official and consensual catalogue of issues; this is the forum in which normative expectations can crystallize and be transformed into (global) laws (Fischer-Lescano, 2005). As a supranational structure, ASEAN (in contrast to the European Union), is in possession of neither regional legal instruments nor institutionalized mechanisms mandated for legally and bindingly addressing HR issues such as the Drug War killings in the Philippines, or the Myanmar Rohingya refugee crisis. Rather than proactively and consensually managing such matters, ASEAN exposes itself to the risk of 'scandalization' if SDG 16 is neglected. Adopting Durkheim's figure of *colére publique* — public temper (Durkheim, 1960, Ch 2, esp. p. 10) — the notion of scandalization describes a formation process of norms or expected behaviour by establishing the visibility of crimes or, in terms of common law, (perceived) injustice through media exposure which leads to public outcry (Dammann, 2012; Fischer-Lescano, 2005, pp. 67ff; Fischer-Lescano & Teubner, 2004). In the world society, public outrage tends to be directed to the economic sphere, where 'soft spots' provide the greatest leverage towards policy change (International Labor Rights Fund, 2007; Romeo & Romeo, 2017).

2 Research Question

Revolving around HR dimensions inherent particularly to SDG 16 (specifically as they pertain to extrajudicial killings), this chapter focuses on two ostensibly different spheres. The first is the *national* sphere — or, with reference to Anderson (2003), 'imagined communities'. Here, the analysis is only interested in passing in the acts and numbers of EJKs in the Philippines: patterns of state involvement are sufficiently documented. A second sphere is the *global*, which is neither the sum nor simply the arena of imagined communities as reflected in the *international* or the *supranational*. Rather, the notion of the global expresses post-Westphalian system formations along processes of functional (out-)differentiation, which radically break with, and tend to disregard, residues or forms of pre-modern system differentiations (Luhmann, 1998b, Ch 4). It is within this sphere that HR and SDGs, with their characteristic universal claims, are being developed and established as heterarchically organized global regimes (Fischer-Lescano, 2005; Fischer-Lescano & Teubner, 2004). In contrast to this postnationality, state involvement constitutes a qualifying circumstance of HR violations (OHCHR, 2000, p. 2). The latter are state infringements along state–citizen interfaces.

Against this backdrop, the chapter focuses on the Philippines as a sovereign state, a signatory to international treaties and a member state of the supranational ASEAN, and addresses the following research questions (RQ). RQ1: how can perceived diverse forms of killings within different societal dynamics (e.g., related to the insurgency/counterinsurgency complex, Duterte's Drug War, or fanatical millenarian vigilante groups) be abstracted in order to reveal common causes or drivers? RQ2: how do/can global regimes respond to HR violations and, given limited international legal venues, how is compliance with global standards being demanded? RQ3: how does SDG 16 relate to the aforementioned issues (RQ1 and RQ2) as they pertain to ASEAN integration processes?

3 Scientific and Social Significance of the Research

The term 'rights' in human rights does not imply that HR as a research object is limited to the exclusive domain of law. Approaches solely oriented towards legal thinking may leave students and advocates of HR mired in debates over jurisdictions or other necessities and intrinsic particularities of law. Other important aspects of HR, such as the societal dimension, may then be obscured and material factors outside the area of legal definitions may be disregarded. Focusing on the Philippines — i.e., the national or domestic sphere — this chapter identifies theory gaps in the prevailing analyses of HR violations. Addressing RQ1, it will distinguish specific conflict lines and factors that are conducive to EJKs. These findings will help to open up narrowly selective perceptions of HR violations. The theory-guided discussion of HR presented here aspires to illuminate issues that are not recognized by other frameworks and to propose them for further discussion.

Defined by, and embodied in, international treaties, human rights are integral to global or postnational legal regimes, yet those same legal regimes lack jurisdiction over alleged perpetrators of HR violations or, applying the chain of command principle, over those responsible for such infringements. Thus, a typical reflex on the part of governments and statesmen accused of massive violations of HR is to deny international or supranational jurisdiction, and not to tolerate interference in domestic affairs. The Government of the Philippines and President Duterte continue to play the sovereignty card in response to international criticism (Duterte, 2017, p. 23; Harry Roque as quoted in Gita, 2017; Macas, 2017). This strategy provokes scandalizations with possible negative effects on the Philippine economy and, collaterally, on ASEAN growth prospects. Regarding the sphere of the global (RQ2), this chapter aims to offer a means of understanding and interpreting general communications pertaining to the Philippine case. The potential role of international treaties and regimes such as SDG 16 in the context of ASEAN integration will be elucidated in the course of this discussion.

4 Methodology

Reviewing HR along the scale from national to postnational contexts, this study is aligned to strands of systems theory discussed by Luhmann (see, e.g., Luhmann, 2012b). As a theoretical discussion, it identifies methodological issues that arise when contrasting this approach with unsophisticated and popular Marxist concepts, and with policy designs and self-descriptions mapped according to Westphalian Order semantics. The theoretical part of the study is based on literature and available online documents relevant to the Philippine HR situation as well as to the theory of HR. The analysis applies critical discourse techniques. The findings are corroborated by the analysis of semi-structured interviews conducted, and data collected, in the context of ongoing research on expressions and origins of violence in the Philippines. Between 2015 and 2018, key informant interviews and follow-up interviews were conducted with six respondents in Metro Manila and during seven research visits to Negros Island. One respondent from Mindanao was interviewed via telephone. The interviews were facilitated by three non-governmental organizations (NGOs) from Manila and Negros Occidental.

5 Literature Review

It is not only since the wave of EJKs in the wake of President Duterte's Drug War that the Philippines' HR record has become an issue of international concern. State-sponsored/tolerated killings and other forms of grave HR violations such as enforced disappearances and torture have been reported and monitored since the rule of President Marcos (Alston, 2007b, 2007c; Amnesty International, 1982, 1994, 2005). McCoy's (2010) edited volume traces violence as a means towards economic and political ends as far back as the colonial period. Based on the existing literature, this section presents the main stakeholders in the Philippine landscape of violence, identifies conflict lines attributed to those stakeholders, and traces a popular, agenda-driven explanatory approach. Alongside this discussion, it identifies blind spots in the Philippine HR discourse, which will be addressed in the subsequent analysis section. As far as possible, I refrain from presenting absolute figures in order not to contribute to sociologically fruitless debates over numbers. Rather, the focus of the section is on patterns and drivers.

5.1 *Landscape of Violence*

Three major stakeholder groups are identified in the Philippine landscape of violence: (1) traditional politicians, (2) state security forces, and (3) instrumental structures of violence — a mélange of heterogeneous armed groups such as paramilitary, vigilante groups and goon squads.

Traditional politicians enter power equations under the notions of bossism or boss-rule (Kreuzer, 2009; Sidel, 1999); cacique democracy (Anderson, 1988); bureaucrat capitalism (Sison, 1971); oligarchic rule (Anderson, 1988; Kreuzer, 2011; McCoy, 2010); warlordism (Creative Associates International, 2012); or strongman rule (Migdal, 1988). These labels describe unconstitutional concentrations of legislative and executive powers in the hands of a family/clan-centred elite stratum. Violence even appears with a folklore tint (McCoy, 2010). Kreuzer (2012, p. 3) describes horizontal and vertical pointers that indicate intra-stratum contests and rivalries on the one hand and violence as an instrument of domination on the other. Elsewhere, he sees a "specific ruling class, made up of families and strongmen, combining political, economic and social power and, if necessary, making use of private or privatized state-violence to defend their position" (Kreuzer, 2009, p. 51).

State security forces such as the Philippine National Police (PNP) and the Armed Forces of the Philippines (AFP) are constantly accused of being involved in HR violations against perceived "enemies of the state" (AFP, n.d.), against the backdrop of insurgency/counterinsurgency dynamics. EJKs have been reported for the martial law period (Amnesty International, 1982), and for all post-martial law administrations (Karapatan, 1997; Leary, Ellis, & Madlener, 1984; NCCP, 1988; US Department of State, 1995, 1997, 1998, 1999, 2000, 2001). Killings reached high peaks during both terms of President Gloria Arroyo (Karapatan, 2010, p. 16). Philippine NGOs, alongside international networks and affiliations, increased the international visibility of the deteriorating HR situation after 2001 through scandalization or the generation of *colére publique mondiale* (Durkheim, 1960; Fischer-Lescano 2005, p. 67ff; Luhmann, 2012a, p. 249). In response to such public outrage, President Arroyo established the so-called Melo Commission (see Melo, Mantaing, Zuño, Gonzales, & Pueblos, 2006), and invited the then UN Special Rapporteur on extrajudicial, summary or arbitrary executions, Phillip Alston, for a country visit in February 2007. The increased visibility also led to a hearing on EJKs in the Philippines at the United States Senate (US Senate, 2008), and to critical statements by the European Parliament and the European Commission (European Parliament, 2008a, 2008b), among others.

In his reports, Alston goes beyond the narrow technical, evidence-focused perspective of the Melo Commission. Looking at patterns rather than court rulings, he rejects the use of the term "unexplained killings" (Alston, 2007b, pp. 2f; 2007c, p. 2; Melo et al., 2007, p. 2) but argues: "one does not have to wait for a court to secure a conviction before one can conclude that human rights violations are occurring. The term 'extrajudicial killings' which has a long pedigree is far more accurate and should be used" (Alston, 2007a, p. 3). Following Alston's mission, incidents of extrajudicial killings declined but remained a persistent matter of concern for the last three years of the Arroyo administration and throughout the 2010–2016 Benigno Aquino III administration (European Parliament, 2012; Karapatan, 2016, p. 10; US Department of State, 2014, 2016).

Since mid-2016, the Philippines has been experiencing an unprecedented surge of killings in the context of Duterte's Drug War, which is generically known as *Oplan Tokhang* (Operation Plan Knock and Plead). Oplan Tokhang is a component of 'Project Double Barrel' (PNP, 2016). Aiming to eradicate drug abuse, the Tokhang leg of the Double Barrel campaign is a direct extension of an anti-crime drive sparked and implemented under Rodrigo Duterte when he was mayor of Davao City, and, according to Human Rights Watch (HRW, 2009), carried out by the so-called 'Davao Death Squad'. The Davao Death Squad bears features of an instrumental structure of violence (see below) which operationally and logistically overlaps with the local PNP (HRW, 2009). The threads come together in the Office of the Mayor (personal communication, identity withheld, November 2017). Although vigilante structures are implicated in the extension of Oplan Tokhang to the national level, the involvement of state organs in the killings is beyond question (PNP, 2017). To illustrate the actual scale of the Drug War killings, taking cases with confirmed PNP involvement only, killings related to Oplan Tokhang (July 2016 to January 2017) outnumber EJKs of dissidents (July to December 2016) by a factor of approximately 134 and by a factor of 373 if we include all cases of killings in the wake of the Drug of War (based on PNP statistics as quoted in Bueza, 2017; Karapatan, 2017, p. 31).

Instrumental structures of violence include paramilitary groups, vigilante groups and goon squads/private armies (Beaulieu, 2010; Espino, 2004; Kraft, 2010; May 1992; Zenarosa, 2010). It is difficult to attribute these groups unambiguously to state or non-state structures or conflict lines. 'Paramilitary' refers to armed groups which are constructed under, and legalized by laws and decrees, and given auxiliary military and/or police functions (Department of Interior and Local Government, 2010; Espino, 2004; Global Security, n.d.a, n.d.b; Kraft, 2010; Zenarosa, 2010). 'Vigilante' are not legally mandated armed structures but are tactically or temporarily tolerated, endorsed or nurtured by the state or branches of the state (May, 1992, p. 50); and/or cater to the interests of traditional politicians (ibid., p. 54). They are visible in several conflict arenas including the insurgency/counterinsurgency theatre (military dimension); disputes over access to (natural) resources (socio-economic dimension); disputes over domination or safeguarding domination (power dimension); and are implicated in the area of drug-related killings. These dimensions may overlap (Kraft, 2010, p. 189). A recurrent feature of vigilante groups is their frequent cultist-chiliastic orientations anchored in religious, cultural, ethnic, or ideological constructs (ibid, p. 199; May, 1992; Van der Kroef, 1986, pp. 5, 6, 8). 'Goon Squads' refers to any armed structures which are deployed to serve the interests of the personal, clan or family agendas of non-state stakeholders, or which operate extralegally at the discretion of public incumbents, for instance in the context of self-administered justice; and which are condoned tacitly by law enforcement and prosecution bodies. Thus, the term 'goon squads' generically covers groups otherwise described as private armies, death squads, and freelance gun-for-hire gangs (Paredes, 2010; Zenarosa, 2010; personal communications, identities withheld, June–August 2015).

5.2 *Conflict Lines*

For the time before the Duterte administration, conflict lines can clearly be attributed to specific stakeholder groups such as security forces and traditional politicians with pronounced vertical (inter-class) and horizontal (intra-class) characteristics (Kreuzer, 2012). Aguilar, Mendoza, and Candelaria (2014) show that killings of journalists tend to be "local events", with local non-state stakeholders acting within local non-state dynamics. Alston (2007b, p. 5) reports on killings along the landlord–peasant conflict line. Preliminary analysis of data on violence and killings in Negros Island corroborate a pattern of non-state involvement if the local elite is seen to be involved in conflicts, typically over land issues (personal communications, identities withheld, 2015, 2016). Hence, such violence occurs along elite conflict lines. State involvement is apparent in the killings of dissidents, especially in the context of large-scale development projects. Here, the military homes in on perceived enemies of the state (AFP, n.d.; Alston, 2008; personal communications, identities withheld, 2015, 2016). Hence, such violence can be said to occur along state conflict lines. We can refer to old state conflict lines if violence is directed against dissidents.

6 Analysis

Prior to the Duterte administration, instrumental structures of violence can be located in the above-mentioned dimensions (military, socio-economic, and domination) of arenas attributed to state stakeholders and non-state elite stakeholders. The distinction between elite and state conflict lines opens theoretical perspectives on the drivers of extrajudicial state violence. Pertaining to the state, we can differentiate between the old conflict line and a new conflict line. The latter becomes apparent in Drug War killings, which, at this level of analysis, cannot be explained by the drivers and dynamics attributed to killings of dissidents. This differentiation merely serves the ordering of the research object. The merits of these two steps — viz., identifying the dimension and differentiating conflict lines — can easily be seen.

6.1 *Law, Morality, and Evil*

Karapatan, the HR voice of the so-called national democratic (ND) Left, observes killings along both elite and (old) state conflict lines, respectively, from agitprop vantage points of vulgar Marxism. Seasoned with cut-and-dried anti-imperialist semantics, these killings are indiscriminately attributed to particular administrations of the "fascist" state (Karapatan, 2017). Mistaking Duterte's unsavory demeanour and invective rhetoric for anti-imperialist sentiment, Drug War killings, on the other hand, are not readily spotted as HR violations: Karapatan's 2016 Year End Report presents a 26-page review of the first six months of Duterte's term which dedicates merely one paragraph to Drug War killings and reports a total of only 19 EJKs (ibid., pp. 30f). This low-key reportage on Drug War-related killings must be seen in the light of Duterte's political appreciation of the ND political spectrum: recognized as a negotiating partner in peace talks and looking forward to cabinet inclusion, the ND Left was elated (ibid., pp. 27, 29). The ingratiation of the underground Left even went so far as to offer help in hunting down drug lords with the use of violence (Communist Party of the Philippines, 2016).

Other HR champions saw the Drug War-related HR crisis looming on the horizon and stood firm (Aktionsbündnis Menschenrechte Philippinen, 2016). Jose Manuel I. Diokno of the Free Legal Assistance Group emphasizes the congruence between the hunt for communists and drug addicts and traders: "The only difference is the branding. The label before was *fear* of communists. The label now is the *fear* of drug addicts and drug pushers" (Diokno quoted in Dizon, 2017, emphasis added). In a nutshell, the hunts are not threat-driven but fear-driven — or, in other words, driven by *threats that are attributed to groups*. Here, 'communist' stands neither for a *weltanschauung* nor for organizational affiliations. Rather, the label stands for traits, characteristics, and threats ascribed from outside to a group. Thus, it is precisely the ascription which is creating the social segregation.

Aiming to explain extreme forms of inter-group violence, Appadurai (1998) focuses on ethnicity, which he understands as "labels" and "frameworks of identity" (p. 226). Attempting to overcome theoretical constraints, Appadurai subsequently abstracts from his initial concept by introducing neutral categories like us/them perceptions ("otherness") and quasi-ethnicity (ibid., pp. 237, 242, 244). Breaking his anthropological mould, Appadurai implicitly recognizes that ethnic and quasi-ethnic (e.g., political) backdrops of killings are first and foremost ethnic or political on the level of the phenomenon.

In the context of counterinsurgency, 'communist' is a semantic or social construct in the same way as ethnicities (with reference to Luhmann; see Bommes, 1995) or nations (Anderson, 1988) are social constructs — and with the same terrible property of being a potential target for exclusion and extermination. The detachment of the 'communist' label from any ideological discourse is reflected by the synonymous use of the 'enemies of the state' formula in reference to the same (now: target) group; it eventually became apparent when the latter label was also employed by Duterte in the Drug War context (Punay & Romeo, 2017). This supports the view that the killings of dissidents on the one hand, and drug users and drug traders on the other, are different expressions of congruent societal segregations. In the same way that the 'communists' become the 'enemies of the state', the 'demonized other', the targets for extermination, the drug users and traders also become the 'enemies of the state', become demonized and, eventually, the targets of 'justified' killing. On an abstract level, we need to explain how the perpetrators — the state or agents of the state — justify extralegality.

Transcending the scope of his originally ethnicity-centred focus, Appadurai anchors his framework in the neighbourhood of Arendt's notion of totalitarianism. Arendt elaborates "central ideological fiction" as a constitutive element in the formation of totalitarian states and movements (Arendt, 2015, p. 766). Ideological fictions crystallize around "totalitarian dichotomies" (p. 772). On one side of this dichotomy — Appadurai's "us" — we have the notion of, for instance, an Aryan national community (*volksgemeinschaft* as in the case of Nazi Germany) or a classless society (as in the case of the Stalinist Soviet Union). This "us" side "is contrasted with Appadurai's 'other'" — a vision of Jewish conspiracy or class enemies (ibid., p. 261ff). Appadurai and Arendt describe extreme expressions of what Luhmann (1998b) identifies as semantics, and which, Arendt (2015) elaborates, are paired with "ideological fictions" — chiliastic notions of redemption by physically exterminating the "other". This millenarian figure can be identified in the context of HR violations in the Philippines. We first encountered it above, in relation to vigilantism against the backdrop of insurgency/counterinsurgency dynamics. Secondly, the figure is echoed by the 'enemies of the state' formula applied to dissidents and drug users and drug traders alike. 'Enemies of the state' are dehumanized, practically outlawed and, hence, subjected to physical extermination: "When you kill criminals, that is not crime against humanity. The criminals have no humanity, goddamn it" (Duterte quoted in Ranada, 2017).

We have seen that seemingly distinct categories of killing campaigns in the Philippines actually share the segregational practice of 'othering'. However, group and trait descriptions per se are not sufficient to explain the proliferation of state-orchestrated systematic killing at the theoretical level. Even the inclusion of Arendt's (2015) "ideological fiction" as a co-factor, although important, fails to fully explain the emergence of such phenomena. From the perspective of the functional systems of law or politics, the introduction of an "ideological fiction" as a policy-guiding momentum in state-driven extralegal killing campaigns implies that these fictions are hampering or practically repealing the systems' basal codes (Luhmann, 2012b, p. 602), i.e., the lawful/unlawful or the government/opposition binary distinction (Luhmann 1998a, p. 70; 1998b, p. 632). This is a grave defect from the perspective of functional differentiation; it indicates a reversed or frustrated autopoiesis (i.e., functionality or system closure). In other words, the functional system of law loses its ability to observe the society indiscriminately and exclusively by means of its original binary. In the case of the political system, the displacement of the government/opposition binary by a morality code, or the equalization of the former with the latter, is a "deathblow" to or the end of functional differentiation (Luhmann, 2012a, p. 259). Morality's good/bad difference colours expected behaviour more or less strongly (ibid., pp. 97, 111).

As far as these expectations are concerned, law is a functional equivalent to morality. A principal difference, however, is that law provides room for discourse and legal action while morality tends to take a hard line and, thus, resides in the orbit of quarrel and violence (ibid., pp. 177, 260). Luhmann (ibid., p. 333) states that morality is being applied across or within different functional systems. However, base codes such as the lawful/unlawful binary of the law system need to be robust vis-à-vis moral pervasions; the good/bad distinction is not a metacode (ibid., p. 333). "Functional codes must be established on a level of higher amorality because both their values must be accessible for the operations of the system" (ibid., p. 259, my translation). Calibrated for confrontation, the good/bad differentiation tends to merge with us/them perceptions, which can be seen, for example, in the negative connotations of terms like "capitalist", "monopoly" (ibid., pp. 302, 40), "US-imperialism", or "finance capital". In this context, values like "freedom", "solidarity", "security", or "health", as well as norms, can be transformed into agents of morality warfare (ibid., pp. 242, 337). Serious problems emerge, however, if an us/them perception is loaded with morality and if the good/bad binary is escalated to the level of good/evil expressions (ibid., pp. 97, 115).

Günther (1994, p. 136) states that the introduction of an "evil" perspective in criminal policy discourses narrows down room for possible courses of action. The attribution of evil, he argues, has the property of pre-empting legal exculpation discourses. Hence, it saturates legal issues with extralegal loads and practically removes jurisdiction from the legal system by surrendering the prerogative of evaluation to the binaries of morality/ethics and/or religion (ibid., p. 153). In effect, "evil becomes a category of exclusion: the enemy as *inimicus humanis generis*. The evil one has made a conscious decision to be the antagonist and thus deserves punishment to the point of (physical) extermination.... Where a concept has already been semantically so firmly defined, the possibility of further political disposition is excluded" (ibid., p. 153; my translation, emphasis added).

"Evil" enters the Philippine discourse on the Drug War at both ends of the debate. It is a recurrent motif in Duterte's sanctification of the Drug War (Duterte, 2017, p. 3). During a Senate Committee meeting, a vocal reformist, Senator Antonio Trillanes, accused his colleague Allan Peter Cayetano, a supporter of Duterte's campaign, of defending evil (Philippine Senate, 2016, p. 20). Although critical of the killing campaign, the clergy can connect to the rhetorical figure of evil. Referring to drug abuse, the Philippine Jesuits state: "The evil that attacks the human with the power of the demonic, should unite us, not divide us ... we must fight this enemy together" (Moreno, 2017).

6.2 *Extrajudicial Killing, Westphalian Obstinacy, and the Global*

This discussion has touched on different dimensions of the legal system. First, it focused on expressions, or self-descriptions of the political system: nation states. Second, it discussed the traditional view of the sphere wherein nation states interrelate: the international. Third, it mentioned formations of superordinate structures such as the EU or ASEAN: the supranational. Lastly, it went on to develop arguments from the perspective of postnationality, which is dominated by functional differentiations: the global.

6.2.1 The Nation State Dimension

The general problem is that national legal systems tend to be inaccessible in instances of systematic and grave HR violations on a large scale. In the Philippines, state-orchestrated extralegal killing campaigns reflect defects within the functional systems of law and politics. Within the functional system of politics and its organizations such as the police and the military, morality's good/evil binary becomes *the* guiding principle vis-à-vis dissidents and alleged drug users and drug traders. The burden of proof for the functionality of the Philippine legal system is on the state. The HR community has kept close tabs on the latest killing campaign against dissidents that started in 2001: not a single perpetrator has been convicted so far. Likewise, the Philippine judiciary and crime investigative bodies are failing dismally in getting the more recent Drug War killings under control. Obviously, remedy for the current HR crisis is not immediately at hand from within Philippine state structures. The political system does not have complete control over the security forces; the Philippine Congress seems incapable of complying with its control function over the administration; the judiciary is unable to bring perpetrators of killing campaigns to justice. Given the decades-long history of extrajudicial killings in the Philippines, death threats against witnesses and actual killings of those who refuse to cooperate in the Drug War campaign (personal communications, identities withheld, June–October 2015, November 2017), confidence in the Philippine legal system seems little more than wishful thinking.

6.2.2 The International Dimension

As far as HR violations are concerned, it is certainly true that the international legal system rarely pierces the sovereignty veil. However, extrajudicial killings are regularly punishable beyond the national level if the state concerned is a State Party to the 2002 Rome Statute of the International Criminal Court (ICC),[1] and if the specific crime falls under the jurisdiction of the ICC which, among others, covers crimes against humanity (Rome Statute, Article 5). The latter include the act of murder "when committed as part of a widespread or systematic attack directed against any civilian population, with knowledge of the attack" (Rome Statute, Article 7.1). The Rome Statute further specifies: "'Attack directed against any civilian population' means a course of conduct involving the multiple commission of acts [such as murder] referred to in paragraph 1 against any civilian population, pursuant to or in furtherance of a State or organizational policy to commit such attack" (Article 7.2.a).

As the Philippines is a State Party to the Rome Statute, ICC Chief Prosecutor Fato Bensouda claimed possible ICC jurisdiction with regard to killings committed in the context of Duterte's Drug War as early as October 2016, pledging her office's keen attention to the situation: "Let me be clear: any person in the Philippines who incites or engages in acts of mass violence including by ordering, requesting, encouraging or contributing, in any other manner, to the commission of crimes with the jurisdiction of the ICC is potentially liable to prosecution before the Court" (ICC, 2016). In February 2018, the ICC Office of the Prosecutor opened a preliminary investigation on that matter (ICC, 2018).

Even before the formation of the ICC, national sovereignty and non-intervention — core doctrines of the Westphalian Order — had ceased to be solid. The Nuremberg Trials, the International Military Tribunal for the Far East and the International Criminal Tribunals for Rwanda and the former Yugoslavia show that the principle of national sovereignty is not a perfect shield against criminal prosecution.

---

1 Rome Statute of the International Criminal Court. UN Ref. A/CONF.183/9. Retrieved from https://www.icc-cpi.int/nr/rdonlyres/ea9aeff7-5752-4f84-be94-0a655eb30e16/0/rome_statute_english.pdf (3 November 2017).

6.2.3 The Supranational Dimension

As a regional body programmatically geared towards supranationality, ASEAN did not develop general superordinate judiciary structures as embodied, for example, in the European Court of Human Rights, the Inter-American Court of Human Rights and the European Court of Justice, which is vested with particularly far-reaching jurisdiction. The most noteworthy ASEAN body exclusively concerned with HR is the ASEAN Intergovernmental Commission on Human Rights (AICHR). The Terms of Reference of the AICHR make it perfectly clear: this body is most tightly aligned to Westphalian doctrines and, thus, precludes any superordinated political or judicial jurisdiction that would put sovereignty into question (ASEAN Intergovernmental Commission on Human Rights, 2017, p. 106). It remains to be seen whether the AICHR resembles, or will eventually evolve into, an arena of discrete HR advocacy that, along the lines of Track 2 or 1.5 diplomacy (Jones, 2015, pp. 26ff), could create margins to deal with serious HR issues. However, ASEAN member states may also "remain stubbornly insistent on its principle of non-interference" (Kvanvig, n.d., p. 1).

Rather than maintaining the principle of national sovereignty, this means trading off potential ASEAN internal legal recourses, policing mechanisms, and judicial remedies on the one hand, against scandalizations and regime collisions in the global, with incalculable effects on, for example, the economy, on the other. Based on international treaties and agreements, supranational unions such as ASEAN or the EU are fundamentally characterized by renouncing (legal) leeway to various degrees and in various fields. Precisely *this* is the structure-forming element of contracts and treaties in the context of functional out-differentiation: reducing complexity (leeway) in order to increase complexity in terms of economics, science, markets etc. under conditions set out by systems that forcefully establish themselves in the global. The increase of complexity as achieved by the EU vis-à-vis the challenges of world society went hand in hand with a considerable narrowing down of spaces of national sovereignty and subordinating principal parts of national judiciaries to European jurisdictions and/or the cognizance defined by treaties and agreements signed between the EU and third parties.

6.2.4 The Global Dimension

Contrary to decided claims on *globality* inherent to operational closures of functional systems, the systems of law and politics show internal secondary differentiations along segmentary lines which become visible, for example, in the self-descriptions of political systems as states or nation states (Luhmann, 1998b, pp. 758, 760f; 1998a, p. 582). Structural couplings between the functional systems of law and politics (e.g., constitutions), however, are not echoed in the global or "world society" (Luhmann, 1998a, p. 582). Neither does the latter provide state-like organs in the form of a world government, and nor has it developed a global equivalent or extension of a hierarchically organized jurisprudence such as that found in national legal systems. This must be viewed historically from the perspective of the Westphalian Order that lastingly impressed the genesic context of the out-differentiation processes of both the functional system of law and politics. The very core doctrines of the Westphalian Order, i.e., the principles of national sovereignty and non-intervention in internal affairs, are precisely the line of defence of the Duterte government vis-à-vis international pressure towards halting the Drug War. But how solid or macerated is this legal retreat?

It is not exclusively the prospect of actual prosecution that generates pressure towards compliance with expected behaviour. Given the considerable legal obstacles to being tried, the process of making HR violations visible is itself a viable HR advocacy instrument (Cabico, 2017). Thus, aside from the Rome Statute as a positive and enforceable law, other positivized as well as non-positivized normative expectations — including legally non-actionable treaties or development goals — beyond being *inter*national agreements or conventions, form parts of *post*national heterarchically organized global legal regimes, which forcefully transcend the concept of national (and supranational or international). To illustrate: analysing regime collisions, Fischer-Lescano and Teubner (2004: pp. 1024ff), among others, refer to the example of Brazil which, facing public health challenges due to the AIDS epidemic around the turn of the millennium, announced that it was considering producing medicine without paying patent royalties. Merck being an affected company, the US government entered the fray and argued on behalf of Merck on the basis of the General Agreement on Tariffs and Trade (GATT). Cursorily reviewed, the issue may appear as a conflict between states — in this case, Brazil and the US. However, a different picture unfolds if we factor in scandalizations conducted by diverse (state and non-state) actors: now the economic global regime (GATT) is being challenged by the World Health Organization. The latter is championed by the NGO-supported SDG regime and, as far as the right to health is concerned, the HR regime.

Thus, it is analytically illuminating not to attribute communications at the face value of their *erscheinungsebene* (level of appearance), but rather to scrutinize specific contexts of functional differentiation within which communication, or discourse for that matter, happens. Keeping this in mind may safeguard one from mistaking critical voices for hypocrisy, and thus snubbing potential HR advocacy allies as Karapatan (2017, pp. 28f) did. We can observe decidedly different contexts of functional differentiation if, for example, distinct foreign state organs, government departments with different portfolios, or congressional/parliamentary committees comment on a given issue (in, nota bene, dissentient manners). Stuck deep in anti-imperialist rhetoric, Karapatan (see above) fails to recognize and exploit the normative power available in postnational regimes such as HR treaties or development goals agendas. As has been argued above, international outrage led to a decrease in the number of dissident killings under President Arroyo. For the case of killings in the wake of the Philippine Drug War, the same mechanisms are at work. Scandalization campaigns and the subsequent mobilization of decision makers generated pressure on President Duterte. Of course, pressure generated at soft spots such as trade and Official Development Assistance directly affects the country's ability to tap its development potential (Katigbak, 2018; Romeo & Romeo, 2017).

In this context, the SDGs — and particularly SDG 16 — are important in the formation process of legal regimes in the global. Viewed as certified benchmarks along which scandalizations crystallize, the SDGs (and HR agendas) can be transcribed into normative expected behaviour, eventually in the sense of a *jus gentium*. However, it remains to be seen if indicators aligned with SDG 16 adequately address the problem of EJKs along the new state conflict line. Indicator 16.1.1 is prone to understating due to the creative relabelling of EJKs. Indicator 16.3.1 does not consider either the number of unreported Drug War-related EJKs in the countryside, or the fact that state-driven violence creates fear, preventing victims or their relatives from reporting such assaults (personal communication, identity withheld, November 2017). Finally, indicator 16.10.1 defines by enumeration groups of victims that qualify for inclusion in SDG 16 statistics — a seemingly exhaustive listing that nevertheless excludes the thousands of EJKs that have occurred in the context of the Philippine Drug War.

7 Summary and Conclusion

The review of EJKs from several perspectives has led us to strands of systems theory that provide the most comprehensive theoretical frameworks for thoroughly and profoundly understanding different expressions of what are, in fact, identical forms of state violence. Searching for common drivers (RQ1, above), we identified the dominance of morality-driven good/evil perceptions as the determining maxim across several spheres of Philippine society. This is a plausible major variable or co-factor contributing to explaining both EJK campaigns under review. Focusing on different dimensions (RQ2, above), we examined the effectiveness of — and instruments, institutions, organizations, and legal processes provided by — the legal system. This line of argument may be seen as an imposition by those who are (ideologically) reluctant to recognize established or establishing regimes, such as HR and SDG treaties, in the global (i.e., the postnational) as heterarchically organized legal systems. Nevertheless, it should be immediately understood and conceded that the same regimes define, and also positivize, expected behaviour and ultimately demand compliance with the same in their eventual expression as global law (Fischer-Lescano, 2005).We might thus summarize the findings as follows:

- the Philippine judiciary, obviously in a derelict state, is incapable of adequately addressing the decades-long problem of EJKs;
- in the international dimension, barriers are high for prosecution before the ICC, as practically the sole regular international institution endowed with (limited) prosecuting power;
- the supranational structure ASEAN, adhering to factually obsolete doctrines of the Westphalian Order, has not established superordinated legal institutions;
- in the (postnational) global, a diversity of actors is engaged in defining and enforcing norms of expected behaviour.

This leads us to the role of SDG 16 in the ASEAN context (RQ3, above). In line with HR and development goals, and despite the aforementioned definitional constraints, attention is drawn to non-compliance with normatively expected behaviour by scandalization campaigns carried out by diverse actors including NGO networks. Due to the lack of institutions endowed with prosecution powers, a common strategy is to identify soft spots and subsequently put pressure on them. Restricting trade, economic leeway, and development potential are powerful and particularly effective measures directed against governments accused of HR violations. In this context, SDG 16 provides an officially accepted address for delivering scandalizations in instances of utter non-compliance. Like any other international treaty or regime, the SDGs, therefore, actually restrict sovereignty. Thus, ASEAN digging in its heels on sovereignty and non-intervention principles eventually narrows the scope for pro-actively approaching humanitarian crises like the Philippine Drug War or the Rohingya refugee issue in Myanmar. ASEAN members' mutual assurance of the validity of Westphalian doctrines applies internally. In the global, it is not respected. With the economy being the main focal point for protests in the global, ASEAN actually trades off economic growth potentials against the myth of national sovereignty.

Acknowledgement

Part of this research work had been conducted under the Research & Creative Work Load Credit Grants programme of the University of the Philippines, Diliman.

References

AFP (n.d.) Knowing the enemy. Power-point presentation. Armed Forces of the Philippines.

Aguilar, F. V. Jr., Mendoza, M. P., & Candelaria, A. L. K. (2014). Keeping the state at bay: The killing of journalists in the Philippines, 1998–2012. *Critical Asian Studies*, 46(4), 649–677. Retrieved from http://dx.doi.org/10.1080/14672715.2014.960719 (23 August 2015).

Aktionsbündnis Menschenrechte — Philippinen (2016). Nach der Wahl in den Philippinen: Sorgen um Menschenrechte. Retrieved from http://www.menschenrechte-philippinen.de/index.php/de/nachrichten-leser/nach-der-wahl-in-den-philippinen-sorge-um-die-menschenrechte-204/ (10 May 2016).

Alston, P. (2007a). Press statement by Professor Phillip Alston, Special Rapporteur of the United Nations Human Rights Council on extrajudicial, summary or arbitrary executions. Manila 21 February. Retrieved from http://pcij.org/blog/wp-docs/alston_press_statement.pdf (20 February 2016).

Alston, P. (2007b). Report of the Special Rapporteur on extrajudicial, summary or arbitrary executions. UN Document A/62/265. New York. Retrieved from https://documents-ddsny.un.org/doc/UNDOC/GEN/N07/461/40/PDF/N0746140.pdf?OpenElement (26 January 2016).

Alston, P. (2007c). Statement by Professor Phillip Alston, Special Rapporteur on extrajudicial, summary or arbitrary executions. United Nations General Assembly, Third Committee, 25 October. New York.

Alston, P. (2008). Report of the Special Rapporteur on extrajudicial, summary or arbitrary executions. UN A/HRC/8/3/Add.2. Retrieved from https://documents-dds-ny.un.org/doc/UNDOC/GEN/G08/130/01/PDF/G0813001.pdf?OpenElement (26 January 2016).

Amnesty International (1982). Human rights violations in the Philippines. New York: Amnesty International.

Amnesty International (1994). "Disappearances" and political killings: Human rights crisis of the 1990s. Retrieved from https://www.amnesty.org/en/documents/act33/001/1994/en/ (5 October 2016).

Amnesty International (2005). Philippines sharp rise in "vigilante" killings as human rights activist's death remains unsolved. Retrieved from https://www.google.com/url?sa=t&rct=j&q=&esrc=s&source=web&cd=5&cad=rja&uact=8&ved=0ahUKEwjL2uHOo8TWAhXLXLwKHQajDbcQFghDMAQ&url=https%3A%2F%2Fwww.amnesty.org%2Fdownload%2FDocuments%2F84000%2Fasa350012005en.pdf&usg=AFQjCNGtO9g-utDaZ-GHU6J3POex-JDsQQ (27 September 2017).

Anderson, B. (1988). Cacique democracy in the Philippines: Origins and dreams. *New Left Review* 169, 3–31.

Anderson, B. (2003). *Imagined communities*. Pasig City: Anvil.

Appadurai, A. (1998). Dead certainty: Ethnic violence in the era of globalization. *Public Culture*, 10(2), 225–247. Retrieved from http://www.arjunappadurai.org/articles/Appadurai_Dead_Certainty_Ethnic_Violence_in_the_Era_of_Globalization.pdf (25 February 2015).

Arendt, H. (2015 [1951]). *Elemente und Ursprünge totaler Herrschaft*. München: Piper.

ASEAN Intergovernmental Commission on Human Rights (2017). Terms of Reference of the ASEAN Intergovernmental Commission on Human Rights. In ASEAN Intergovernmental Commission on Human Rights, *What you need to know: ASEAN Intergovernmental Commission on Human Rights. ASEAN 50th Anniversary Edition, A Compendium* (pp. 103–116). Jakarta: ASEAN Secretariat. Retrieved from aichr.org/?dl_name=AICHR_What_You_Need_to_Know_Compendium_FINAL_min.pdf (12 January 2018).

Beaulieu, J. A. (2010). Protracted state insurgencies: Counterinsurgency lessons from Sri Lanka, Indonesia, the Philippines, and Colombia. Master's thesis, Georgetown University, Washington, DC. Retrieved from https://repository.library.georgetown.edu/bitstream/handle/10822/553445/beaulieujames.pdf?sequence=1&isAllowed=y (5 September 2015).

Bommes, M. (1995). Migration and ethnicity in the national welfare state. In M. Martiniello (Ed.), *Migration, citizenship and ethno-national identities in the European Union* (pp. 120–143). Aldershot: Avebury.

Bueza, M. (2017, April 23). In numbers: The Philippines' "war on drugs". *Rappler*. Retrieved from https://www.rappler.com/newsbreak/iq/145814-numbers-statistics-philippines-war-drugs (17 August 2017).

Cabico, G. K. (2017, October 13). Duterte "finally feeling the heat" — Human Rights Watch. *philstar.com*. Retrieved from http://old.philstar.com/headlines/2017/10/13/1748430/duterte-finally-feeling-heat-human-rights-watch (28 November 2017).

Communist Party of the Philippines (2016). Response to President Duterte's call for anti-drug cooperation. 2 July. Retrieved from https://www.ndfp.org/response-president-dutertes-call-anti-drug-cooperation/ (28 September 2017).

Creative Associates International (2012). The Philippines electoral security framework: Electoral security assessment. Creative Associates for USAID. Retrieved from http://www.creativeassociatesinternational.com/wp-content/uploads/2003/01/Creative-Philippines-Electoral-Security-Assessment-Final.pdf (22 November 2016).

Dammann, K. (2012). Human rights against collective and organized violence. *Revista Filosófica de Coimbra*, 21(41), 247–264. Retrieved from http://www.uc.pt/fluc/dfci/public_/publicacoes/vol_21_n_41/human_rights (9 May 2018).

Department of Interior and Local Government (2010) Memorandum circular 2010–94. Basic information on Police Auxiliary Units (PAUs). Quezon City: Department of Interior and Local Government. Retrieved from http://www.dilg.gov.ph/PDF_File/issuances/memo_circulars/DILG-Memo_Circular-2010920-277a5d99a9.pdf (9 November 2016).

Dizon, N. (2017, February 23). "Fight the fear" under Duterte rule, FLAG's Diogno urges Filipinos. *Philippine Daily Inquirer*. Retrieved from http://newsinfo.inquirer.net/874712/fight-thefear-under-duterte-rule-flags-diokno-urgesfilipinos (28 September 2017).

Durkheim, E. (1960 [1933]). *The division of labor in society*. Glencoe, IL: Free Press of Glencoe. Retrieved from http://fs2.american.edu/dfagel/www/Class%20Readings/Durkheim/Division%20Of%20Labor%20Final%20Version.pdf (10 May 2018).

Duterte, R. (2015). National platform of government thrust and priority programs. Retrieved from https://i2.wp.com/www.manilalivewire.com/wp-content/uploads/2015/11/11254574_1651101205151450_9194550690694276661_n.jpg (23 February 2018).

Duterte, R. (2017). State of the Nation address 2017. Retrieved from http://www.nwpc.dole.gov.ph/pages/download/messages/2017/President-Duterte-2017-SONA.pdf (27 February 2018).

Espino, I. C. (2004). Counterinsurgency: The role of the paramilitaries. Master's thesis, Naval Postgraduate School, Monterey, CA. Retrieved from http://www.dtic.mil/dtic/tr/fulltext/u2/a429863.pdf (8 September 2015).

European Parliament (2008a). Human rights situation in the Philippines. Re. Pl6 TA(2007)0171 in C 74E/788. Retrieved from http://eur lex.europa.eu/legalcontent/EN/TXT/PDF/?uri=CELEX:52007IP0171&from=EN (18 October 2016).

European Parliament (2008b). Note on the Republic of the Philippines. Re. DGExPo/PolDep/Note/2007_197. Retrieved from http://euinformationservice.rsconsulting.com/Policy%20Department%20for%20External%20Relations/Countries%20and%20Regions/Asia/2.%20Country%20Notes/Philippines/Current%20Political%20Situation,%20Human%20Rights,%20Environment%20Situation.PDF (18 October 2016).

European Parliament (2012). Cases of impunity in the Philippines. Re. P7_TA(2012)0264 and (2013/C 332 E/20). Official Journal of the European Union C 332 E/99. Brussels. Retrieved from http://eur-lex.europa.eu/legal-content/EN/TXT/?uri=uriserv:OJ.CE.2013.332.01.0099.01.ENG&toc=OJ:C:2013:332E:TOC (18 October 2016).

Fischer-Lescano, A. (2005). *Globalverfassung*. Weilerswist: Velbrück Wissenschaft.

Fischer-Lescano, A., & Teubner, G. (2004). Regime-collisions: The vain search for legal unity in the fragmentation of global law. *Michigan Journal of International Law*, 25(4), 999–1046. Retrieved from https://www.jura.uni-frankfurt.de/42852885/regimecollisions.pdf (15 November 2016).

Gita, R. A. (2017, November 2). Roque on Duterte's remarks: "Get used to it". *SunStar*. Retrieved from http://www.sunstar.com.ph/manila/local-news/2017/11/02/roque-dutertes-profanity-laced-remarks-get-used-it-572649 (27 February 2018).

Global Security (n.d. a). Civilian Armed Forces Geographical Units (CAFGU). Retrieved from http://www.globalsecurity.org/military/world/philippines/cafgu.htm (9 November 2016).

Global Security (n.d. b). CVO — Civilian Volunteer Organization. Retrieved from http://www.globalsecurity.org/intell/world/philippines/cvo.htm (9 November 2016).

Günther, K. (1994). Kampf gegen das Böse? Zehn Thesen wider die ethische Aufrüstung der Krimilapolitik. *Kritische Justiz*, 27, 135–157. Retrieved from https://www.nomos-elibrary.de/10.5771/0023-4834-1994-2-135.pdf (11 November 2017).

HRW (2009). You can die any time. New York, NY: Human Rights Watch. Retrieved from https://www.hrw.org/report/2009/04/06/you-can-die-any-time/death-squad-killings-mindanao (6 December 2014).

ICC (2016). Statement of the Prosecutor of the International Criminal Court, Fatou Bensouda concerning the situation in the Republic of the Philippines. International Criminal Court Office of the Prosecutor. Retrieved from https://www.icc-cpi.int/pages/item.aspx?name=161013-otp-stat-php (3 November 2017).

ICC (2018). Statement of the Prosecutor of the International Criminal Court, Mrs Fatou Bensouda, on opening preliminary examinations into the situations in the Philippines and in Venezuela. International Criminal Court Office of the Prosecutor. Retrieved from https://www.icc-cpi.int/Pages/item.aspx?name=180208-otp-stat (1 March 2018).

International Labor Rights Fund (2007). Request for the review of the GSP status of the Republic of the Philippines for violation of workers' rights. Washington, DC. Retrieved from https://laborrights.org/sites/default/files/publications-and-resources/GSPPhilippines.pdf (1 March 2017).

Jones, P. (2015). Track two diplomacy in theory and practice. Stanford, CA: Stanford University Press.

Karapatan (1997). The Ramos presidency and human rights. Quezon City: KARAPATAN (Alliance for the Advancement of People's Rights). Retrieved from http://www.omct.org/files/2003/09/2437/stateviolence_philippines_03_eng.pdf (12 November 2016).

Karapatan (2010). 2010 Year-end report on the human rights situation in the Philippines. Quezon City: KARAPATAN (Alliance for the Advancement of People's Rights). Retrieved from http://www.chrp.org.uk/wp-content/uploads/2010/12/2010-HR-Report-1Dec2010.pdf (5 November 2016).

Karapatan (2016). Jail BS Aquino!. Quezon City: KARAPATAN (Alliance for the Advancement of People's Rights). Retrieved from http://www.karapatan.org/print/1374 (30 August 2016).

Karapatan (2017). 2016 Year-end report. Quezon City: KARAPATAN (Alliance for the Advancement of People's Rights). Retrieved from http://www.karapatan.org/files/2016%20Karapatan%20YEReport.pdf (25 May 2017).

Katigbak, J. (2018, February 1). US on EJKs in PHL: Encountering signs of decline. *Philippine Star*.

Kraft, H. J. S. (2010). The foibles of an armed citizenry: Armed auxiliaries of the state and private armed groups in the Philippines (overview). In D. Rodriguez (Ed.), *Primed and purposeful: Armed groups and human security efforts in the Philippines* (pp. 185–215). Geneva: Small Arms Survey.

Kreuzer, P. (2009). Private political violence and boss-rule in the Philippines. *Behemoth. A Journal on Civilisation*, 2(1), 47–63. Retrieved from https://ojs.ub.uni-freiburg.de/behemoth/article/view/721/647 (8 September 2015).

Kreuzer, P. (2011). Domination in Negros Occidental: Variants on a ruling oligarchy. PRIF Report No. 112. Frankfurt am Main: Peace Research Institute Frankfurt. Retrieved from https://www.hsfk.de/fileadmin/HSFK/hsfk_downloads/prif112.pdf (19 March 2015).

Kreuzer, P. (2012). The Philippines: Mafia style domination in an oligarchic polity. Draft paper prepared for IPSA's XXIInd World Congress of Political Science, Reshaping power, shifting boundaries, Madrid 2012. Retrieved from http://paperroom.ipsa.org/papers/paper_11398.pdf (22 July 2017).

Kvanvig, G. (n.d.). ASEAN, sovereignty and human rights. Oslo: Norwegian Center for Human Rights, Faculty of Law, University of Oslo. Retrieved from http://www.jus.uio.no/smr/english/about/programmes/vietnam/docs/asean-sovereignty-and-human-rights---gisle-kvanvig.pdf (12 January 2018).

Leary, V., Ellis, A. A., & Madlener, K. (1984). The Philippines: Human rights after martial law. Geneva: The International Commission of Jurists. Retrieved from http://icj.wpengine.netdna-cdn.com/wp-content/uploads/1984/01/Philippines-human-rights-mission-report-1984-eng.pdf (11 October 2016).

Luhmann, N. (1998a). *Das Recht der Gesellschaft*. Frankfurt: Suhrkamp.

Luhmann, N. (1998b). *Die Gesellschaft der Gesellschaft Teilband 1 und 2*. Frankfurt: Surhkamp.

Luhmann, N. (2012a). *Die Moral der Gesellschaft*. Frankfurt: Suhrkamp.

Luhmann, N. (2012b). *Soziale Systeme*. Frankfurt: Suhrkamp.

Macas, T. (2017, October 19). Duterte dares Western countries to take over PHL drug war. *GMA News*. Retrieved from http://www.gmanetwork.com/news/news/nation/630170/duterte-dares-western-countriesto-take-over-phl-drug-war/story/ (24 February 2018).

May, R. J. (1992). Vigilantes in the Philippines. Philippine Studies Occasional Paper No. 12. Honolulu, HI: Center for Philippine Studies, School of Hawaiian, Asian and Pacific Studies, University of Hawai'i at Manoa. Retrieved from http://citeseerx.ist.psu.edu/viewdoc/download?doi=10.1.1.732.2471&rep=rep1&type=pdf (8 March 2015).

McCoy, A. W. (Ed.). (2010). *An anarchy of families*. Quezon City: Ateneo de Manila University Press.

Melo, J. A., Mantaing, N. M., Zuño, J. R., Gonzales, N. T, & de Dios M. Pueblos, J. (2006). Report of the independent commission to address media and activist killings. Metro Manila. Retrieved from http://pcij.org/blog/wp-docs/melo_commission_report.pdf.

Migdal, J. S. (1988). *Strong societies and weak states: State–society relations and state capabilities in the third world*. Princeton, NJ: Princeton University Press.

Moreno, A. (2017). Statement of the Philippine Province Jesuits on fighting the evil of illegal drugs. Retrieved from https://www.phjesuits.org/portal/coming-together-in-the-power-of-the-spirit-statement/ (26 February 2018).

NCCP (1988). Betrayed hopes — "... human rights violations continue ...". Occasional papers 1988–2. Quezon City: Human Rights Desk, National Council of Churches in the Philippines.

OHCHR (2000). Human rights: A basic handbook for UN staff. Geneva: Office of the High Commissioner for Human Rights.

Paredes, A. K. R. (2010). Definitions: Warlords, private armies, and partisan armed groups. In D. Rodriguez (Ed.), *Primed and purposeful: Armed groups and human security efforts in the Philippines* (p. 203). Geneva: Small Arms Survey.

Philippine Senate (2016). Report 18, Committee on Justice and Human Rights jointly with the Committee on Public Order and Dangerous Drugs. Pasay City: Office of the Secretary. https://www.senate.gov.ph/lisdata/2518121701!.pdf.

PNP (2016). PNP anti-illegal drugs campaign plan — Project 'Double Barrel'. Command Memorandum Circular No. 16-2016. Philippine National Police. Retrieved from http://didm.pnp.gov.ph/Command%20Memorandum%20Circulars/CMC%20 2016-16%20PNP%20ANTI-ILLEGAL%20DRUGS%20CAMPAIGN%20PLAN%20 %E2%80%93%20PROJECT%20DOUBLE%20BARREL.pdf (13 September 2017).

PNP (2017). #RealNumbersPH update: The anti-drug campaign in the Philippines. Philippine National Police. Retrieved from http://www.pnp.gov.ph/images/2017 News/realnumbers/rn_82917.pdf (14 September 2017).

Punay, E., & Romeo, A. (2017, September 9). Rody meets with Carl's parents. *Philippine Star*.

Ranada, P. (2017, March 3). Duterte on war on drugs: "More killings to come". *Rappler*. Retrieved from https://www.rappler.com/nation/163067-duterte-more-killings-drug-war (28 November 2017).

Romeo, A., & Romeo, P. (2017, April 3). Palace to EU: Don't link trade to EJK. *Philippine Star.*

Sidel, J. T. (1999). *Capital, coercion, and crime bossism in the Philippines.* Stanford, CA: Stanford University Press.

Sison, J. M. (1971). *Philippine society and revolution.* Manila: Pulang Tala.

UN (2015). Transforming the world: The 2013 agenda for sustainable development. A/RES/70/1. New York, NY: United Nations. Retrieved from https://sustainable development.un.org/content/documents/21252030%20Agenda%20for%20Sustainable%20Development%20web.pdf (10 February 2018).

UN (n.d.). Sustainable Development Goal 16. Sustainable Development Platform. New York, NY: United Nations. Retrieved from https://sustainabledevelopment.un.org/sdg16 (9 February 2018).

US Department of State (1995). The Philippines human rights practices, 1994. Retrieved from http://dosfan.lib.uic.edu/ERC/democracy/1995_hrp_report/95hrp_report_eap/Philippines.html (1 September 2016).

US Department of State (1997). Country report on human rights practices 1996: Philippines. Retrieved from http://www.refworld.org/cgi-bin/texis/vtx/rwmain?page=printdoc&docid=3ae6aa310 (1 September 2016).

US Department of State (1998). The Philippines country report on human rights practices for 1997. Retrieved from https://1997-2001.state.gov/www/global/human_rights/1997_hrp_report/philippi.html (1 September 2016).

US Department of State (1999). Philippines country report on human rights practices for 1998. Retrieved from http://www.state.gov/www/global/human_rights/1998_hrp_report/philippi.html (1 September 2016).

US Department of State (2000). 1999 country reports on human rights practices: Philippines. Retrieved from http://www.state.gov/www/global/human_rights/1999_hrp_report/philippi.html (1 September 2016).

US Department of State (2001). Philippines country reports on human rights practices, 2000. Retrieved from http://www.state.gov/j/drl/rls/hrrpt/2000/eap/764.htm# (1 September 2016).

US Department of State (2014). Philippines 2013 human rights report. Retrieved from http://www.state.gov/documents/organization/220436.pdf (1 September 2016).

US Department of State (2016). Philippines 2015 human rights report. Retrieved from http://www.state.gov/documents/organization/253005.pdf (1 September 2016).

US Senate (2008). Extrajudicial killings in the Philippines: Strategies to end the violence. Hearing before the subcommittee on East Asian and Pacific Affairs of the Committee on Foreign Relations, March 14, 2007. Washington, DC. Retrieved x/21/2016 from https://www.gpo.gov/fdsys/pkg/CHRG-110shrg40811/pdf/CHRG-110shrg40811.pdf (21 October 2016).

Van der Kroef, J. M. (1986). Private armies and extrajudicial violence in the Philippines. *Asian Affairs*, 13(4), 1–21. Retrieved from URL: http://www.jstor.org/stable/30172017 (5 September 2015).

Villamordec, F. (2016, December 22). Rodrigo Duterte of Philippines calls U.N. human rights chief an "Idiot". *New York Times*. Retrieved from https://www.nytimes.com/2016/12/22/world/asia/rodrigo-duterte-philippines-zeid-raad-al-hussein.html (23 February 2018).

Zenarosa, M. A. (2010). Executive summary of "A journey towards H.O.P.E." The Independent Commission against Private Armies report to the President. Memorandum for Leandro R. Mendoza (Executive Secretary, Office of the President). (Unpublished). Manila: Office of the President.

PART 3

SDGs and Progress on the Social Agenda in Middle-Income and Developing Countries in Southeast Asia

CHAPTER 6

Sustainable Development Goals and Capacity Building in Higher Education in Malaysia and ASEAN

Azirah Hashim and Aliyyah Nuha Faiqah Azman Firdaus

1 Introduction

This chapter examines higher education and the Sustainable Development Goals (SDGs) in the ASEAN region by focusing on the contribution of higher education and research in striving to achieve not only SDG 4, for quality education, but also other SDGs in the economic, social, and environmental areas. It studies the targets of ASEAN member states in general, and Malaysia in particular, in the education sphere and related areas, as well as plans for implementation. This involves a review of multi-level governance and regional mechanisms, local government, stakeholder involvement, and inter-regional cooperation with the European Union (EU). The role of the ASEAN Secretariat, the ASEAN University Network, the Southeast Asian Ministers of Education Organization Regional Centre for Higher Education and Development (SEAMEO RIHED), and the SHARE EU ASEAN project will be discussed. For Malaysia, the chapter will focus on the Declaration on Higher Education in ASEAN, the CLMV programme and mobility programmes. Cross-national engagement initiatives to address the development divide through knowledge exchange, engagement, and collaboration for the academic community will also be evaluated, in particular in Cambodia and Laos. Although differing in terms of development, these two countries are emerging as higher education players in the region. Spurred on by the ASEAN Economic Community since 2015 and inspired by the ASEAN Community Vision 2025, both nations are aware of the need for highly educated and skilled people if socio-economic development is to be achieved.

2 Higher Education and the Sustainable Development Goals

Building upon the success of the Millennium Development Goals (MDGs), the SDGs aim to improve life opportunities and outcomes for the less privileged parts of society. The role of education in achieving several of the SDGs is crucial. Education is claimed to be "the bedrock of sustainable development,

 | DOI:10.1163/9789004391949_007

contributing to its social, economic and environmental dimensions, and underpins peace and security" as well as "creating a lasting impact on health and gender equality, and leading to safer, more resilient and stable societies" (UNICEF, 2013, p. 2).

This chapter starts by delineating how SDG 4 for quality education ("Ensure inclusive and equitable quality education and promote lifelong learning opportunities for all")[1] directly contributes to other SDGs in the social, economic, and environmental spheres. A recent study by Educate A Child and FHI 360 (2016) offers a useful exploration of the connections between education and the other SDGs. For example, SDG 4 impacts in several ways on SDG 8 for Decent Work and Economic Growth ("Promote sustained, inclusive and sustainable economic growth, full and productive employment and decent work for all"). It is anticipated that the post-2015 era will see education playing a crucial part in addressing the global development challenges implicit in SDG 8: "Demographic trends indicate that as disadvantaged groups such as women and rural populations gained access to education, they shifted from a choice between agricultural or household-oriented work to an increased supply of labor choices" (Educate A Child & FHI 360, 2016, p. 15). The study notes that such a shift from agricultural to industrial or technological work is likely to result in turn in higher levels of education; in a kind of virtuous cycle, the type of work available can impact on the content of educational curricula (ibid., p. 16). On the other side of the balance sheet, the report points out that child labour — which may be a feature of economic growth — can act as a barrier to education (ibid.).

Similarly, the goal of quality education is likely to impact on SDG 16 ("Promote peaceful and inclusive societies for sustainable development, provide access to justice for all and build effective, accountable and inclusive institutions at all levels"). The Educate A Child and FHI 360 study cites a survey carried out in 36 low- and middle-income countries, which found a link between education and levels of voting, especially in countries with lower levels of education: "In India, education also had a positive effect on the probability of campaigning, discussing electoral issues, attending rallies and establishing contacts with local government officials in the states of Madhya Pradesh and Rajasthan (UNESCO 2014)" (Educate A Child & FHI 360, 2016, p. 26). A link was also found between education and (in)tolerance:

> In Latin America, those with secondary education were 47 percent less likely to express intolerance of other racial groups, and in Arab States

1 For the text of all 17 SDGs, see https://sustainabledevelopment.un.org/sdgs.

> people with secondary education were 14 percent less likely to express intolerance towards other religions (both compared to primary).... In Sub-Saharan Africa, those with secondary school completion were 23 percent less likely to express intolerance towards those with HIV and in Central/ Eastern Europe those with secondary education were 16 percent less likely to express intolerance towards immigrant groups (UNESCO 2014). (ibid.)

The Educate A Child and FHI study also reports on research which indicates that "countries with high rates of education inequality [in 1960–2010] were twice as likely to experience conflict in the following years. A subsequent study found that conflict exacerbated education inequalities by gender, wealth and other dimensions. For example, conflict worsened gender equality in educational attainment for females by 5.4 percent in fragile countries" (ibid.).

In relation to SDG 15 ("Protect, restore and promote sustainable use of terrestrial ecosystems, sustainably manage forests, combat desertification, and halt and reverse land degradation and halt biodiversity loss"), the study highlights the powerful impact of education on awareness of environmental issues: "respondents with more education were more likely to have signed a petition, given money or taken part in a protest or demonstration, in relation to the environment, over the past five years" (ibid., p. 25). In some cases, the likelihood of such political action jumped from 12% amongst those with no more than primary education, to 26% for those with secondary education (ibid.).

3 Malaysia, Higher Education and the Sustainable Development Goals

With this snapshot of the relevance of education to all the SDGs as background, we now turn to the Malaysian case. In May 2015, at its gathering in Incheon, Republic of Korea, the World Education Forum adopted the Incheon Declaration/Education 2030 Framework for Action, which reaffirmed earlier commitments to Education for All (EFA). The Incheon Declaration/Education 2030 was incorporated into the Sustainable Development Goals as SDG 4 and adopted by the UN General Assembly in September 2015. "SDG4/Education 2030 was developed through a broad consultative process driven and owned by Member States and facilitated by UNESCO as well as other partners and guided by the EFA Steering Committee" (UNESCO, 2015, p. 22). Each SDG has specific targets to achieve. SDG 4 has seven targets which can be summarized as follows:

- Target 1: Primary and secondary education for all, leading to relevant and effective learning outcomes
- Target 2: Early childhood development, care and pre-primary education leading to school-readiness
- Target 3: Equal access to technical, vocational, and tertiary education, including university
- Target 4: Skills for employment, decent jobs and entrepreneurship
- Target 5: Gender equality, equity and inclusion, focusing on vulnerable groups
- Target 6: Youth and adult literacy and numeracy
- Target 7: Knowledge and skills to promote sustainable development and global citizenship.[2]

As argued above, SDG 4 (quality education) is crucial for the success of all the SDGs. According to Wang (2016, p. 10), SDG 4 focuses on "increased and expanded access, inclusion and equity, quality and learning outcomes at all levels, within a lifelong learning approach". The use of ICTs is encouraged, while gender equality is championed through a rights-based approach stressing "not only access [to education] and completion, but also empowerment for girls and women" (ibid.). UNESCO has been given the mandate to lead and coordinate SDG 4/Education 2030 through "advocacy for political commitment and resource mobilization; capacity development; policy dialogue and advice, knowledge-sharing, standard-setting; facilitating South–South and triangular cooperation", as well as for monitoring progress towards the goals and targets, through the UNESCO Institute for Statistics (UIS) and the Global Education Monitoring (GEM) Report (ibid., p. 21).

Malaysia's foray into sustainable development began in the 1970s with the introduction of the New Economic Policy (NEP) to eliminate poverty and address societal imbalance. Every five years, it produces a national development plan: since the time of the NEP, these plans have incorporated elements of sustainable development such as "sustainable economic growth, access to basic infrastructure and utilities, access to education and healthcare services and mainstreaming of environmental conservation" (Government of Malaysia, 2017, p. iv).

In 2009, Malaysia developed the New Economic Model (NEM) based on the three pillars of high income, inclusivity, and sustainability. These pillars are the equivalent of the three main components of the SDGs, namely economy, society, and environment. The current national development plan — the Eleventh

2 See Wang (2016, p. 7); also see UNESCO (2017, p. 1). For the full text of the SDG targets, see https://sustainabledevelopment.un.org/sdgs.

Malaysia Plan (2016–2020) — is based on the three NEM pillars and has the theme "Anchoring Growth on People" (Government of Malaysia, 2015). This stresses a people-centred approach, ensuring that the development process is fully inclusive (ibid.), and again echoes one of the themes of the Sustainable Development Goals, "Leaving no one behind".

By adopting the 2030 Agenda for Sustainable Development, Malaysia indicated its commitment to the broad goal of a more sustainable, resilient, and inclusive development. In terms of SDG 4, two important regional follow-up meetings were hosted by Malaysia after the adoption of the Education 2030 document, namely the Asia-Pacific Regional Conference on Education and Training: Making Skills Development Work for the Future (Kuala Lumpur, 3–5 August 2015) and the 2nd Asia-Pacific Regional Policy Forum on Early Childhood Care and Education (Putrajaya, 19–21 July 2016). SDG 4 is addressed in the Eleventh Malaysia Plan under its Strategic Thrust 3 (Accelerating human capital development for an advanced nation) which is to be pursued through a range of policies and plans including the Malaysia Education Blueprint 2013–2025, the Malaysia Education Blueprint (Higher Education) 2015–2025, and the National Higher Education Strategic Plan, Blueprint on Enculturation of Lifelong Learning for Malaysia 2011–2020. As Wang points out, "the SDG4/ Education 2030 can serve as a reference document for the national review of the Blueprint implementation" (Wang, 2016, p. 22).

Importantly, Malaysia is taking a leading role in the region when it comes to SDG 4-related matters, and is emerging as an important hub for higher education. It has "become a donor country by establishing a Funds-in-Trust Agreement with UNESCO, which is benefiting developing and least developed countries and contributing to South–South cooperation in delivering the SDG4/Education 2030 agenda" (ibid., p. 23). It is also considering ratifying the 2011 Tokyo Convention on the Recognition of Higher Education Qualifications, which could increase student mobility across borders (ibid.).

According to the Malaysian Education Act of 1996 (Act 550), "Education in Malaysia is an ongoing effort towards further developing the potential of individuals in a holistic and integrated manner so as to produce individuals who are intellectually, spiritually, emotionally and physically balanced and harmonious, based on a firm belief in and devotion to God" (Loke & Hoon, 2011, p. 100, quoted in ASEAN, 2014, p. 59). For the period 2012–2025, the Malaysia Education Blueprint (Higher Education) has set out five aspirations as illustrated in Figure 6.1.

The Ministry of Education (MoE) aims to achieve universal access to education and 100% enrolment from pre-school to upper secondary school by 2020. The quality of education must be comparable to international standards and

FIGURE 6.1 Five system aspirations
SOURCE: PEREIRA (2016, P. 14)

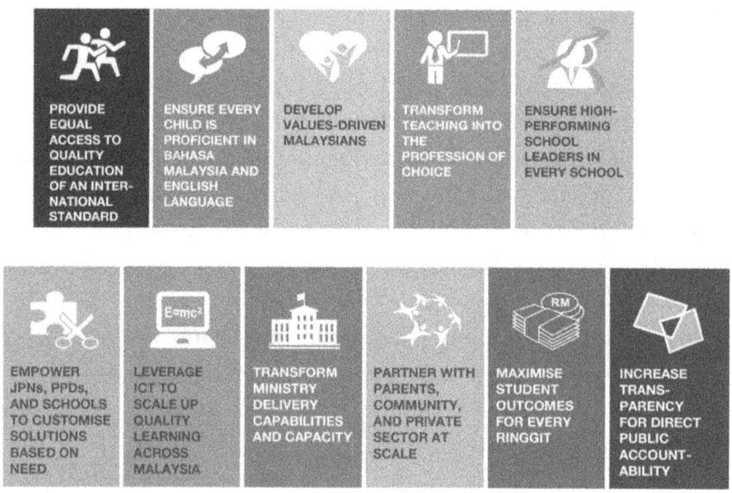

FIGURE 6.2 Eleven shifts to transform the Malaysian education system
SOURCE: MINISTRY OF EDUCATION, 2012, REPRODUCED IN PEREIRA (2016, PP. 14–15)

adopt best practices. Malaysia also aspires to be in the "top third of countries" for international benchmarks such as Programme for International Student Assessment (PISA) and Trends in International Mathematics and Science Study (TIMSS), within 15 years (Pereira, 2016, p. 14). By 2020, equity goals will be achieved by reducing the urban–rural, socio-economic, and gender achievement gaps by half. To accomplish these five aspirations, the MoE proposes eleven shifts (changes) (see Figure 6.2). These are intended to transform the Malaysian education system in three waves (time periods):

> In Wave 1 (2013 to 2015) the focus is on strengthening the vocational education and creating alternative pathways and ways of improving opportunities for special needs students. In Wave 2 (2016–2020) the MoE will focus on scaling up programmes piloted under the first Wave; and in the final Wave (2021–2025) refining of individual pathways and greater involvement of the private sector are envisioned. (Education for All, 2015, p. 14)

4 ASEAN Community and the Sustainable Development Goals

The post-2015 development agenda for ASEAN integration in the field of higher education comes with challenges to achieve new sustainable development goals to ensure inclusive, equitable, quality education and sustainable education capacities as well as promote lifelong learning and opportunities for all.

The Preamble to the 2007 ASEAN Charter states that the ASEAN nations are "Resolved to ensure sustainable development for the benefit of present and future generations, and to place the well-being, livelihood and welfare of the peoples at the centre of the ASEAN community building process" (ASEAN, 2007, p. 2). While Southeast Asia's performance in achieving the MDGs has been mixed, it can count among its successes a reduction in extreme poverty, improved school enrolment, the eradication of a number of diseases including tuberculosis, the protection of some biodiversity hot spots, and the provision of safe drinking water (UN ESCAP, 2017, p. ii). H. E. Vongthep Arthakaivalvatee, ASEAN Deputy Secretary-General for Socio-Cultural Affairs reiterated that ASEAN recognizes the importance of creating synergy between the mutually-reinforcing ASEAN 2025 Blueprints and the SDGs within the framework of ASEAN–UN cooperation (ASEAN, 2016). It is clear that "ASEAN is the subregion's mechanism that boosts the implementation of the 2030 Agenda to attain the Sustainable Development Goals" (UN ESCAP, 2017, p. ii). The collaboration between ASEAN and the UN appears to be a smart partnership, as it offers a joint effort towards executing both the global and the subregional agenda, with programmes and initiatives put into place to tackle the challenges of implementation.

4.1 *Complementarity between ASEAN Community Vision 2025 and the SDGs*

Southeast Asia is in the unique position of having two closely related routes to achieving the SDGs, namely, the 2030 Agenda and the ASEAN Community Vision 2025, subtitled "Forging ahead together" (ASEAN, 2015). Both agendas aim to achieve the level of well-being that any individual can reasonably expect during his/her lifetime. The creation of an ASEAN Community is realized in Vision 2025 through its three main pillars: the ASEAN Political Security Community (APSC), ASEAN Economic Community (AEC) and the ASEAN Socio-Cultural Community (ASCC). Each pillar has its own blueprints; together they will achieve the ASEAN vision by 2025 of "a peaceful, stable, and resilient Community with enhanced capacity to respond to challenges"; "an outward-looking region within a global community of nations, while maintaining ASEAN centrality"; "vibrant, sustainable, and highly integrated economies, enhanced ASEAN connectivity as well as strengthened efforts in narrowing the development gap"; "an ASEAN empowered with capabilities to seize opportunities and address challenges in the coming decade" (UN ESCAP, 2017, p. 27). ASEAN's people-centred goals are most clearly expressed in the ASCC Blueprint, which spells out a vision of "a community that engages and benefits the peoples and is inclusive, sustainable, resilient and dynamic" (ibid.).

The UN ESCAP (2017) document goes on to highlight the complementarities between the ASEAN Vision 2025 goals and the SDGs. First, they both have a people-centric focus, reflected in the theme "leaving no one behind" of the 2030 Agenda, and the "one vision, one identity, one community" motif of Vision 2025 (ibid., p. 28). Second, "there is clear alignment at the sectoral outcome level as all the elements in ASEAN Community blueprints respond to or support the SDGs.... Conversely, all the SDGs have equivalent goals in the three Community blueprints" (ibid.). This mutual alignment is illustrated by a Table, which is reproduced here (Table 6.1).

TABLE 6.1 Alignment of 2030 agenda and ASEAN vision 2025

ASEAN community pillar blueprint elements	**Aligned/supported SDGS**
ASEAN Political Security Blueprint Elements	
A. Rules-based, people-oriented, people-centred community	16
B. Resilient community in a peaceful, secure and stable region	16, 17
C. ASEAN centrality in a dynamic and outward looking region	17
D. Strengthened ASEAN institutional capacity and presence	17
ASEAN Economic Community Blueprint Elements	
A. A highly integrated and cohesive economy	8, 9, 17
B. A competitive, innovative, and dynamic ASEAN	9, 12, 17
C. Enhanced connectivity and sectoral cooperation	2, 3, 4, 5, 7, 9, 17

TABLE 6.1 Alignment of 2030 agenda and ASEAN vision 2025 (*cont.*)

ASEAN community pillar blueprint elements	Aligned/supported SDGs
D. A resilient, inclusive, people-oriented and people-centred ASEAN	1, 4, 5, 8, 10, 17
E. A global ASEAN	16, 17
ASEAN Socio-cultural Community Blueprint Elements	
A. A committed, participative, and socially-responsible community (engages and benefits the people)	1, 2, 16, 17
B. An inclusive community that promotes high quality of life, equitable access to opportunities for all and promotes and protects human rights	4, 5, 10, 16
C. A sustainable community that promotes social development and environmental protection	6, 9, 11, 12, 13, 14, 15
D. A resilient community with enhanced capacity and capability to adapt and respond to social and economic vulnerabilities, disasters and climate change	3, 5, 8, 9, 12, 16, 17
E. A dynamic and harmonious community that is aware and proud of its identity, culture and heritage	4, 8, 16, 17

SOURCE: UN ESCAP (2017, P. 29)

According to UN ESCAP (2017), there are also distinct differences between Vision 2025 and 2030 Agenda, which in fact accentuate their complementarity. UN ESCAP (2017, p. 28) summarizes the differences as follows: Vision 2025 is a "framework of aspirational goals that drive the process of regional community and nation building", whereas the SDGs are "indivisible aspirational goals that can strengthen and transform nations and peoples so that these may contribute to the betterment of the global community and the planet". Thus, "Vision 2025 focuses on the regional community of states while the 2030 Agenda focuses on the global community and its states".

5 Multi-level Governance and Regional and Inter-regional Mechanisms

The pursuit of equitable, quality education is a universal effort, constantly on the agenda of governments in both developing and developed nations. The then Director of SEAMEO RIHED, Arief Sadiman noted that "the promotion of a quality and equitable education is common educational policy across Southeast Asia" (quoted in Pereira, 2016, p. 7). The cultural diversity present in Southeast Asia establishes the national agenda for education — to build social cohesion.

In order to meet the challenges of the global environment and knowledge-based economy, there is an increasing demand for skilled human capital, as reflected in various ASEAN strategy documents. Priority 2 in the ASEAN State of Education Report 2013 stresses "the need to increase access to quality primary and secondary education across the ASEAN region", as well as "the need for an improved quality of education, including higher performance standards, more opportunities for lifelong education, and the more widespread provision of professional development support" (ASEAN, 2014, p. 10). The ASEAN Work Plan on Education (2016–2020) focuses on eight areas: those related to higher education are education for sustainable development (ESD) and higher education development and quality assurance mechanisms. The ASEAN University Network (AUN) has five strategic goals — G1, quality gap narrowed; G2, emerging priorities engaged; G3, exposure expanded; G4, in-depth awareness; and G5, teaching and learning. Although the Kuala Lumpur Declaration on Higher Education mentions nine points, the majority of these can be seen to align with the AUN's five, as indicated below:

(1) quality education (G1)
(2) intra-ASEAN mobility of students and scholars (G3, G4)
(3) university–government–industry–community cooperation (G2)
(4) diversify Higher Education for innovation-driven ASEAN community (G2, G5)
(5) contribute to and foster sustainable local community development (G2, G4)
(6) increase visibility of ASEAN Higher Education globally (G1, G2, G3, G4). (cited in Morshidi Sirat, 2017, p. 19)==

Since the establishment of the ASEAN Economic Community (AEC) in 2015, higher education has become an enabler in driving the free flow of people (academics, students, and others) as well as the free flow of information and knowledge in the region. This resonates with the aspirations of the "Europe of knowledge" or the "knowledge economy" (Pol, 2015, p. 14) within the EU's Europe 2020 Strategy. In the ASEAN's post-2015 vision on education, the AEC vision and key elements of education emphasize the "widening of educational cooperation" for quality, inclusive education, and lifelong learning, which feeds into the outline of the Vision 2025 (ASEAN, 2015). The three pillars of the ASEAN Community are tasked with promoting "an innovative ASEAN approach to higher education" (ibid., p. 117) through people-to-people connectivity and mobility; free flow of ideas, knowledge, expertise and skills; and strong regional and global cooperation in higher education. The ASEAN 5-Year Work Plan on Education (2016–2020) under the ASCC 2025 contains key elements such as higher education quality assurance, university–industry and global partnership, capacity building for educators, education for sustainable development and others. These are to fulfil the vision of a dynamic ASEAN higher education area as envisioned in the blueprint.

Two bodies formed by the ASEAN Secretariat within the Education, Youth and Sports Division — Senior Officials Meeting on Education (SOM-ED) and ASEAN Education Ministers Meeting (ASED) — carried out a series of programmes to promote ASEAN cooperation in higher education. Working with them are other key players including:

- the ASEAN University Network (AUN)
- the ASEAN Quality Assurance Network (AQAN)
- the Southeast Asia Ministers of Education Organization (SEAMEO) and SEAMEO-RIHED (Regional Centre for Higher Education and Development)
- individual ASEAN member states (education ministries, institutions, and bodies)
- various dialogue partners of ASEAN+3 (Japan, China, and Korea), East Asia Summit nations, the EU, etc.
- global institutions such as UNESCO and UNICEF.

The programmes include:

- international mobility programmes (such as the AUN mobility programmes, the AIMS programme, ASEAN–China and ASEAN–ROK academic exchange programmes, AUN/SEED-NET education network with the Japanese government, and others)
- projects related to the harmonization of higher education in ASEAN (such as AQAN, the ASEAN Qualification Reference Framework or AQRF, the AUN-QA programme)
- credit transfer systems including AUN-ACTS (ASEAN Credit Transfer System), UMAP-UCTS by the University Mobility in Asia-Pacific, the ACTFA (Academic Credit Transfer Framework) etc.
- intra-ASEAN cooperation in higher education (scholarships via bilateral and subregional agreements, leadership-building programmes, research collaboration, TVET training and so on).

In terms of ASEAN–EU cooperation in higher education, several initiatives are underway stemming from a paradigm shift in both European and ASEAN higher education, partly driven by the external dimension of the Europe 2020 strategy for education. Programmes include EU-SHARE, Erasmus+, Horizon

2020, Newton Fund, and ASEAN-QA. EU-SHARE has carried out a number of capacity-building activities in ASEAN since 2015 to strengthen regional cooperation, and enhance the quality, competitiveness, and internationalization of ASEAN higher education institutions. EU-SHARE comprises a consortium of EU partners including the British Council, ENQA, DAAD, EUA, Nuffic, and Campus France, and ASEAN players (AUN, AQAN, SEAMEO, etc.) including the Asia-Europe Meeting (ASEM), and other relevant national-level target beneficiaries and stakeholders (government bodies, university leaders and managers, higher education experts, quality assurance agencies, students, and industrial employers).

6 Cross-national Engagement Initiatives

In line with UNESCO's Education for All movement, Cambodia and Laos are committed to strengthening and developing their higher education systems to meet the learning needs of the academic community. Both countries have expanded their academic systems/institutions to meet a growing demand for higher education across Asia (Altbach & Umakoshi, 2004; Wiedman, 1995, both cited in Chapman, 2009, pp. 102–103). This expansion was grounded in the economic rationale that education plays a role in improving the socio-economic development and competitiveness of a nation and its communities.

A substantial increase in investments in higher education systems has led to the expansion of HE institutions and universities especially in the capital cities of Vientiane (Laos) and Phnom Penh (Cambodia). This has led to some concerns about, for example, lack of quality higher education services, unskilled human resources, poor supply of instructional materials, lack of research expertise, and inadequate access to education for some disadvantaged communities. Furthermore, tremendous pressures have been placed on the countries' research capacity, which is needed to strengthen the education systems and ultimately to prepare the two nations for regional integration, in line with the AEC's call in 2015 for regional mobility and cooperation between neighbouring countries in the Southeast Asian region.

In this context, it is informative to look at the CLMV (Cambodia, Laos, Myanmar and Vietnam) research programme, a cross-national initiative sponsored by the Ministry of Education in Malaysia. The focus of the project is on addressing the developmental divide through knowledge exchange, engagement and collaboration, and capacity building in research development between Malaysia and the other countries. Here we will examine the challenges and understanding of key issues or concerns in higher education development in the areas of English language teaching/learning (ELT) and education-based research, looking at Cambodia and Laos.

The project aimed to help empower Cambodian and Laotian academic communities through knowledge sharing, skills development, and participation in research, training, and exchange programmes between higher education institutions. A team from the University of Malaya (UM) was involved in the CLMV programme in ELT and education from 2013 to 2017. This programme operated in tandem with Malaysia's National Higher Education Strategic Plan (PSPTN Phase 2, 2011–2015) for global outreach and engagement, especially in the ASEAN region. The programme also aligned with the goals of the Initiative for ASEAN Integration (IAI) Strategic Framework to narrow the development gap in the CLMV countries, and the objectives of the AUN for regional mobility and cooperation in terms of teaching, research collaboration, and training and development. The project aimed:

> To explore, understand and help address the challenges and issues faced by the academic community regarding English language teaching/learning and research related to tertiary education.
>
> To develop capacity building activities to enable potential candidates to become leaders/trainers at their institutions and to ensure the continuation and sustainability of training and development programmes.
>
> To explore strategic sites/areas for cooperation between the academics in Malaysia and Cambodia/Laos to facilitate the sharing of knowledge, expertise and ideas across borders — academic development, international staff/student exchanges, and quality of teaching/learning curriculum e.g., in tourism-based education.
>
> AZIRAH HASHIM, YEE, & ALIYYAH NUHA FAIQAH, 2015

The project consisted of a series of training and outreach activities that were conducted by the ELT/education team from UM and associates from other universities. These activities included research and teaching seminars, involvement in a World Bank research project, co-writing of a research paper on higher education in Cambodia (published in an ISI-ranked journal; see (Azirah Hashim, Yee, & Pheak, 2014), international exchanges of staff and academic visits between Malaysia, Cambodia, and Laos, curriculum development workshops on tourism and hospitality, and so on. The series of seminars and workshops focused on topics such as a basic introduction to research, research writing, research methodologies, quantitative analysis, and research skills and techniques. These were presented at several universities in Cambodia and Laos, mainly at the Institute of Foreign Languages, Royal University of Phnom Penh, the University of South-East Asia (Siem Reap), the National University of Laos (Vientiane) and the Souphanouvong University (Luang Prabang), Laos. The purpose of these activities was to empower Cambodian and Laotian academics by giving them the necessary research knowledge and skills in conducting English language teaching (ELT) and education-based research.

Trained academics may become potential leaders/trainers at their respective institutions and be in a position to continue similar capacity-building activities in a sustainable way that is appropriate to their local context through knowledge sharing, training, and skill-development programmes. By engaging these participants in research discussions, it was hoped that there would be better understanding about the challenges and issues faced by the academic communities in Cambodia and Laos, particularly on ELT and their research needs in higher education. Through research discussions and seminar question and answer sessions, the UM team developed closer ties and friendships with Cambodian and Laotian counterparts and will continue to collaborate on future joint projects and co-writing of papers for publication. This exchange thus brought benefits to all parties in terms of research networking, intercultural understanding, capacity building, promotion of Malaysia as a potential hub for higher education, and narrowing of the development gap among these ASEAN countries.

As a follow-up to CLMV Phase I (2011–2012), UM in collaboration with the University of Southeast Asia (USEA) also successfully secured a World Bank grant to develop a curriculum for the tourism and hospitality programme that corresponds to the needs of the AEC and the demands of the industry as ascertained through employer surveys and graduate feedback. Two intensive workshops on curriculum development and teaching methodologies were conducted successfully with the help of UM experts and tourism consultants.

This project involved other university partners from Thailand (Ubon Ratchathani University and Dhurakit Pundit University) and China (Liaoning Technical University), as well as Sokha Angkor Hotel (an employer in Cambodia) and the Siem Reap Department of Tourism. In May 2014, a full week of research training seminars and workshops were conducted at UM for 10 invited delegates from Laotian universities. Academic representatives from Laos were given lectures and ideas about conducting research which they can hopefully impart to colleagues and students in their home institutions. Field trips and visits to the Main Library, Institute of Graduate Studies, IT Centre, Museum of Asian Arts, and the Art Gallery of UM were also arranged to promote intercultural learning between Laotians and Malaysians.

In all the interactions described above, small communication barriers in English speaking and listening were observed, partly because of the low to average level of English proficiency in the CLMV community and the distinct features of different variations of English (e.g., the deletion of the phonemes /f/ and /g/ in spoken Cambodian English). Some difficulties were experienced in the comprehension of specific terminologies, and some words were mispronounced. Very few research/studies on Cambodia and Laos are published in English-language journals as they are usually written in the Khmer or Lao language (the exceptions being articles written by non-Cambodian and non-Laotian scholars). This scenario poses some challenges for Cambodian and Laotian academic discourse.

A number of other problems can arise in these cross-national engagements, mostly at the practical/organizational level. For example, the academics involved in collaborations between international partners may not have the necessary authority to approve the projects, with such decisions usually requiring the consent of the president/rector of the educational institution. Another challenge that has been observed is the generation of ideas for joint activities; in the initial stages, such discussions tend to lead to silence or non-committal responses. Other challenges include the possible lack of interest and participation in activities across borders, and late communication on non-participation, making it impossible for a replacement to be organized. Communications in general can be difficult; this was the case, for instance, regarding the World Bank project between UM and collaborators from the Cambodian university. In Laos, there are different challenges: there are very few universities and private institutions/colleges for the team to collaborate with, and networking with multiple universities is a challenge because they are scattered through different regions of Laos (north and south). Hence, extra travel time and more effective network linkages are necessary to reach potential collaborators from these areas. Some improvements are also needed in the promotion of research/teaching seminars, including better advertising and posters, and better communication about the CLMV seminars with the upper management of the institutions involved.

In spite of the challenges, however, more than 30 research seminars were conducted at various HE institutions in Cambodia and Laos as part of the project. As a result of the goodwill and trust built through CLMV initiatives, UM was later able to collaborate with its Cambodian and Laotian contacts to embark on a Training of Trainers programme to develop a sustainable approach towards human resource development and capacity building. This should help prepare these nations for regional and global demands and facilitate regional mobility in terms of teaching, learning, and research in higher education. The CLMV project also promoted a better understanding of different ASEAN cultures (Malaysian, Cambodian, and Laotian in particular) between the collaborators. Most importantly, a network of potential research collaborators has been formed which can work together on existing and future joint projects or the co-writing of research articles for publication.

7 Conclusion

This chapter has examined higher education and the SDGs, looking particularly at SDG 4 and how that impacts on the other Goals. It has focused on Malaysia and the wider ASEAN region, highlighting the frameworks and policies that align with the SDGs, the targets set by Malaysia and ASEAN in the education sphere and related areas, and plans for implementation. The chapter has reviewed multi-level governance and regional mechanisms, local government and stakeholder involvement, and inter-regional cooperation with the EU. It has described the role of the ASEAN Secretariat, the ASEAN University Network and SEAMEO RIHED as well as the SHARE EU ASEAN project. It has reported on the cross-national CLMV programme and one of its projects. Although different from Malaysia in terms of development, the other two countries involved in the project are emerging as higher education players in the region. Spurred on by the AEC blueprint and inspired by the ASEAN Community Vision 2025, both Cambodia and Laos are aware of the need for highly educated and skilled people for socio-economic development. This project has also played an important role in regional mobility and integration. The SDG 4 goal of quality education impacts on numerous other SDGs such as gender equality, end to poverty, decent work, and economic growth. A clear understanding of the connections and interactions between the SDGs will therefore be needed to develop future actions and policies for the nation and the region.

References

Altbach, P., & Umakoshi, T. (Eds) (2004). *Asian universities: Historical perspectives and contemporary challenges*. Baltimore, MD: Johns Hopkins University.

ASEAN (2007). ASEAN charter. Jakarta: The ASEAN Secretariat. Retrieved from http://asean.org/storage/images/archive/publications/ASEAN-Charter.pdf.

ASEAN (2014). *ASEAN state of education report 2013*. Jakarta: The ASEAN Secretariat. Retrieved from http://www.asean.org/storage/images/resources/2014/Oct/ASEAN%20State%20of%20Education%20Report%202013.pdf.

ASEAN (2015). ASEAN 2025: Forging ahead together. Jakarta: The ASEAN Secretariat. Retrieved from http://asean.org.storage/2015/11/67.-December-2015-ASEAN-2025-Forging-Ahead-Together-2nd-Reprint.pdf (accessed 26 January 2018).

ASEAN (2016). ASEAN taps on Vision 2025 to support SDGs. ASEAN Secretariat News. Retrieved from http://asean.org/asean-taps-on-vision-2025-to-support-sdgs-2/ (1 August 2018).

Azirah, H., Yee Chee Leong, & Pheak Tra Pich (2014). English in higher education in Cambodia. *World Englishes*, 33(4), 498–511.

Azirah, H., Yee Chee Leong, & Aliyyah Nuha Faiqah Azman Firdaus (2015). Empowering the academic community with knowledge and skills in ELT and education-based research: Capacity building in selected public and private higher institutions in Cambodia/Laos. Paper presented at workshop, The success story of Malaysia's global reach and CLMV programs, Penang, 2 September.

Chapman, D. W. (2009). Education reforms and capacity development in higher education. In Y. Hirosato & Y. Kitamura (Eds), *The political economy of educational reforms and capacity development in Southeast Asia: Cases of Cambodia, Laos and Vietnam* (pp. 91–109). Berlin: Springer Science + Business Media B.V.

Educate A Child & FHI 360 (2016). Education and the SDGs. Occasional paper #2 EAC and FHI 360. Doha: Education Above All. Retrieved from https://educationaboveall.org/uploads/library/file/2a8e15847d.pdf (26 January 2018).

Education for All (2015). Education for All 2015 national review report: Malaysia. Retrieved from http://unesdoc.unesco.org/images/0022/002297/229719E.pdf (26 January 2018).

Government of Malaysia (2015). *Eleventh Malaysia plan (2016–2020): Anchoring growth on people*. Kuala Lumpur: Economic Planning Unit. Retrieved from https://www.talentcorp.com.my/clients/TalentCorp_2016_7A6571AE-D9D0-4175-B35D-99EC514F2D24/contentms/img/publication/RMKe-11%20Book.pdf.

Government of Malaysia (2017). Malaysia: Sustainable Development Goals Voluntary National Review. Kuala Lumpur: Economic Planning Unit. Retrieved from https://sustainabledevelopment.un.org/content/documents/15881Malaysia.pdf (26 January 2018).

Loke, S. H., & Hoon, C. L. (2011). Education in Malaysia: Development and transformations. In C. Brock & L. P. E. Symaco (Eds), *Education in South-East Asia* (pp. 95–119). Oxford: Symposium Books.

Ministry of Education (2012). *Preliminary report: Malaysian Education Blueprint 2013–2025*. Putrajaya: Ministry of Education.

Morshidi Sirat (2017). From MDG to SDG: Update on ASEAN's quality of higher education for social transformation [Keynote Presentation slides]. Presented at the International Conference on Social Transformation, Community and Sustainable Development (ICSTCSD), 7–9 November, Promenade Hotel, Kota Kinabalu, Sabah, Malaysia.

Pereira, J. D. (2016, July). THF Literature Review. Equity, access and educational quality in three South-east Asian Countries: The case of Indonesia, Malaysia and Vietnam. Singapore: The HEAD Foundation.

Pol, P. (2015). The Bologna Process and the European Higher Education Area from consolidation to the search of a new impetus. In A. Hashim (Ed.), *The ASEAN Higher Education Forum (AHEF) 2015* (pp. 12–24). AEI Occasional Paper 28. Kuala Lumpur: Asia–Europe Institute.

UN ESCAP (2017). Southeast Asia subregion challenges and priorities for SDG implementation. Economic and Social Commission for Asia and the Pacific. Retrieved from http://www.unescap.org/resources/southeast-asia-subregion-challenges-and-priorities-sdg-implementation (25 January 2018).

UNESCO (2014). Sustainable development begins with education: How education can contribute to the proposed post-2015 goals. Paris: UNESCO. Retrieved from http://unesdoc.unesco.org/images/0023/002305/230508e.pdf.

UNESCO (2015). Education 2030 Incheon Declaration and framework for action: Towards inclusive and equitable quality education and lifelong learning for all. World Education Forum 2015. Paris: UNESCO.

UNESCO (2017). Education 2030 Brief. Volume 1: Integrating SDG 4 in national education planning. Retrieved from http://unesdoc.unesco.org/images/0024/002472/247239e.pdf.

UNICEF (2013). Envisioning education in the post-2015 development agenda: Executive summary. Global thematic consultation on education in the post-2015 development agenda. Retrieved from http://en.unesco.org/post2015/sites/post2015/files/Post-2015_EN_web.pdf. (accessed 25 January 2018).

Wang, L. (2016). Aligning national education planning with education 2030: Perspectives from UNESCO Bangkok [PowerPoint slides]. Presented at the Education 2030 launch and symposium: Connecting the dots between national priorities and global targets, Kuala Lumpur, 23 August. Retrieved from https://www.moe.gov.my/images/Terbitan/Rujukan-Akademik/Presentation-Education-2030-Launch-Symposium-23rd-August-2016-Hotel-Istana-Ballroom-Kuala-Lumpur/Libing-Wang-Chief-of-EISD-UNESCO-Bangkok/Libing%20Wang.pdf (accessed 25 January 2018).

Weidman, J. (1995). Reform of higher education in the Lao People's Democratic Republic. Paper presented at the 1995 Annual Meeting of the Association for the Study of Higher Education, Orlando, FL.

CHAPTER 7

Healthcare System Reform and Governance for Sustainable Development under Indonesia's Health Insurance (JKN) Policy

Laksono Trisnantoro

1 Introduction

The effective implementation of Indonesia's policy on health and social security, which aims at universal coverage in health services, began in 2014. The policy is based on two laws, one concerning the National Social Security System (Law no. 40/2004, the SJSN Law), enacted in 2004, and the second concerning the Social Insurance Administration Organization (Badan Penyelenggara Jaminan Sosial — BPJS), Law no. 24/2011, enacted in 2011. BPJS administers the government's National Health Insurance scheme (Jaminan Kesehatan Nasional — JKN). The BPJS is divided into two groups: BPJS Employment and BPJS Health. The objective of the BPJS is to improve the equity of access to high quality healthcare and social security.

The two laws aim to reform the healthcare system by changing the financing mechanism. Basically, the poorest citizens should be funded by central and local government budgets, while middle- and higher-income groups are required to pay a premium to the BPJS, albeit at a relatively low price. For this purpose, BPJS Health acts as a health insurance agent and has the authority to collect these premiums.

1.1 *Research Question*

In the context of these new laws, the principal question to be answered is whether the National Health Insurance scheme (JKN) is truly a reform policy, capable of achieving its objective of economic and geographical equity of healthcare in Indonesia. The answer to that question has both a scientific and a social significance. Examined scientifically, a number of factors come into play, such as benefit package rates, premium rates, supply-side readiness, governance, and family economic status. Defining equity in different ways will result in different evaluations of the JKN's success. The social significance of the question lies in the fact that the gap between rich and poor, and between cities and remote areas, must be carefully managed in order to promote social justice.

 | DOI:10.1163/9789004391949_008

1.2 *Methodology*

In this chapter, an assessment of whether JKN can reform the Indonesian healthcare system will be conducted through policy analysis, using the policy triangle framework introduced by Walt and Gilson (1994). The content, actors, and policy-making and implementation processes of the JKN will be elaborated in the first part of the chapter, while a more in-depth analysis of governance and sustainability will be explored in the second part, before drawing some conclusions.

2 A Policy Analysis of JKN

2.1 *Content Analysis*

Thus far, scepticism and doubt still surround the health system reform in Indonesia as stipulated by the SJSN Law of 2004. There is a question as to whether the SJSN Law aims to govern social security in broad outline, as mandated by the relevant Convention of the International Labour Organization (ILO), or whether it aims to also organize operational matters. It should be understood that implementation of the SJSN Law requires the BPJS Law (2011) to work effectively — a process which it is anticipated could take up to seven years.

One of the uncertainties surrounding the SJSN Law is its role in regulating the function of government in the healthcare system. The law does not establish a clear division of roles between provincial and district/city governments. While it refers to the central government and the state budget (Anggaran Pendapatan Belanja Negara — APBN), in reality, local-level governments also play an important role in the healthcare system. This uncertainty triggered a judicial review in the Constitutional Court and remains controversial, in terms of the relations between the BPJS and the provincial and district/city governments.

Although the SJSN Law aimed to reform the healthcare system, relatively few of its articles are directly related to healthcare (only Articles 19 to 28). Even these 10 articles are not sufficiently detailed, given that health insurance in Indonesia is a technically complex issue. In the light of this complexity, it is all the more important that the law should establish clear regulations, for example in relation to the Regional Health Insurance (Jamkesda), private insurance companies, service quality, contractual relationships, and the roles of relevant stakeholders such as the Ministry of Health, the Health Office, and hospitals. In terms of the objective of equity, there is an article in the SJSN

Law — Article 23, Section 3 — that complicates the operational implementation of this ideal: "Where a region is not yet equipped with healthcare facilities sufficient to meet the needs of citizens, the Social Security Organization Body (BPJS) shall be required to provide compensation" (author's translation).

The main problem with the implementation of this article lies in the uneven distribution of health workers and facilities in Indonesia. As an illustration, at the time of writing, in most of East Nusa Tenggara Province, there are no full-time anaesthesiologists. If Article 23, Section 3 were to be fully implemented, BPJS should provide compensation to the people in East Nusa Tenggara for this shortfall. But in what form should the compensation be given? Should the BPJS compensate individuals for their transportation costs to the nearest city, such as Bali, for proper treatment and medicine? This is just one example of the technical difficulties in implementing the SJSN Law, particularly in remote areas.

When the BPJS Law was first formulated in 2011, it contained many ambiguities. As noted above, it covers two groups, namely BPJS Employment and BPJS Health. As a result of these two groups being brought together under the BPJS Law, additional regulations were required at various levels — governmental, ministerial, and presidential. By 2017, more than 80 regulations related to JKN had been issued. This indicates that the contents of the SJSN and BPJS Laws, and their accompanying regulatory frameworks, were far from comprehensive.

In assessing the content of the two laws, it is informative to compare them with similar legal forms in other countries. In terms of country size, economic system, and the division of governance between central and local governments, Indonesia and the United States have many similarities. In the USA, ILO Convention 102 (1952) was used as a reference point for social security measures introduced in 1953, and for several amendments which followed. These measures and amendments established regulations for federal old age benefits, survivors' insurance, disability insurance, unemployment benefits, temporary assistance for needy families, health insurance for the aged and disabled (Medicare), grants to States for the medical assistance programme (Medicaid), the State Children's Health Insurance Program (SCHIP), and supplemental security income (SSI). More specific issues related to health insurance were regulated by subsequent measures, including the Health Insurance Portability and Accountability Act of 1996, the Medicare Prescription Drug, Improvement, and Modernization Act, 2003, and the Patient Protection and Affordable Care Act, 2010, which came to be known as Obamacare. This brief overview gives an indication of the depth and breadth of regulations governing health and social security insurance in the USA.

In contrast, Indonesia still needs a body of law that comprehensively deals with health insurance. Judging from their content, the SJSN Law of 2004 and the BPJS Law of 2011 are not sufficient to regulate Indonesia's very complex health insurance system. The failure of the two laws to cover many of the technical issues related to insurance has proved to be a problem in the process of implementation so far. At the time of writing, the inevitable conclusion is that the SJSN and BPJS laws are inadequate and will not have the desired impact in reforming the healthcare system.

2.2 *The Context of the Law on SJSN and the Law on BPJS*

It is important to note that the SJSN Law in Indonesia was formulated during the transition period from centralization to decentralization. In 1999, a Law on Governmental Decentralization was passed, but it was not yet effectively implemented when the SJSN Law was enacted. As the result, there was still some confusion about the functions of the central and local governments, which also affected the health insurance and social security sector. As mentioned above, this led to a lawsuit from a regional health administrator and a review in the Constitutional Court of the roles of provincial and city governments in SJSN Law. The result is still a matter of dispute. Against this background, some tough negotiations took place during the formulation of the later BPJS Law. At the formulation stage, there were several policy options; two of them were as follows:

1) The SJSN law should remain exactly as enacted in 2004. The likely implications of this option were prolonged conflict over political, economic, and technical interests, and recurring implementation bottlenecks.
2) The SJSN Law should be amended, and laws on health insurance and on social security should be formulated. Clearly this option would require a lot more effort, and many of the relevant parties were reluctant to engage with such a process.

The second option adopted the principle of taking a step back in order to move forward. The SJSN Law (2004) would be amended by strengthening its focus on social security: it would function as social security law, as in the USA. Health and labour issues would then be governed by two different laws, with health insurance regulated by a special law (noting that a special law on health insurance in the future should not overlap or interfere with Law no. 32/2004 on the issue of health decentralization). However, in the process of formulating the BPJS Law, a proposal explaining the importance of a special law on health insurance did not feature in the policy agenda.

The decision of the central government and the House of Representatives was for the formulation of a BPJS Law that was divided into two: BPJS

Employment and BPJS Health. This decision had the merit of being pragmatic. Nevertheless, it had two fundamental weaknesses: it did not have the capacity to regulate all the relevant technical matters; and it missed an opportunity to coordinate with laws governing decentralization, including in the health sector.

In the context of the JKN, there are three conditions which should ideally be met if the JKN is to provide equitable and quality healthcare. The first is the availability of adequate healthcare infrastructure, such as hospitals, clinics, doctors, paramedics, and others, in sufficient quantities and proportional to the size and geographical spread of the population. Prior to the BPJS, Indonesia's health infrastructure was very uneven, in quality and distribution, and this was not addressed during the policy formulation phase. Second, the JKN programme makes the state responsible for providing subsidies and protection to needy and poor communities. Hence, the availability of adequate and sustainable budgets is vital. Third, the JKN programme requires an integrated and orderly administration in order to produce a more targeted service — something which was lacking both before and after the enactment of the SJSN and BPJS laws. It is thus clear that there are a number of problems that reduce the capacity of the SJSN Law (2004) and BPJS Law (2011) to reform the healthcare system in Indonesia.

2.3 *The Actors*

The formulation and implementation process of the SJSN and BPJS laws used up enormous financial resources, and involved a wide array of stakeholders, including beneficiary communities, central, provincial, and district authorities, the House of Representatives, the Regional Representatives Council, the private sector, various companies that manage social security, associations and organizations of healthcare providers, professional organizations, and others. Obviously, such a disparate range of stakeholders will hold differing or even contradictory political views on many aspects, including financial and economic interests. On the question of social security, many of the actors involved hold positions that are difficult to manage collaboratively.

Indonesia has chosen to follow a 'single pool' principle with BPJS as the sole organization to manage the health insurance system. Comparisons with other countries which are similar to Indonesia in terms of economic structure and system, and population, as well as associations between central and local government, suggest that this is a risky choice. Thailand, for instance, has three pools in its health/social security system, reinforcing doubts as to the likely effectiveness of the Indonesian system (Trisnantoro, 2014). From the outset, technical matters were neglected in the political debate; the main focus was on differences in economic interests among implementers, including in the context of a single pool option. This is evident in the current form and composition of BPJS.

After nearly seven years in the pipeline, the BPJS bill finally became Law no. 24/2011. In spite of this long process, the debates around the SJSN and BPJS laws have not been silenced. At least three groups of protagonists can be identified: 1) those in favour of the SJSN and BPJS laws; (2) those who oppose the concept of premiums which is fundamental to the SJSN Law; and (3) those who are reluctant to accept the current formulations of the SJSN and BPJS laws. This debate occurs not only in parliament or in academic circles, but also on the streets. This raises the question: is this simply a pragmatic disagreement, or is there an ideological argument at play?

2.4 *The Formulation Process*

The formulation process of the SJSN and BPJS laws was neither quick nor easy. In the late 1990s, Universitas Gadjah Mada (UGM) had drafted similar regulations related to a national health insurance system in collaboration with PT Askes Indonesia. However, the process was discontinued in that form, and converted instead into the drafting of the SJSN bill. The development of the SJSN bill did not run smoothly. In the end, the SJSN Law not only covers health insurance, but also combines it with other social securities advocated by the ILO. The process of formulation went through many difficulties and compromises. The eventual enactment of the law was announced on the last day of the administration of President Megawati.

For several years, the SJSN Law could not be properly applied since its implementation required the supporting BPJS legislation. In the second government of President Susilo Bambang Yudhoyono, various groups and political parties strove to pass the BPJS bill, so that the law could be enacted during his presidency. A study by Pisani, Kok, and Nugroho (2017) suggests that the journey to establish universal health coverage in Indonesia is likely to have been influenced by political rather than technical considerations. Stakeholders were swayed by the political and economic value of reaching an agreement to develop such a policy in Indonesia, to the point that technical matters were subordinated to political factors.

After a number of years of implementation, the goal of reforming the health system in terms of improved equity has not been achieved. Growth in the number of hospitals, including those with advanced medical facilities and educational capacities, as well as access to specialist doctors, is concentrated in certain areas, like Java. In addition, due to the nature of medical payments that require the use of claims systems, most of the budget is absorbed in Java.

3 Governance Issues of JKN

This section addresses the governance aspects of universal health coverage. The term 'governance' in the context of Indonesia's JKN refers to how the process of solving various problems in health security in Indonesia is conducted and whether decisions made are being implemented. For this, it is necessary to focus on formal and informal actors who contribute to decision making and implementation. In the implementation of the two laws, there is a recurring question: who regulates and operates the financing system? It is important to understand that governance in the health system plays a key role in supporting a universal health coverage policy — represented in Indonesia by JKN — especially in terms of strategic decision making (Greer & Mendez, 2015). The underlying issue here is how the health system is governed and whether the organization is managed properly (Fattore & Tediosi, 2013).

In examining the role of government in JKN, the term that is often used in such contexts is stewardship, which means that one party is trusted by various actors to carefully manage the programme. The concept of stewardship is derived from the theory of principal–agent relationships (Waterman & Meier, 1998; see also Honda, McIntrye, Hanson, & Tangcharoesathien, 2016), which sees interactions in terms of a contract to be carried out. The purchaser or commissioner in the contractual relationship is called the principal, while the party that provides the service is called the agent. The relationship between principal and agent is regulated by a contract that sets out what the agent should do and what the principal must do by way of compensation.

Applying the principal–agent theory to the Indonesian case, BPJS plays the role of purchaser, serving as the principal for healthcare institutions. As a principal, BPJS uses various tools such as contract systems, financing, and regulation, and implements monitoring mechanisms to ensure that healthcare institutions that serve as the agents provide good-quality service within the agreed tariff. It also acts as an agent appointed by the government under the SJSN Act (2004) and BPJS Act (2011). The government, in this case, has a stewardship role, ensuring the equity and quality of service specified in the purchase contract. Stewardship roles depend upon trust. Hence, it is important to determine which government units are the principals of BPJS. Does the office of the President fill this role? Or is perhaps the Ministry of Health, the Ministry of Finance, or some other? One of the noteworthy aspects of the JKN is the position of BPJS, as summarized by Situmorang (2017), former head of the National Social Security Council (DJSN): "The SJSN Law and BPJS Law in general terms do not authorize any single ministry to be responsible over the administration of social security except for BPJS under the supervision of

DJSN. Even in the BPJS Law, it is mentioned that BPJS is a public corporation, and is only responsible to the President" (Situmorang, 2017; translation by the author).

In the structure of Indonesia's health insurance sector, the two arms of BPJS (Employment and Health) have come about through the transformation of PT Askes Indonesia and PT Jamsostek. Under the old structures, the two agencies each fell under the authority of a ministerial department. Since the restructuring, BPJS is directly under the authority of the President. Information obtained since the implementation of the new system suggests that the Ministry of Health — responsible for regulating the health sector — has found it difficult to make transparent and accountable decisions about the national healthcare system, and about the utilization of BPJS data. Even now, the Ministry of Health seems to have problems accessing the data necessary to plan and to evaluate the health programme. The DJSN also has difficulties accessing BPJS data. At the same time, falling under the direct responsibility of the President brings its own concerns, not least in fighting for space in the President's busy agenda. In the light of these problems, how can decisions be implemented? Has good governance been applied in Indonesia's health insurance sector?

3.1 *The Role of Central Government*

There are several institutions related to JKN: the Ministry of Health, the Office of the President, the Office of the Vice President, DJSN, and the Coordinating Ministry of Human Resources. The role of the central government in funding, formulating JKN policies, maintaining the quality of health services, and implementing healthcare systems is, in fact, unlimited. In terms of financing, the government's role is to ensure a budget that is adequate for JKN policies. Through the state budget (APBN), which is modified annually, the government and the House of Representatives determine the budget allocation for the health sector. In addition, the central government works with local governments on managing investments and financing health security.

Despite the principle of equitable access, and the provision of health facilities through BPJS, Indonesia's poor still have difficulty accessing health services. The main problem is that, as noted, the majority of facilities are concentrated in certain areas, which implies expenses such as transportation and accommodation for those who need to travel to the facilities — costs which are not covered by BPJS. To reduce this burden, the central and regional governments must increase health facilities, especially in remote areas. However, data from 2014 and 2015 show that the central government undertook little investment for the regions. It is clear that investment to increase health facilities, infrastructure, and human resources in support of JKN in these years fell short because of the limited fiscal ability of central government.

What about the role of the Ministry of Health as part of the central government in JKN services? Here we return to a point made earlier: because the Director of BPJS is directly responsible to the President, there is no hierarchical relationship between the Ministry of Health and BPJS. The Ministry of Health has difficulty accessing BPJS data for purposes of quality improvement, planning, and evaluating the health system, which leads to problems in optimizing the organization of health services under BPJS financing.

3.2 *The Role of Regional Government*

The governors and mayors/regents are the local government officials who represent the President in various matters in the regions, especially in the context of Indonesia's current move to decentralization. BPJS is responsible to the President; therefore, in the regions, it is logical that the local branch of BPJS is responsible to the governor or mayor/regent. However, mirroring the relationship between BPJS and the Ministry of Health at the central level, relationships at the local level have also been problematic. Different BPJS structures at the central, provincial, and city/regency levels further complicate the relationship. There has never been any analysis of BPJS health and financial at the provincial and city/regency level. An examination of the period 2014–17 shows that the BPJS system and local governance system are differently organized and structured, are governed by different laws, and are not well interlinked. This means that there are constant difficulties in trying to make connections between the BPJS and public health programmes managed by local governments.

3.3 *Is JKN an Example of "Good Governance"?*

Good governance has eight key characteristics: active participation, respect for the law, transparency, responsiveness, consensus-oriented action, effectiveness and efficiency, equity and inclusiveness, and accountability (see, e.g., UNESCAP, n.d.). In the context of JKN, it is hoped that, with these characteristics in place, mismanagement of health services and misuse of funds can be reduced, and marginalized and disadvantaged groups in society will get the same treatment as the rest of the population. In order to assess whether JKN has benefited from good governance in formulation and implementation, this section will examine each of the eight characteristics in turn.

3.3.1 Active Participation

Active participation in important decisions related to JKN is expected. Indeed, participation in the decision-making process is cultivated, with all parties — BPJS, central government, local government, professional organizations, hospital associations etc. — encouraged to participate and have an influence in the proceedings. During the formulation of the SJSN and BPJS, the parties most actively involved were donor organizations in the formulation of relevant policies and the BPJS Law of 2011. The participation of non-governmental groups such as the Indonesian Hospital Association (Perhimpunan Rumah Sakit Seluruh Indonesia — PERSI) and the Indonesian Doctors Association (Ikatan Dokter Indonesia — IDI) was limited in the formulation process, but began to increase in the implementation phase. Nevertheless, lobbying capacity and participatory mechanisms in the decision-making process remain weak. By the end of 2017, hospital associations such as PERSI were just beginning to participate more actively in policy formulation. This received a positive response from the government, which always invites PERSI representatives to attend various conferences and meetings as it prepares amendments to the Presidential Regulation.

3.3.2 Respect for the Law

Good governance requires a good legal framework, especially in order to protect minority groups and those living in remote areas. BPJS has tried to stay within the existing law, but it has met numerous obstacles during its implementation, including in financial matters. These financial constraints have made it difficult for BPJS to comply fully with the law. For example, the "compulsory" compensation policy, which forms part of the SJSN Law, Governmental Regulation, Health Minister's Regulation, and BPJS regulation, has not yet been implemented. Similarly, the wording of the SJSN law includes a very weak statement regarding government assistance for health insurance that is directed to those in need — the Beneficiary Health Insurance Contributions (Penerima Bantuan Iuran — PBI). This weak wording means that there has been no real effort to ensure compliance with the rules, and leaves the door open for widespread violations in the future. It is important that this matter is resolved as quickly as possible.

3.3.3 Transparency

The characteristic of transparency implies that decisions are formulated based on known rules and regulations, and that the information needed for any decision is freely and easily accessible for relevant stakeholders and those who will be impacted by the decision. It also means that information is available to the media. However, the experience of implementing the JKN thus far shows that there has been little transparency in the making or implementation of decisions. Data available at BPJS have not been readily provided or used as the basis of local decision making (where, for instance, data on increased rates of cancer or heart disease could lead to particular investment decisions). This chapter has already mentioned the surprising fact that, as of 2017, the central government, in the form of the Ministry of Health and other focal institutions such as the DJSN, found it difficult to access data from BPJS. Another example of the lack of transparency is the determination of the size of premiums for payrolled workers (Pekerja Penerima Upah — PPU) and for non-payrolled workers (Pekerjan Bukan Penerima Upah — PBPU). To date, the basis for determining these premium rates is unclear. Recognizing the problem of data transparency as a key issue, at the end of 2017 DJSN established a team of academics to work with the BPJS data. It is too early to assess the outcome of this.

3.3.4 Responsiveness

The characteristic of responsiveness implies that a quick and adequate service should be provided that meets the needs of the parties concerned. In the context of JKN, the main purpose of the policy is to provide access to quality health services. This should be the remit of service providers, whose goal is to serve wholeheartedly. However, financial difficulties in BPJS over the past four years indicate problems with the level of tariffs set by BPJS for First Level Health Facilities (Fasilitas Kesehatan Tingkat Pertama — FKTP) and Referral Health Facilities (Fasilitas Kesehatan Tingkat Lanjutan — FKTL). This has resulted in complaints about poor responses and inadequate treatment from healthcare providers and BPJS. The problem of responsiveness has been further exacerbated by the difficult relations between the government, BPJS, and service providers, which culminated in a deficit that affected the cash flow to hospitals in 2017.

3.3.5 Consensus-Oriented Actions

Good governance requires mediation from various parties to produce the best outcome for the whole community. In the health insurance sector there are various actors who need to work together and consensually to achieve goals. So far, however, this has not been a feature of Indonesia's JKN. Communication among the actors has been poor, including between the Ministry of Health and BPJS, as well as between BPJS and the hospitals, especially private hospitals where BPJS has disagreements about fund management. The problems faced by JKN will require a long-term process to resolve, including revision to the SJSN and BPJS laws. As already indicated, relations with the DJSN are strained, which makes it all the more difficult for the DJSN to perform its function as mediator in talks and negotiations. There is much work to be done here: achieving consensus will remain difficult until all parties are able to understand the historical, cultural, and social context of the formulation of Indonesia's health security policy over the past 25 years.

3.3.6 Effectiveness and Efficiency

Good governance implies that processes and institutions deliver products or services that meet the needs of society effectively and efficiently. Efficiency also means sustainability — that goods and services can be provided over the long term. In the context of JKN policy implementation, it is too early to judge whether it fully meets this criterion. So far, there are both positives and negatives to report. On the positive side, BPJS has undoubtedly provided protection to millions of Indonesian citizens who would otherwise not have had any health insurance cover. On the negative side, BPJS is suffering from a serious financial deficit which may indicate that it has not been efficient and might threaten its long-term sustainability. There may also be significant potential for fraud in the system, although to date no court cases for fraud have been brought. Another negative point is the failure to control the spread of various diseases; for instance, there has been an increase in tuberculosis cases and an outbreak of diphtheria. These highlight some of the weak points of JKN in controlling and preventing disease.

3.3.7 Equity and Inclusiveness

One of the most important tests of good governance is the way that marginalized groups are treated, and whether opportunities to improve the lot of minority or marginalized groups are taken. In the context of the SJSN and BPJS laws, and their implementation over recent years, the aim of resolving the uneven distribution of health services by 2019 seems unlikely to be achieved. The main problem is how to sustainably improve the level of services to those who live in remote geographical areas or belong to marginalized social groups. To date, the JKN system, as mandated by BPJS, has struggled with uneven distribution issues, both on the supply side and in the processing of claims, which have overwhelmed the principle of fairness.

3.3.8 Accountability

If good governance is to be achieved, every agency should be responsible and accountable for its actions. In the legislation behind JKN, however, the accountability of relevant agencies is unclear. What are the responsibilities of the central government, especially the Ministry of Health, and what are the responsibilities of local governments and private health facilities? Is BPJS actually capable of managing a system that provides access to quality and fair health services for the entire community? Given the lack of clarity around the authority of each level of government in the laws, it can be said that JKN does not meet the criterion of accountability.

The above analysis suggests that the implementation of JKN policy cannot at this point be said to demonstrate good governance. There is still much to be done by the actors involved in JKN to improve governance.

4 How Sustain Is JKN?

The long-term sustainability of JKN as a policy will be heavily dependent on financial support from the government. This means that the political dynamics of government must be taken into account. There are several issues to be considered. First, politically, after the reform era of 1997, there was a strong desire to transform Indonesia into a welfare state. The 1945 Constitution includes the concept of the welfare state explicitly, and indicates the need for differentiated treatment by the state between the poor and those who are better off. The 1945 Constitution and the 2004 SJSN Law both mention specific points of attention for the state in order to regulate the financial system for the entire population.

A successful welfare state uses funding derived from taxes and non-tax revenues to help the poor. In principle, those in a secure financial position are required to pay for their own healthcare through various mechanisms such as buying commercial health insurance, while the state uses its revenue to protect the poor. In practice, however, the situation in Indonesia is such that PBI funds that should be used by the poor, are actually being used by those in the middle and upper classes who could afford to pay for themselves. This situation ultimately raises the issue of social justice.

To resolve this issue, there needs to be a change in the financing component of the health system; this in turn requires a change in the country's system of taxation. A substantial and reliable flow of revenue from personal income tax is the foundation of the welfare state. The tax mechanism is used to redistribute income in the form of benefits. Ultimately, the rich help those in need through the payment of income tax. In the case of healthcare, tax funds are used to finance provision for those who need it and cannot afford it — an example of mutual cooperation through the tax mechanism.

The personal income tax paid in Indonesia is relatively unprogressive (up to a top rate of 30%) and the amount of revenue generated for the state from income tax is less than the amount generated by sales taxes. The ratio of tax to GDP is low, at about 11% — far below the OECD countries, which have a ratio of up to 25%. There are many non-compliant personal taxpayers, or citizens who should pay taxes but do not register as taxpayers. The state also derives funds from indirect taxes, for instance in the form of tobacco excise and sales taxes. Such indirect taxes are regressive, as people on low incomes spend proportionally more of their income than the better off. If JKN relies for most of its funding on indirect taxes, there will be no real redistribution of income: rather, funds that should be used to provide services for the poor will in fact be used by the relatively well off. This is contrary to the 1945 Constitution which states that the government should take care of the poor first. This is, therefore, a political issue: how to ensure that government subsidies are mainly used by the poor.

The second issue to be addressed regarding the sustainability of the JKN is Indonesia's fiscal situation. Since the financial crisis of 2008, much of the world economy has slowed down. Indonesia, in contrast, has maintained an average economic growth rate of 4–5% per year. According to World Bank national accounts data, Indonesia's GDP exceeded US$932 billion in 2016. However, this economic growth is not reflected in the government's fiscal space. As already indicated, Indonesia's average ratio of tax revenues to GDP is around 11%, even dropping from 11% in 2013 to 10.3% in 2015. It appears that, while GDP is growing, tax revenues are flat. This causes a squeeze in the fiscal space for the funding of public programmes. A widening of this fiscal space is important in the context of public programme funding. The commitment to provide 5% of the state budget for health funding was achieved by 2016, after struggling at about 3.5–4% in previous years. However, after reaching 5%, the trend is flattening. This raises an ideological question: should rich people who pay insufficient income tax be entitled to benefit from BPJS funds which are intended to be used by the poor? It may well be impossible to apply welfare-state principles in Indonesia at the present time, given the weak tax situation and the high rate of utilization of BPJS benefit packages by the middle and upper classes of society. A practical solution may be for people in good financial circumstances to pay more, but not through the tax mechanism: rather, they should buy commercial health insurance inside or outside the BPJS system, or share costs in the BPJS system.

One of the most important — and surprising — lessons learned from JKN implementation to date, with BPJS as the managing institution, concerns the change in financing. There has been a radical change, shifting from a market mechanism to a managed system. The managed system relies on BPJS with all its attendant regulations, which are mostly related to BPJS's main budgetary resource, the PBI for the poor. At present, some of the budget intended for the poor is used by better-off economic groups. This reliance of BPJS on PBI — a government funding source — is one of its most serious weaknesses. Most poor patients have limited access to medical services as a result of cultural, economic, and geographic factors. This means that wealthier citizens enjoy more benefits from BPJS than the majority of poor people: in other words, the wealthier 'customers' are using up a high proportion of BPJS's financial resources. If this situation is to change, it will be important to find ways of getting wealthier citizens to contribute more to health funding. Some options might include out-of-pocket payments for the better off; commercial health insurance premiums; or increases in premiums for Class 1 independent PBPU. If improvements in the funding of healthcare are implemented effectively, this could create an environment in which healthy people subsidize the sick, the rich subsidize the poor, the working-age population subsidizes the elderly, and health services employ strategies to promote the efficient use of resources (Duran et al., 2014).

With regard to the Sustainable Development Goals, if the majority of the current budget is used to finance medical care for the rich, there will be no resources left to fund SDG programmes in other sectors, and the achievement of the SDGs will be at risk. This raises the question of whether funding can be found for the health sector, beyond that already regulated under BPJS — that is, from sources outside of taxation? If we look at health sector expenditure as a percentage of GDP, the current figure stands at about 3.1%. The equivalent figure for Thailand is approximately 4.6% as reported by World Bank (Ernst and Young, 2015). This suggests that Indonesians could be encouraged to spend an additional 1.5% of GDPs on health. Taking the 2016 total GDP of US$932 billion mentioned above, 1.5% of GDP would mean approximately US$14 billion per year which could be spent on healthcare. The challenge for the health sector will be to compete for these additional funds with other claimants. But without serious efforts from all parties, ensuring the sustainability of the JKN will be hard.

5 Conclusion

In assessing Indonesia's JKN, this chapter has reached a number of conclusions. The policy is limited to funding and finance; there is no coordination with other healthcare policies such as reducing smoking, improving health infrastructure, human resources for health production and deployment, improving health sector governance and policies. If these aspects of healthcare are not addressed, Indonesia is unlikely to see the lifestyle improvements that are the target of SDG 3 ("Good health and well-being"), or the reductions in inequality within and between regions which are the aim of SDG 10 ("Reduced inequalities").

In terms of governance in the health sector, there is confusion over the functioning of government units at different levels in Indonesia's decentralized system, and over the roles of the private sector and civil society. To date, JKN has not been managed according to the principles of good governance. Moreover, affluent members of the population enjoy medical benefits at the expense of the poor, yet pay very low premiums. Current levels of financial support for universal health coverage seem inadequate to guarantee its sustainability. More funds are needed from non-tax sources from middle- and upper-class citizens to reduce inequality within and between the regions in Indonesia. There is a serious risk that the curative care burden will increase sharply and reduce the budget for sustainable development programmes.

The conclusion of this chapter is, therefore, that radical changes are necessary, perhaps implementing the principles of healthcare system reforms described by Roberts, Hsiao, Berman, and Reich (2004). Policies to promote healthier lifestyles, behaviour, and environments should be introduced. Health sector governance in Indonesia's decentralized political system should be strengthened and the use of tax revenue to finance medical treatment for affluent members of society should be reduced to ensure a healthy life and improved welfare for people of all ages and income levels.

References

Duran, A., Kutzin, J., & Menabde, N. (2014). Universal coverage challenges require health system approaches: The case of India. *Health Policy*, 114(2–3), 269–277. DOI: 10.1016/j.healthpol.2013.11.009.

Ernst & Young (2015). Ripe for investment: The Indonesian health care industry post introduction of universal health coverage. Retrieved from https://www.ey.com/Publication/vwLUAssets/EY-ripe-for-investment-the-indonesian-health-care-industry-post-introduction-of-universal-health-coverage/%24FILE/EY-ripe-for-investment-the-indonesian-health-care-industry-post-introduction-of-universal-health-coverage.pdf.

Fattore, G., & Tediosi, F. (2013). The importance of values in shaping how health systems governance and management can support universal health coverage. *Value in Health*, 16(1), S19–S23.

Greer, S. L., & Mendez, C. A. (2015). Universal health coverage: A political struggle and governance challenge. *American Journal of Public Health*, 105(S5), S637–S639.

Honda, A., McIntrye, D., Hanson, K., & Tangcharoesathien, V. (Eds) (2016). Strategic purchasing in China, Indonesia and the Philippines. WHO Comparative Country Studies 2(1). Geneva: World Health Organization. Retrieved from http://apps.who.int/iris/bitstream/handle/10665/252763/9789290617686-eng.pdf;jsessionid=6035C06AC04B8F771EDE51A7E03BA935?sequence=1.

Ministry of Health (2016). Minister of Health's keynote address to Public Health Association Conference in Makassar.

Pisani, E., Kok, M., & Nugroho, K. (2017). Indonesia's road to universal health coverage: A political journey. *Health Policy Plan*, 32(2), 267–276.

Roberts, M. J., Hsiao, W., Berman, P., & Reich, M. R. (2004). *Getting health reform right: A guide to improving performance and equity*. Oxford and New York: Oxford University Press.

Situmorang, C. (2017, October 13). Eksistensi BPJS sebagai badan hukum public. *Jurnal Social Security*. Retrieved from http://www.jurnalsocialsecurity.com/opini/eksistensi-bpjs-sebagai-badan-hukum-publik.html.

Trisnantoro, L. (2014). Financial risk of BPJS. Paper presented at Indonesian Health Economics Association meeting, Bandung (24–25 January).

UNESCAP (n.d.). What is good governance? Bangkok: United Nations Economic and Social Commission for Asia and the Pacific. Retrieved from https://www.unescap.org/sites/default/files/good-governance.pdf.

Walt, G., & Gilson, L. (1994). Reforming the health sector in developing countries: The central role of policy analysis. *Health Policy and Planning*, 9(4), 353–370. https://doi.org/10.1093/heapol/9.4.353.

Waterman, R., & Meier, K. (1998). Principal–agent models: An expansion? *Journal of Public Administration Research and Theory*, 8(2), 173–202. DOI: 10.1093/oxfordjournals.jpart.a024377.

PART 4

SDGs and the New Urban Agenda, Cities, and Transport

CHAPTER 8

Urban Transformation in Indonesia, the SDGs, and Habitat III: Political Will, Capacity Building, and Knowledge Production

Bakti Setiawan

1 Introduction

Urban transformation[1] in Indonesia is facing unprecedented challenges in the next 20 years. The transformation is occurring not only in terms of demographic and spatial changes, but also along more complex social, economic, and cultural dimensions. Given a number of factors including the anticipated demographic dividend,[2] ongoing democratic consolidation and decentralization, the strengthening of community capacity, social capital, and civil society, and the prospect of economic growth in the future, Indonesia is expected to become a country that is not only prosperous and just, but also has the ability to compete and to play a more positive role in the global arena. Urbanization and city transformation will be important in driving the growth, development, and future of Indonesia.

In 2015, the Sustainable Development Goals (SDGs) were agreed by the global community in an attempt to realize the shared vision of development and the 'global village'. The SDGs comprise 17 objectives that are elaborated through detailed indicators of performance. There is much optimism surrounding the SDGs, strengthened by the experience of implementing and localizing the previous development agenda — the Millennium Development Goals (MDGs). This chapter focuses on SDG 11, which aims to "make cities and human settlements inclusive, safe, resilient and sustainable". In this context, delegates to the United Nations Conference on Housing and Sustainable Urban Development in Quito, Ecuador, in October 2016, known as Habitat III, adopted the New Urban Agenda (the NUA) — a new framework that lays out how cities should

1 In this chapter, urban transformation is defined in general as the growth, development, conservation, and regeneration of a given urban area; it covers spatial/physical, economic, social, and cultural dimensions.

2 Also known as the demographic bonus, this refers to a shift in the age structure of a nation's population, so that a high percentage are of working/productive age, relative to those of non-productive age.

 | DOI:10.1163/9789004391949_009

be planned and managed to best promote sustainable urbanization. The NUA is a roadmap for building cities that can serve as engines of prosperity and centres of cultural and social well-being while protecting the environment. It also offers guidance for achieving the Sustainable Development Goals and provides the underpinning for actions to address climate change (UN, 2017a).[3] Although Indonesia has a long history as a UN member state which has participated actively and positively in many UN commitments and agreements, it faces challenges in adopting, elaborating, implementing, and localizing all the points contained in the UN agreements. There are many complex dynamics to be taken into account, as well as internal social, economic, and political factors within Indonesia: these will contribute, directly and indirectly, to the implementation and localization of the SDGs and the NUA in the country.

1.1 *Research Question*

This chapter thus assesses Indonesia's prospects of successfully implementing the international agenda at national and local levels, through a review of its adoption and implementation of the NUA. It examines how Indonesia views, perceives, and responds to the NUA, and argues that without a clear, strong, and consistent political commitment, along with comprehensive and systematic capacity building and knowledge production, it will be difficult to effectively and successfully implement and localize the NUA. It also demonstrates that coordination is required within the framework of multi-level governance to ensure the implementation and localization of the international agenda at national and subnational levels.

1.2 *Scientific and Social Significance of the Research*

The particular scientific relevance of the chapter is to enhance our understanding of the relationship between global and local forces, which has been perceived as a process by which global forces will determine and dictate the activities of local actors and agencies. This perspective sees local parties as passive agents that simply follow global influences and directives. This chapter examines the role of local actors and agencies in Indonesia, and concludes that — in contrast to previous assumptions — they in fact represent an active and sovereign body which assesses, filters, and integrates global influences and

3 It is clearly stated in paragraph 9 of the NUA that "The implementation of the New Urban Agenda contributes to the implementation and localization of the 2030 Agenda for Sustainable Development in an integrated manner and to the achievement of the Sustainable Development Goals and targets, including Goal 11 of making cities and human settlements inclusive, safe, resilient, and sustainable" (UN, 2017a, p. 4).

agreements strategically with Indonesia's local conditions, context, and interests. Finally, the chapter proposes some policy recommendations to ensure that the international agenda can be integrated and localized in Indonesia, without sacrificing national and local interests and agendas.

1.3 *Methodology*

The research is a descriptive exploratory type of study, mainly making use of secondary data and information. Some interviews with stakeholders at the national, provincial, and local levels were also conducted, particularly on issues related to government and civil society responses to the NUA. The analysis is qualitative, while simple quantitative data are used to support discussions.

2 Literature Review: Global–Local Dynamics and the Politics and Leadership of Urban Governance

Within the Indonesian urban literature, it is interesting to discover that there are relatively few studies on the local–global dynamics of urban development processes or, more specifically, the social-political creation of urban space and the role of government in that process.[4] Several studies have focused more on the demographic aspect of urbanization,[5] on environmental problems, on the urban community,[6] or on particular aspects of the urban sector such as housing, land, and infrastructure. Recently, a number of studies have been conducted with the particular aim of improving the management capacity of local governments (e.g., Holzhacker, Wittek, & Woltjer, 2016).

All these studies play an important role as they provide us with useful information on the trajectory and the consequences of urban development and transition in Indonesia. Yet, we know very little about the processes through which urban spaces have been created, or the local–global relations between various actors or agencies in the urban development and transformation process, such as communities, developers, investors, politicians, the private sector, and government, not only at the national and provincial but also at the local levels. How are relationships formed among actors and agencies? Who benefits and who loses in this development process? And, importantly for our

---

4 Although Dutch scholars did some work on this in the 1920s, very few comprehensive studies have been done since then. Among these few are studies by Cobban (1971), Geertz (1965), and Nas (1985).

5 See, for example, Hugo, Hull, Hull, and Jones. (1987), Milone (1966), and Rutz (1987).

6 See Guiness (1986), Jellinek (1991), Krausse (1975), and Sullivan (1992).

purposes, what is the role of international agreements and the global agenda in this process? How are such international or global agreements perceived, treated, and integrated into local contexts, interests, and agendas? Do the multiple levels of the governmental system in Indonesia support or hinder such a process?

With increasing urbanization and global markets, as well as global agreements, agendas, and influences, and with the role of cities being seen as an engine of the global market, states and governments could be caught in a difficult situation. They are in a position to support urbanization, the urban development process, and the global agenda, but they also need to moderate its possible adverse externalities, to protect national, regional, and local needs and interests. It therefore becomes important to document a country's attitude towards urban transformation, especially in the context of global markets, interests, and influences.

It can be argued that, under unprecedented levels of globalization and global agreements, the state tends to be the passive agent of capital and global forces, leaving urban development to the mercy of market forces and unable to protect national and local needs and interests. This implies that urbanization and urban transformation, as both engine and symptom of globalization, will follow the same process and take on the same form in all countries under the same global influences. Yet, the fact is that many cities around the world are able to acquire and retain unique characteristics or take on very different forms: this is an indication of the ability of national and local variables or human institutions to adapt or change global influences (Flanagan, 1993). As Harrison (1988) has argued, those who view national and local society as pale reflections of globalization and external influences have overlooked the dynamic capacities of society as a group of free human agents.

In the context of the democratization and decentralization trends that have transformed Indonesia since 1998, another important question to be answered is whether democratization and decentralization bring improvement and benefit to all societies. This question becomes even more crucial in the context of the unprecedented urban transformation process that is happening in Indonesia today. Are democratization and decentralization in Indonesia bringing development and transformation for a better, more prosperous, inclusive, and sustainable city?

While acknowledging that there has been much criticism and pessimism over decentralization and democratization in Indonesia, this chapter takes a more optimistic view and argues that, after just 20 years, these two processes are not yet fully consolidated; while there is room for further improvement, decentralization and democratization have brought some progress and benefits.

As Holzhacker et al. (2016, p. 6) argue: "at the increasingly remote yet coordinative national level, at the local district level with more or less independent and diverging 'local kings', and at the regional gap in between … A wide variety of innovations in governance are emerging at all of these levels". These innovations are occurring, not in a vacuum, but in the dynamic and complex political culture of Indonesia. It is our role, as scholars, to continue to monitor, evaluate, and provide suggestions to ensure that such processes have a positive impact on the welfare and sustainability of the country.

Nowadays, global agendas such as the SDGs and the NUA bring both opportunities and pressures to national, sub-national, and local governments. These agendas lead, directly and indirectly, to a shift in the existing urban governance and management. As van Dijk, Edelenbos, and van Rooijen (2017) argue, urban governance is now facing a complex environment, in which stakeholders from different organizations and domains — governments, private actors, non-governmental organizations (NGOs), and citizens — are engaged. The implication is that city governments are no longer in control and have lost the ability to direct events in the urban context. To face these challenges, leadership and bureaucracy are important in shaping the character and quality of public services in urban areas. Leaders must provide appropriate services to the community to ensure the sustainability of the city and the environment. As the world becomes increasingly urban, leadership in city planning and governance can make a difference (Barber, 2013).

3 Urbanization and Urban Transformation in Indonesia

Renowned as the world's largest maritime nation with its rich biodiversity and natural beauty, many people forget that, by 2015, Indonesia had become an urbanized country — that is, a country in which more than 50% of the population lives in urban areas. By 2030, the end year for both the SDGs and the NUA, about 63% of Indonesia's population are expected to be living in urban areas, slightly more than the projected average rate of urbanization in the world (see Table 8.1).

Currently, there are 381 cities spread across Indonesia, including one megapolis (Jakarta), 10 metropolitan cities, 11 big cities, 112 medium cities, and 247 small cities. Of the total, 90 cities (24%) are located in coastal areas and the rest are inland; 93 cities (25%) have become municipalities with autonomous governmental power over legislation.

Beyond these demographic facts and figures, the issues and challenges of urbanization and urban growth in Indonesia require special attention and

TABLE 8.1 Indonesia's urban population (percentage) according to different sources

Year	UN data[1]	BPS, BAPPENAS, UNFPA Indonesia[2]	BPS[3]
2005	45.9	48.4	NA
2010	49.9	54.1	49.8
2015	53.7	59.3	53.3
2020	57.2	63.7	56.7
2025	60.3	67.5	60.0
2030	63.0	NA	63.4
2035	65.2	NA	66.6
2040	67.2	NA	NA
2045	69.1	NA	NA
2050	70.9	NA	NA

Note: NA indicates data not available

SOURCES: [1] UN (2014); [2] BPS (2008); [3] BPS (2013)

understanding, because of Indonesia's unique spatial and environmental characteristics. There are at least five such characteristics that merit listing here. First, urbanization takes place against the setting of Indonesia's enormously diverse environment; it can boast the second richest biodiversity in the world, after Brazil, with an extraordinary variety of flora and fauna, many of which are under threat. It is thus vital that the process of urbanization should be undertaken with minimum disruption to the preservation of Indonesian biodiversity. Second, Indonesia also has the second-largest tropical forest in the world. Tropical forests contribute greatly to the preservation of the world's environment, which again implies that careful monitoring of urban growth is necessary.[7]

Third, as already mentioned, Indonesia is the world's largest maritime nation, with 17,508 islands and 54,716 km of coastline. This has implications for the pattern of distribution and development, and the form of cities in Indonesia, in terms of the number of coastal cities, including those which may

7 According to the FAO (2010), 52.1% (about 94,432,000 ha) of Indonesia is forested. Of this, 50% (47,236,000 ha) is classified as primary forest, the most biodiverse and carbon-dense form of forest. Between 1990 and 2010, Indonesia lost an average of 1,205,650 ha or 1.02% of its forest per year. In total, between 1990 and 2010, 20.3% of its forest cover, or around 24,113,000 ha, was lost.

be vulnerable to subsidence and sea level rises;[8] the number of cities located in remote areas with poor accessibility (for instance, on small and far-flung islands); and the opportunities for coastal cities to take advantage of existing marine resources. Indonesia's maritime setting is thus relevant in the formulation of policies and strategies of urban development in Indonesia.

Fourth, Indonesia is vulnerable to various types of natural disasters, including volcanic eruptions, earthquakes, and tsunamis. This of course affects its cities, many of which have a high level of vulnerability or natural disaster risk.[9] It is clear that policies and strategies related to urban development in Indonesia must pay attention to this risk.

Fifth, urbanization and urban development can lead to an increase in a wide range of social and economic problems, including poverty, social imbalance, social intolerance and conflict, terrorism, crime, and corruption. At the same time, from an economic perspective, urbanization is not usually accompanied by significant increases in local income. Most cities in Indonesia still rely on development financing in the allocation of funds from the central government and few, if any, have seen an increase in productivity. Moreover, in terms of institutions and governments, Indonesian cities have not been particularly effective or innovative, largely due to insufficient or incompetent human and government resources, exacerbated by inadequate local leadership capacity.

In the midst of these challenging environments, the transformation of cities in Indonesia is also accompanied by positive opportunities that need to be optimized, such as the demographic bonus.[10] If competent and competitive human resources can be marshalled, Indonesia will become a country with great production and consumption potential. Furthermore, the consolidation of democracy and decentralization which is already underway, and which is expected to increase, will provide opportunities for improved public services and regional development progress which will eventually reduce regional gaps and strengthen Indonesia in the arena of global competition. Equally importantly,

8 The study by Chaussard, Amelung, and Abidin (2012) shows that five major cities in Indonesia (Lhokseumawe, Medan, Jakarta, Semarang, and Surabaya) are undergoing land subsidence at vertical rates of up to 24 cm per year.

9 Over the last 30 years, there have been an average of 289 significant natural disasters per year involving an average annual death toll of approximately 8,000. Furthermore, rising sea levels and changing weather patterns may lead to increased uncertainty in water availability and food production, as well as disruptions to transport, commerce, and urban development (GFDRR, 2016).

10 In 2020–2030, Indonesia will have the chance to benefit from a demographic bonus, as 43% of the total population will be in the productive age group (BPS, 2014). However, according to BPS (2015), only 10% of the productive age group are university graduates, and only 3.8% of them are skilled workers.

Indonesia enjoys a great diversity of people, from Sabang to Merauke; over the course of its history, Indonesia has proven itself as a republic in which the principles of unity and diversity act as major forces, pushing Indonesia to be more competitive at the global level. Its economic and development potential offer great opportunities that must be seized if positive and prosperous urbanization and city transformations are to be achieved.

3.1 *Indonesia's Responses to SDGs*

As an active member of the United Nations, with an open foreign policy with regard to the notions of equality and democracy, Indonesia has long been involved and fully compliant in pursuing global agreements. For example, Indonesia was actively engaged in the process of formulating the MDGs, as well as implementation and achievement efforts. Indonesia's compliance with this global agreement was then optimized by integrating items from the MDG agenda into the country's central, provincial, and local development planning system.

As a developing country, Indonesia assumes that only through a rational, systematic, and gradual development planning process, will the state achieve its development goals. The development process in Indonesia has made significant progress since the days of the Repelita model,[11] which was established at the beginning of the New Order era under Soeharto's authoritarian regime. The process of reform and democratization which started in 1998 brought significant changes in the political system in Indonesia while the country continued to pursue the concept and model of rational development planning.

This process is specifically embodied in what are known as the National Long-term Development Plan (RPJP), which covers a 20-year span, and the National Medium-term Development Plan (RPJMN), which covers five years, as well as the annual action plan (RKT). In a rational, comprehensive, and gradual development planning system, global agreements and agendas, such as the SDGs and NUA, are screened, adapted, and integrated within the RPJP and RPJMN, especially at the central level. In certain cases, to ensure and demonstrate Indonesia's commitment to the global community, Indonesia has also established institutional and regulatory support to ensure the implementation of such agreements and agendas.

In the context of the SDGs agreed in 2015, Indonesia was able to directly integrate this agreement into its RPJMN. The National Planning Agency

11 Repelita is abbreviation from Rencana Pembangunan Lima Tahun — Five (5) Year Development Planning. Repelita was used by the Government of Indonesia for setting the long-term goals of development planning.

(BAPPENAS) managed to compile, critique, filter, adjust, and then integrate the agenda items of the SDGs into the 2015–2019 RPJMN. Indonesia's seriousness in integrating the SDGs into the RPJMN is evident in the presence of 169 indicators in the 2015–2019 RPJMN adopted from 244 SDG indicators.

Furthermore, in a sign of Indonesia's long-term commitment, the government issued Presidential Decree No. 59, 2017 (Republic of Indonesia, 2017). This decree is very significant, as it initiated and legally guarantees the implementation of the SDGs in Indonesia. Six particularly important points in this presidential regulation are worth highlighting: 1) it integrates the SDGs into RPJP and RPJMN, both at national and provincial levels; 2) it guarantees the government's political commitment to the SDGs, not least by including the President and Vice-president of Indonesia at the head of the country's SDG steering board; 3) it sets out Indonesia's obligation to develop a roadmap for SDG implementation for the period 2017–2030, as well as a National Action Plan for SDGs and Regional Action Plan/SDG RAD in each province; 4) it establishes a clear institutional framework to ensure the implementation of the SDGs in Indonesia, in particular in the form of a National Coordinating Team involving representatives of all stakeholders; 5) it asserts that the SDGs can only be achieved through the cooperation of all stakeholders, including government, community, private sector, philanthropic organizations, and academia; and 6) it emphasizes the importance of monitoring and evaluation systems to reduce possible deviations from implementation of the SDGs.

In brief, it can be concluded that Indonesian's response to the SDGs has been very positive:[12] its political support is based on the rational and tangible commitment of national leaders, as well as a clear institutional and legal system. With its recent and extensive experience of implementing the MDGs, Indonesia is in a position to ensure the effective implementation of the SDGs, even in the context of different political, economic, and social systems.

3.2 *Indonesia's Responses to the New Urban Agenda*

Although on a different scale from the SDGs, Indonesia's involvement with and responses to the New Urban Agenda (NUA) have also been relatively positive, at least in the early stages of NUA formulation. Indonesia has been involved in the whole NUA process: it was active at the Habitat I conference in Vancouver in 1976, and at Habitat II in Istanbul in 1996. In the lead-up to Habitat III,

12 During a high-level meeting at the UN in 2017, the Minister for the National Planning Agency of Indonesia stated that "for Indonesia, SDGs is not just to accommodate the UN agreement, but also crucial as a platform for Indonesia to be a developed country" (BAPPENAS, 2017).

which gave birth to the NUA treaty, Indonesia played a key role, especially in hosting the preparatory committee, PrepCom 3, in Surabaya in July 2016; this was the final drafting phase of the NUA document which was agreed at the Habitat III conference, in Quito, in October 2016. As illustrated in Table 8.2, several key events demonstrate Indonesia's active level of engagement, from the preparation of its own National Report to its participation in the Habitat III conference.

In terms of substance, too, Indonesia's contribution to the formulation of the NUA has been clear and important. Since 2014, Indonesia has been a member of the council and of the working group to draft a policy paper in preparation for a draft agreement at Habitat III. Through a National Expert Team for Habitat established by the government, it was also involved in formulating points of agreement from various Asia-Pacific countries through the Jakarta Declaration in 2015. The Jakarta Declaration revealed the concerns of the Asia-Pacific countries in the context of the NUA and included some 55 proposals.

Indonesia was not only active and successful in hosting PrepCom 3 in Surabaya; it also had some substantial input in formulating the last draft of NUA. There were at least five important issues of concern to Indonesia which were raised in the process of drafting the NUA in Surabaya. These were: 1) the right to the city; 2) the LGBT community; 3) urban–rural linkages; 4) island cities; and 5) urban heritage and culture. Although not all of the Indonesian inputs were fully adopted, the attention drawn to these five issues received a positive response; they remain under discussion among academics and observers.

In spite of its active role in drafting the agreement, Indonesia's response to the NUA can be critiqued on at least four points. First, the Indonesian government has failed to analyse and assess the NUA comprehensively: it tends to look at urban issues from a purely physical point of view and has therefore mandated the Ministry of Public Works (MOPW) as solely responsible for organizing national resources in response to the NUA. Second, the Indonesian government has failed to develop a National Urban Policy (NUP), although this is a requirement clearly specified in the NUA.[13] This failure is unfortunate because, in the absence of a NUP, Indonesia may face difficulties not only in implementing but also in localizing the NUA into the context and perspectives of Indonesia. Third, in contrast to the case of the SDGs, the government has failed to develop clear legal and institutional support to ensure the implementation and localization of the NUA. At the time of writing — mid-2018, almost a year and a half after the NUA declaration — Indonesia has no clear and strong institutional

13 Specific references to national urban policies can be found in the NUA in paragraphs 15(b), 15(c), 21, 87, 89, 130, and 149 (see UN, 2017a).

TABLE 8.2 Towards the NUA: list of events by Indonesian Government and Civil Society

No.	Date	Events/activities	Notes
1	2014	National Secretariat for Habitat formed to prepare the NUA Indonesia, together with the formation of National Expert Team	Initiated and funded by the Ministry of Public Works (MOPW)
2	2014–2015	Preparation of Indonesian National Report	Initiated and funded by the MOPW, involving Expert Team
3	2015–2016	Indonesia joins the Bureau Members for Habitat III; continuous involvement from then on in drafting of the NUA	Ministry of Foreign Affairs (MOFA)
4	2015	Indonesia hosts the Asia-Pacific regional meeting (Jakarta Declaration 21–22 October 2015)	Organized by the MOPW involving academics and NGOs
5	2015	3rd Urban Social Forum held in Surabaya on 19–21 December 2015 and also in Solo, 2016	Initiated by civil society with little support from the government
6	2015–2016	ASPI (Association of Indonesian Planning Schools) holds seminars for Habitat III	Initiated by ASPI with little support from the government
7	July 2016	Indonesia hosts PrepCom 3 in Surabaya (leading to the final draft of the NUA): 25–27 July 2016	Organized by the MOPW and supported by the Ministry of Foreign Affairs
8	October 2016	Indonesia sends a 'small/modest' delegation to Quito for Habitat III (October 2016).	Led by the Minister of Public Works
9	October 2017	Indonesians celebrate Habitat Day with the NUA as the main theme of the celebration	Largely ceremonial activity
10	2017	Preparation of practical guidelines for the implementation of the NUA in Indonesia	No National Urban Policy (NUP) has been drawn up
11	February 2018	Indonesian delegates attend the World Urban Forum in Kuala Lumpur to discuss the implementation of the SDGs and the NUA	Largely ceremonial/ formal activity
12	March 2018	MUO between government and 20 universities in Indonesia to use the NUA as an agenda for "community service programme"	No clear agreement on implementation and funding mechanisms

framework that specifically addresses the NUA. The National Secretariat, established in 2014, is just a small unit responsible to the Directorate of Human Settlements within the MOPW. The secretariat has limited authority, financial and human resources, and is mostly restricted to managing 'ceremonial' activities under the NUA banner. Finally, Indonesia has also failed to secure the participation of the private sector in implementing the NUA.[14] Although some private sector actors were invited to provide data, information, and views on the problems of urbanization and housing in Indonesia during the preparation of the National Report from 2014 to 2016, there is no clear roadmap for how to involve the private sector more fully in the coming years. Cities in Indonesia need to identify innovative strategies to improve stakeholder engagement and increase investment, including from the private sector.

These concerns notwithstanding, there are some reasons for optimism. In February 2018, the Indonesian government participated in the World Urban Forum 9 in Kuala Lumpur, Malaysia. A large delegation consisting of bureaucrats, academics, NGOs, the private sector, and youth groups were all involved in programmes to monitor and ensure the implementation of the NUA. This can be seen as a demonstration of the commitment of the Indonesian government and people to implement the NUA. Discussions of the Indonesian delegates during the event reflect their perception that, so far, the involvement of Indonesia in the NUA has been largely ceremonial, and their concern over the lack of a clear substantive agenda of how to implement and localize the NUA for the benefit of Indonesians.[15]

In brief, almost one and a half years after the adoption of the NUA, Indonesia's effectiveness in implementing and localizing the agreement is still very limited. The prospects for its success will partly depend on whether Indonesia is able to intelligently and strategically link the NUA with the SDGs. The following section discusses this in more detail.

4 Linking the SDGs and the NUA: Indonesia's Challenges

As two documents produced by the United Nations, there is ongoing debate over the relevance of the SDGs and the NUA. A common view, especially in Indonesia, is that the NUA is a more detailed translation of SDG 11, which

14 The important role of the private sector in urban development is emphasized in many paragraphs of the NUA, including 53, 82, 132, 133, 141, 149 (UN, 2017a).

15 Documentation about this event can be found at Kementerian Pekerjaan Umum dan Perumahan Rakyat RI (2018), and PWK ITS (2018).

explicitly emphasizes the importance of achieving inclusive, safe, resilient, and sustainable cities. This view is true, but it is also incomplete: the NUA is a lot more than a translation of SDG 11. While it builds on Goal 11, it also addresses a range of urbanization and human settlement issues.

According to UN Habitat (2016), there are at least five reasons for associating the NUA with the SDGs. First, the NUA provides a spatial framework or location for the delivery of the SDGs. This kind of locational framework is important because it can also be used to coordinate other sectors discussed in the SDGs. In the Indonesian context, the spatial framework or location for the implementation of the SDGs is crucial because of the large geographical area or scope. Although Indonesia has relatively stable macroeconomic foundations, it is still overshadowed by the disparity between its territories; the sheer size and topography of the country impede the uniform distribution of economic growth and the resulting prosperity. Against this backdrop, the implementation of the SDGs in Indonesia could exacerbate spatial inequities and imbalances.

Second, it is anticipated that by 2030, 60% of the world's population will reside in urban areas; it is therefore reasonable to target urban areas in the implementation of the SDGs. Furthermore, since cities account for 70% of the world's GDP and contribute between 39% and 49% of the world's greenhouse gases, then economically and environmentally, it is hard not to mention cities as vehicles in achieving SDG goals. The situation in Indonesia reflects the global trend, with cities expected to play an increasingly significant role in national development. Indonesia's future and sustainability will thus depend to a considerable extent on whether the government can guide urban development in an optimal way, yet avoid urban bias in its policies.

Third, the NUA represents not only a strategic spatial framework, but also an important governance framework: such a framework will be essential for the successful implementation of the SDGs. The national urban development policies, legislation, spatial planning, and local financial framework recommended in the NUA, for example, could guide and facilitate the implementation of the SDGs. Although, at present, Indonesia is focusing its development on remote borders and rural areas, in the medium to long term, more productive, efficient urban–rural relationships will be crucial for the country's sustainability.

Fourth, the NUA provides more detailed targets and strategies for several SDG objectives, especially in urban basic services. For example, transportation is described in more depth in the NUA than in any SDG. The NUA contains various recommendations on sustainable transportation and mobility, including sustainable transport infrastructure and services, rural–urban relations, travel-demand management, road safety, climate change, air quality, energy efficiency, equity, and inclusion. In Indonesia, such detailed guidance is highly relevant and important because municipalities face major problems in this area. The kind of detailed, clear targets and strategies for achieving better urban services offered in the NUA will also be useful for Indonesia in implementing and localizing the SDGs.

Finally, the NUA addresses the importance of implementation at the local level and the role of urban local government, while the SDGs emphasize implementation at the national and regional levels. Such differences in emphasis could be collaboratively linked and synergized. A focus by local governments on the NUA could ensure the realization of services to the people, while a focus at national and regional levels on the SDGs could ensure the equity and distribution of services at those levels. In Indonesia, this kind of synergy is very important. The national targets of the SDGs can be achieved only if there is effective collaboration between national, regional, and local levels (see below for more on the issue of multi-level governance).

Cities are key to the implementation of many of the SDGs, and a successful NUA will create an opportunity to enhance the SDGs' effectiveness. Although the 17 SDGs and their 169 targets are often described as aspirational, it is in cities that they become tangible to regular citizens. SDG 11 aims to make "cities and human settlements inclusive, safe, resilient, and sustainable". Achieving this Goal depends in part on the level of engagement of local stakeholders, regional governments, community-based organizations, academics, and the business sector, as well as on adequate synergies between national and local policies.

Given current conditions in the country, Indonesia faces some serious challenges in achieving SDG 11, as shown in Table A1 in the Appendix to this chapter. Table A1 shows Indonesia's position vis-à-vis 15 indicators related to SDG 11. There have been some achievements; moreover, Indonesia is pursuing a number of programmes and activities to ensure a positive outcome on all indicators over a longer time frame. It is important to recognize, however, that this interpretation of objectives through 15 indicators is too limited, not only for SDG 11, but also in the context of the NUA. There are many dimensions and aspects of the NUA that are not covered by these 15 indicators. This is why the development of a National Urban Policy is so important for Indonesia — to act as a framework and an umbrella for the implementation and localization of the NUA in a more complete and comprehensive form.

4.1 *Implementing and Localizing the NUA: Politics and Leadership in Multi-level Governance*

As a country with a highly centralized and even authoritarian governance history during the New Order (1965–1998), Indonesia is still learning how to improve the quality, effectiveness, and benefits of the decentralization and democratization initiated since 1998. Some critical observers argue that the process of consolidation is very slow and has so far failed to yield results; others are quite optimistic that the process will bring positive change to Indonesia (see the discussions in Holzhacker et al., 2016). While recognizing that the government's efforts to date have delivered few tangible benefits to society, this chapter shares the more optimistic view of the prospects for this process of consolidating decentralization and democratization in Indonesia. Nevertheless, the government will need to be aware of a number of requirements as it goes forward.

First, as emphasized by Budiono (2018), the central government must have clarity of vision regarding long-term development, as well as a clear sense of direction and measurable targets. Second, it needs consistency, in the sense that all the policies, programmes, and development strategies that have been formulated in RPJP and RPJMN are consistently implemented, in accordance with changing conditions, resources, constraints, and opportunities. Third, all policy formulations and policies must be clearly based on proper, rational considerations. Fourth, coherence is needed, in terms of coordination, harmony, and synergy. The issue of coherence is especially relevant in the context of multi-level governance, and concerns relationships and collaborations between central, provincial, and local (district and city) governments.

In the context of implementing and localizing the NUA, there are three major issues to be resolved: 1) sectoral ego; 2) regional/territorial ego; and 3) the weakness of local leadership. Sectoral ego refers to the tendency for different sectors or agencies to formulate and implement programmes and activities in accordance with their own specific mandates or institutions, rather than collaborating and cooperating with related sectors to ensure more effective and synergistic results. In the context of the NUA, this phenomenon of sectoral ego is linked to the fact that, from the very beginning of Indonesian government involvement in the NUA, all activities were delegated to the MOPW, while other very relevant ministries (such as BAPPENAS, Ministry of Agriculture and Spatial Planning, Ministry of Internal Affairs, Ministry of Life and Forestry) were not involved.[16]

16 In informal discussions with senior staff at the MOPW, I was informed that, through a MOPW programme called 100-0-100 (standing for 100% clean water, zero slums, 100% sanitation) under the RPJMN 2015–2019, Indonesia has implemented the NUA. This view of Indonesia's coverage of the NUA illustrates the phenomenon of sectoral ego: because the MOPW mandate covers only the physical aspects of the NUA, it missed other important non-physical aspects, such as the right to the city, inclusiveness, and safety issues.

Regional/territorial ego is the tendency of each region to focus solely on the problems of that region/territory, without cooperating, either vertically with the provincial and central government,[17] or horizontally with the surrounding area. In the context of urban development, this tendency is highly visible and forms a serious impediment to the efficiency and even sustainability of urban services.[18] Some regions have developed forms of inter-regional cooperation, but the effectiveness and the extent of this kind of cooperation need to be developed much further.[19] Among the most immediate negative impacts of this lack of coordination among regions are the ongoing large-scale land conversions from agricultural land to non-agricultural, and from productive land to residential areas, particularly in the urban fringes, outside the administrative boundary of a given city. Central and local governments need to change their approach, reduce their sectoral and territorial ego, and start collaborating on inter-regional issues such as water resources, regional waste management, flood control, and disaster management.

Finally, the success or failure of regional development in this era of decentralization and democratization will ultimately depend on the leadership qualities of the regional leaders, in this case at the level of the general and the regent. In the Indonesian context of multi-level governance, the existence of unclear bureaucratic mechanisms and tendencies towards sectoral or territorial ego will require the presence of innovative and well-qualified local leaders. If local leaders have a sincere commitment to serve, and involve their officials and institutions in a meaningful way in decision-making processes, the result will be an improvement in living conditions in Indonesia's cities.

4.2 *The SDGs and the NUA: Implications for Capacity Building and Knowledge Production*

No matter how robust the commitment of goverments at national, provincial, and local levels to implement and localize the NUA, supported by a rational

17 Again, the case of the 100-0-100 programme is illustrative, this time revealing a lack of coordination between the national and the local level. Interviews with some local government officials indicated that the zero slum target, especially, was difficult to achieve, since no significant resources were given to local governments to resolve their slum issues.

18 This is one of the problems caused by the lack of a strong national urban policy, leading to lack of coordination and imbalanced growth of metropolises, big, medium, and small cities.

19 Several studies have concluded that inter-regional collaborations have not been able to deliver better public services due to the lack of: consistent commitment among regions; clear goals, targets and indicators; clear and binding legal frameworks; and private sector involvement (e.g., Hanggoro, 2008; Lokantara, 2016; Pambudi, 2014; Utami & Ika, 2015; Wahyono, 2015).

and integrated planning system, the effectiveness of that implementation will ultimately depend on the willingness and the capacity not only of the bureaucratic apparatus of government, but of all the stakeholders involved, especially the community and civil society. Moreover, the process of urban transformation should not only lead to an improvement in the city environment and a better life for the inhabitants, but should also maintain and promote the unique characteristics and identity of Indonesia. With this in mind, it is important to explore the process of knowledge production in the shaping of Indonesia's policies and development strategies for the urban environment. Capacity building and knowledge production are two prerequisites not only to effectively implement the NUA, but also to localize it, adapt it, and make it more appropriate for the Indonesian context.

Capacity building has many dimensions, especially when we talk about capacity building for the whole community in a developing country. In order to realize the agenda contained in the NUA, a comprehensive and continuous capacity development is required for all stakeholders involved. Indeed, in the NUA itself, at least 10 paragraphs are dedicated to capacity development for stakeholders involved in the implementation of the agreement.[20] In Indonesia, the condition of capacity building can be illustrated by the case of planning education, which began formally in 1959 and continues to develop. As I have indicated elsewhere (Setiawan, 2018), planning education in Indonesia has tended to serve the needs of government planning practices and has therefore failed to cast a critical eye over those practices. It has turned planning into a largely technocratic exercise and planning professionals into technocrats working in the office to support and to justify development goals decided solely by government. Thus, despite its growing availability and standing,[21] planning education in Indonesia has not been able to inspire planning practices which provide innovative approaches for complex urban and regional problems.

In light of the problems faced by planning education in Indonesia, I have suggested some reforms (ibid.). As an important element in capacity building for the development of the sustainable city, planning education in Indonesia needs to refocus its approach and its orientation. In this context, the most fundamental consideration is whether it should step outside of its traditional way of transferring knowledge and skills on planning in order to inspire students

20 Several paragraphs that are related directly to the important need of capacity development are 81,120, 126, 129, 147, 149, 151, 152, 155, 156 (see UN, 2017a).

21 At the time of writing, in 2018, there are 60 planning schools in Indonesia and all are members of the Indonesian Planning Schools Association (ASPI), formed in 2001 (Setiawan, 2018).

and young planners to adopt a more critical perspective — one that is no longer driven solely by government and market forces. For example, it could move beyond its predominant focus on spatial planning and the physical aspects of planning, and engage in more complex social, economic, cultural, and political dimensions of urban planning and development.

This example of planning education illustrates the importance of a comprehensive and systematic capacity-development programme in Indonesia in order for the NUA to be effectively implemented. In this context, it is particularly important to ensure that such capacity development includes those who are marginalized and neglected, such as women and girls, children and youth, older persons and persons with disabilities, indigenous peoples and local communities, as well as persons in vulnerable situations.[22]

Parallel to their role in capacity development, academics must also continue to explore and understand the dynamic processes of urbanization and urban transformation in Indonesia, and how such processes can result in a better urban environment and living conditions. For this, it is important to establish a research agenda not only to monitor and evaluate the implementation of the SDGs and the NUA,[23] but also to further explore how such global agreements and influences will affect and shape future urbanism in Indonesia. This research agenda should focus on what makes each city unique in its response to global forces and on understanding the impacts of urbanization and urban transformation on broad societal issues such as equity, welfare, rights, and power relations.

A central problem of urban and housing studies in Indonesia is their tendency to maintain a myopic and narrow focus on specific aspects of urban and housing policies. These studies view urban and housing issues merely in terms of their spatial-physical dimensions, analysing them from a simple economic supply and demand model, and neglecting more complex socio-political issues. Within a broader perspective, urban studies in Indonesia should be directed to answer three fundamental questions: 1) how do local variations (within both the state and civil society) respond to global modernization and urbanization processes and who benefits and loses in this process; 2) what is the role of the state in the process of urbanization and urban transformation; and 3) what are the effects of urbanization and urban transformation on the relationship between the state and civil society; do urbanization and urban transformation promote the development of a more just, inclusive, safe, resilient, and

22 See further detail in Paragraph 155 of the NUA (UN, 2017a).

23 The importance of knowledge in the NUA is stated in several paragraphs including 56, 117, 125, 157, 159, and 160 (UN, 2017a).

sustainable society? Only by explicitly relating urban issues to broader societal problems can urban studies become more comprehensive and useful.

As Indonesia becomes increasingly urbanized, a comprehensive and innovative research agenda leading to knowledge production would help not only in achieving more effective policies and strategies to implement the SDGs and the NUA but, more importantly, in localizing the universal SDG and NUA goals to meet more relevant and contextual needs. With its unique, rich, and diverse cultures and traditions, Indonesia could provide other forms of urbanization and urban development.

5 Conclusion

The main conclusions and recommendations of this chapter can be summarized as follows. First, urbanization and urban transformation in Indonesia are taking place in a period of great flux and uncertainty. Indonesia's success in guiding this urban transformation process will be an important — if not the most important — determining factor for a more prosperous, just, and sustainable future. This process is full of challenges and opportunities that must be approached positively and optimistically, in accordance with the spirit of the NUA; that is, how to maximize the potential benefits while addressing the possible negative impacts of urbanization and city transformation. Some of the positive factors that Indonesia can use to its advantage are: 1) the predicted demographic bonus; 2) the ongoing consolidation of democracy and decentralization; 3) community empowerment and social capital; and 4) steady economic growth. Indonesia's success in implementing and localizing the NUA will ensure its success in also achieving the SDGs and the country's long-term goals, as contained in the RPJP.

Second, if Indonesia is to succeed in implementing and localizing the NUA, four measures will be key: 1) critically mapping, adjusting, and defining substantive indicators in the NUA that are relevant in the Indonesian context, as has been attempted with the SDGs; 2) building clear and robust legal and institutional support, to ensure continuity and consistency of implementation; 3) securing strong political and leadership commitment and effective coordination both horizontally and vertically at the national, provincial, and local/city levels; and 4) developing comprehensive and innovative capacity-building programmes, especially aimed at young people, communities, and vulnerable groups, so that they can engage with the process and maximize the opportunities and benefits of urbanization. The strengthening of these networks and capacities should be coupled with knowledge production efforts to better understand the direction, process, and shape of urbanization and urban development in Indonesia.

This is an opportunity for Indonesia to renew its political engagement with the urban agenda, and to empower the people to act as its partners in achieving sustainable urban development. It will be an important part of Indonesia's contribution not only to the implementation of the SDGs and the NUA, but also to developing cultural diversity in this global village.

Appendix

TABLE A1 The link between SDG 11 targets and indicators, and Indonesian conditions

SDG 11 target	Indicator	Indonesian conditions
1.1 By 2030, ensure access for all to adequate, safe and affordable *housing* and basic services and upgrade *slums*	11.1.1 The proportion of urban population living in slums, informal settlements, or inadequate housing	1. The housing backlog in Indonesia was estimated at 7.6 million in 2014. This means that a lot has to be achieved by RPJMN 2015–2019. According to PUPR (2015), the housing backlog can be addressed by government through several types of programmes. 2. PUPR (2015) suggests that percentage of the population with access to appropriate sanitation in Indonesia in 2015 was 60.9%. WHO gives the percentages of Indonesians with access to sanitation as: rural 47.5%; urban 72.3%; total 60.8% (TPRP & Rekan, 2017). This leaves 40% still in need of access to proper sanitation. 3. PUPR (2015) indicates that 68.87% of the total population had access to safe drinking water in 2015. WHO gives figures as: rural 79.5%; urban 94.2%; total 87.4% (TPRP & Rekan, 2017). 4. PUPR (2017) estimates total slum settlement area in Indonesia at 38,431 ha located in 34 provinces spread over 269 regencies/cities in 11,067 villages. At the end of 2017, the slum area that had already been improved with basic infrastructure was 5,603 ha.

TABLE A1 The link between SDG 11 Targets and Indicators (*cont.*)

SDG 11 target	Indicator	Indonesian conditions
11.2 By 2030, provide access to safe, affordable, accessible and sustainable *transport systems* for all, improving road safety, notably by expanding public transport, with special attention to the needs of those in vulnerable situations, women, children, persons with disabilities and older persons	11.2.1 The proportion of the population that has convenient access to public transport, by sex, age, and persons with disabilities	1. Average number of accidents in Indonesia in the period 1992–2016 was 49,781. For a graphic representation see BPS (2018). 2. The accessibility of Indonesian transportation in a whole-world ranking is as follows: ranked 72 in the world (overall); 72 (roads); 42 (railways); 72 (ports); 64 (airports) (TPRP & Rekan, 2017).
11.3 By 2030, enhance inclusive and sustainable urbanization and capacity for participatory, integrated and sustainable human settlement planning and management in all countries	11.3.1 The ratio of land consumption rate to population growth rate	No data available.
	11.3.2 The proportion of cities with a direct participation structure of civil society in urban planning and management that operates regularly and democratically	The Spatial Planning Act No. 26 2007 requires that all spatial planning should be open for community/people's participation (Republic of Indonesia, 2007b).
11.4 Strengthen efforts to protect and safeguard the world's cultural and natural heritage	11.4.1 Total expenditure (public and private) per capita spent on the preservation, protection and conservation of all cultural and natural heritage, by type of heritage (cultural, natural, mixed and World Heritage Centre designation), level of government (national, regional and local/municipal), type of expenditure (operating expenditure/investment) and type of private funding (donations in kind, private non-profit sector and sponsorship)	1. Indonesia has one World Heritage in Danger site: tropical rainforest in Sumatra (see: https://whc.unesco.org/en/soc/3804). 2. Indonesia has four World Natural Heritage sites; four Cultural Heritage sites; and six Intangible Cultural Heritage sites (see: https://whc.unesco.org/en/statesparties/id). 3. Annual budget for environmental protection in Indonesia is still considered very low compared to environmental problems faced by the country. In 2012, the budget for the Ministry for the Environment was only 0.001 per cent of the total annual budget. In 2018, the budget for the Ministry for the Environment and Forestry has increased to 6.77 trillion Rupiah (about 0.00305 per cent of the total budget) (Kementerian Keuangan RI, 2018; Nainggolan, 2016).

TABLE A1 The link between SDG 11 Targets and Indicators (*cont.*)

SDG 11 target	Indicator	Indonesian conditions
11.5 By 2030, significantly reduce the number of deaths and the number of people affected and substantially decrease the *direct economic losses relative to the global gross domestic product caused by disasters*, including water-related disasters, with a focus on protecting the poor and people in vulnerable situations	11.5.1 Number of deaths, missing persons and directly affected persons attributed to disasters per 100,000 population 11.5.2 The direct economic loss in relation to global GDP, damage to critical infrastructure and number of disruptions to basic services, attributed to disasters	1. In August 2016, the number of disasters in Indonesia for that year was estimated at 1,495, with 257 deaths, 303 injured, 2.1 million affected or experiencing displacement (temporary or permanent), and thousands of houses damaged to varying degrees (BNPB, 2016a). Based on data from BNPB (2018), the number of disasters in Indonesia was: 1,674 (2013), 1,967 (2014), 1,732 (2015), 2,342 (2016), 2,341 (2017), and 1,134 (2018, up to 26 May). 2. Annual average of damage and loss due to disasters in Indonesia is around 30 trillion Rupiah (BNPB, 2016a). 3. Government measures related to disaster risk are the Disaster Management Act No. 24 2007, National Action for Disaster Management, and National Action for Disaster Risk Reduction (BNPB, 2014).
11.6 By 2030, reduce the adverse per capita environmental impact of cities, including by paying special attention to air quality and municipal and other waste management	11.6.1 The proportion of urban solid waste regularly collected and with adequate final discharge out of total urban solid waste generated, by cities	1. In general, only 60–70 % of solid waste in Indonesia cities can be collected by the municipal governments (BPS, 2017). 2. Regulations include the Waste Management Act No. 18 2008, and follow-up PP No. 81 2012, The Management of Household Waste and Similar Types of Household Waste (Republic of Indonesia, 2008, 2012).
	11.6.2 Annual mean levels of fine particulate matter (e.g., PM2.5 and PM10) in cities (population-weighted)	No data available.
11.7 By 2030, provide universal access to *safe, inclusive and accessible, green and public spaces*, in particular for women and children, older persons and persons with disabilities	11.7.1 The average share of the built-up area of cities that is open space for public use for all, by sex, age, and persons with disabilities 11.7.2 The proportion of persons victim of physical or sexual harassment, by sex, age, disability status and place of occurrence, in the previous 12 months	1. According to the Spatial Planning Act No. 26 2007, all cities in Indonesia should have 30% of open green spaces and 20% of this should be open to the public (Republic of Indonesia, 2007b). 2. Urban crimes figures fluctuate in Indonesian cities. In terms of spatial distribution, crime tends to occur more frequently in urban fringe areas (Setiawan, 2017).

TABLE A1 The link between SDG 11 Targets and Indicators (*cont.*)

SDG 11 target	Indicator	Indonesian conditions
11.a Support positive economic, social and environmental links between urban, peri-urban and rural areas by strengthening national and regional development planning	11.a.1 The proportion of the population living in cities that implement urban and regional development plans integrating population projections and resource needs, by the size of the city	1. According to the 2015 census, 53.14% of the Indonesian population live in urban areas (BPS, 2015). 2. In 2018, the Government of Indonesia issued a new regulation, Guidelines for Arranging Spatial Planning for Province, Region, and City Level, including the need to follow population projections (BPN, 2018).
11.b By 2020, substantially increase the number of cities and human settlements adopting and implementing integrated policies and plans towards inclusion, resource efficiency, mitigation and adaptation to climate change, resilience to disasters, and develop and implement, in line with the Sendai Framework for Disaster Risk Reduction 2015–2030, holistic disaster risk management at all levels	11.b.1 Number of countries that adopt and implement national disaster risk reduction strategies in line with the Sendai Framework for Disaster Risk Reduction 2015–2030	Indonesia already has the Disaster Management Act No. 24 2007, the National Action for Disaster Management, and the National Action for Disaster Risk Reduction (BNPB, 2014).
	11.b.2 The proportion of local governments that adopt and implement local disaster risk reduction strategies in line with national disaster risk reduction strategies	1. According to the Disaster Management Act No. 24 2007, all local governments in Indonesia must prepare Local Action Plans for Disaster Management and for Disaster Risk Reduction (Republic of Indonesia, 2007a). 2. BNPB annual report 2016 indicates that 555 regions/municipalities were targeted for designing of local Risk Reduction plans. It is possible that the number of regions/cities achieving this could exceed the targeted number (BNPB, 2016b).
11.c Support least developed countries, including through financial and technical assistance, in building sustainable and resilient buildings utilizing local materials	11.c.1 The proportion of financial support to the least developed countries that is allocated to the construction and retrofitting of sustainable, resilient and resource efficient buildings utilizing local materials	Based on National Long-term Development Plan (RPJPN), National Medium-term Development Plan (RPJMN), and the Strategic Plan of the Ministry of Public Works and Housing, Indonesia has a target to eradicate slum areas. Government programmes include Kotaku–Kota tanpa Kumuh ('City without Slums'), and the 100-0-100 programme which targets 100% access to clean water, 0% slum settlement areas, and 100% access to appropriate sanitation facilities (PUPR, 2015, 2017; Kementerian Pekerjaan Umum (2014).

SOURCES: TARGETS AND INDICATORS FROM UN (2017B); OTHER SOURCES AS SPECIFIED IN THE TABLE.

References

BAPPENAS (2017). Indonesia's SDGs Voluntary National Review 2017. Jakarta: Ministry of National Development Planning (BAPPENAS). Retrieved from https://sustainabledevelopment.un.org/content/documents/25469INDONESIA_VNR_PPT.pdf (accessed 19 August 2018).

Barber, B. (2013). *If mayors ruled the world: Dysfunctional nations, rising cities*. New Haven, CT: Yale University Press.

BNPB (2014). Rencana Nasional Penanggulangan Bencana 2015–2019 [National action for disaster management 2015–2019]. Jakarta: BNPB.

BNPB (2016a). Manajemen Bencana di Indonesia [Disaster management in Indonesia]. Jakarta: BNPB. Retrieved from http://sibima.pu.go.id/pluginfile.php/8140/mod_resource/content/1/201610-CPD%20Ahli%20Arsitektur-03-03-Manajemen%20Bencana%20di%20Indonesia.pdf.

BNPB (2016b). Laporan Kinerja BNPB (Badan Nasional Penanggulangan Bencana) Tahun 2016 [Annual performance report Indonesian National Board for Disaster Management (BNPB) 2016]. Jakarta: BNPB.

BNPB (2018) Data Bencana Indonesia [Disaster data in Indonesia]. Retrieved from http://dibi.bnpb.go.id/dibi/ (accessed 22 August 2018).

BPN (2018). Peraturan Menteri Agraria dan Tata Ruang/Kepala Badan Pertanahan Nasional Republik Indonesia No. 1 Tahun 2018 Tentang Pedoman Penyusunan Rencana Tata Ruang Wilayah Provinsi, Kabupaten dan Kota [Guidelines for arranging the spatial planning for province, region, and city level]. Jakarta: BPN.

BPS (2008). Proyeksi Penduduk Indonesia 2005–2025 [Indonesia population projection 2005–2025]. Jakarta: UNFPA Indonesia/BAPPENAS.

BPS (2013). Proyeksi Penduduk Indonesia 2010–2035 [Indonesia population projection 2010–2035]. Jakarta: Statistics Indonesia.

BPS (2014). Proyeksi Penduduk, Mercusuar Pembangunan Negara [Population projection, a signal for national development]. Retrieved from https://www.bps.go.id/news/2014/03/26/85/proyeksi-penduduk--mercusuar-pembangunan-negara.html (accessed 17 August 2018).

BPS (2015). Penduduk Indonesia Hasil Survei Penduduk Antar Sensus 2015 [Population of Indonesia: Result of the 2015 intercensal population census]. Jakarta: Statistics Indonesia.

BPS (2017). Statistik Lingkungan Hidup Indonesia [Environment statistics of Indonesia]. Jakarta: Statistics Indonesia.

BPS (2018). Jumlah Kecelakaan, Koban Mati, Luka Berat, Luka Ringan, dan Kerugian Materi yang Diderita Tahun 1992–2016 [The number of accidents, death, major and minor injuries, and loss & damage 1992–2016]. Retrieved from https://www.bps.go.id/linkTableDinamis/view/id/1134 (accessed 22 August 2018).

Budiono (2018). Factors for effective decentralization in Indonesia. Paper presented at the opening ceremony of Laporan Perekonomian Indonesia 2017 oleh Bank Indonesia. Jakarta, 28 March.

Chaussard, E., Amelung, F., & Abidin, H. Z. (2012). Sinking cities in Indonesia: Space-geodetic evidence of the rates and spatial distribution of land subsidence. Proceedings of the FRINGE 2011 Workshop, Frascati, Italy, 19–23 September.

Cobban, J. L. (1971). The city of Java: An essay in historical geography. (Doctoral dissertation.) Berkeley, CA: University of California.

van Dijk, M. P., Edelenbos, J., & van Rooijen, K. (2017). *Urban governance in the realm of complexity*. Rugby: Practical Action Publishing Ltd.

FAO (2010). *Global forest resources assessment and the state of the world's forests*. Rome: Food and Agriculture Organization of the United Nations.

Flanagan, W. G. (1993). *Contemporary urban sociology*. Cambridge: Cambridge University Press.

Geertz, C. (1965). *The social history of an Indonesian town*. Cambridge, MA: MIT Press.

GFDRR (2016). Country Profile: Indonesia. Global Facility for Disaster Reduction and Recovery (GFDRR).

Guiness, P. (1986). *Harmony and hierarchy in a Javanese kampung*. Singapore: Oxford University Press.

Hanggoro, D. (2008). The effectiveness of regional cooperation 'Pawonsari': Case study of education services in cross-border areas. (Unpublished master's thesis.) Yogyakarta: Universitas Gadjah Mada.

Harrison, D. (1988). *The sociology of modernization and development*. London: Unwin Hyman.

Holzhacker, R. L., Wittek, R., & Woltjer, J. (Eds) (2016). *Decentralization and governance in Indonesia*. New York and Dordrecht: Springer International.

Hugo, G., Hull, T. H., Hull, V. J., & Jones, G. W. (1987). *The demographic dimension in Indonesian development*. Singapore and London: Oxford University Press.

Jellinek, L. (1991). *The wheel of fortune: The history of a poor community in Indonesia*. Sydney: Allen and Unwin.

Kementerian Keuangan RI (2018). Rancangan APBN 2018. Retrieved from https://www.kemenkeu.go.id/rapbn2018 (accessed 22 August 2018).

Kementerian Pekerjaan Umum (2014, June 25). Retrieved from http://ciptakarya.pu.go.id/csr/home/?p=100.

Kementerian Pekerjaan Umum dan Perumahan Rakyat RI (2018). Buletin Cipta Karya Edisi 02/Tahun XVI/Februari 2018. Retrieved from http://ciptakarya.pu.go.id/dok/bulletin/buletin_2017/pdf/1526446374-Buletin%20CK_Feb%202018.pdf (accessed 20 August 2018).

Krausse, G. H. (1975). The kampongs of Jakarta, Indonesia: A study of spatial patterns in urban poverty. (Doctoral dissertation.) Pittsburgh, PA: University of Pittsburgh.

Lokantara, I. G. W. (2016). Kerjasama Kawasan Perkotaan Sarbagita dalam Penyelenggaraan Pelayanan Publik [Regional cooperations in public services: The case of Sarbagita, Bali]. (Unpublished master's thesis.) Yogyakarta: Universitas Gadjah Mada.

Milone, P. D. (1966). Urban areas in Indonesia: Administrative and census concepts. Institute of International Studies, research series no. 10. Berkeley, CA: University of California.

Nainggolan, L. (2016). Kebijakan Fiskal Lingkungan Hidup. Retrieved from https://lukasbn.wordpress.com/tag/ekonomi-fiskal-lingkungan-apbn-kemenkeu/ (accessed 22 August 2018).

Nas, P. J. (Ed.) (1985). *The Indonesian city*. Cinnaminson, NJ: Foris Publications.

Pambudi, W. (2014). Decentralized regional development by inter-municipal cooperation in Indonesia. Case study of Barlingmascakeb, Javapromo, and Kartamantul regional cooperation. (Unpublished master's thesis.) Yogyakarta: Universitas Gadjah Mada.

PUPR (2015). Peraturan Menteri Pekerjaan Umum dan Perumahan Rakyat Republik Indonesia No. 13.1/PRT/M/2015 tentang Rencana Strategis Kementerian Pekerjaan Umum dan Perumahan Rakyat Tahun 2015–2019. Jakarta: Kementerian PUPR.

PUPR (2017). Pelaksanaan Penanganan Permukiman Kumuh untuk Mendukung Terwujudnya Permukiman Layak Huni dan Berkelanjutan. Retrieved from http://kotaku.pu.go.id:8081/pustaka/files/PokjaPKP/06._Kasubdit_Perencanaan_Teknis_PKP-Pelaksanaan_Penanganan_Permukiman_Kumuh.pdf (accessed 22 August 2018).

PWK ITS (2018). World Urban Forum 9, Kuala Lumpur, Implementasi New Urban Agenda Menuju Cities 2030. Retrieved from: https://pwk.its.ac.id/world-urban-forum-wuf-9-kuala-lumpur-implementasi-new-urban-agenda-menuju-cities-2030/ (accessed 20 August 2018).

Republic of Indonesia (2007a) Undang-undang No. 24 Tahun 2007 tentang Penanggulangan Bencana [Disaster management]. Jakarta: Pemerintah Republik Indonesia.

Republic of Indonesia (2007b) Undang-undang No. 26 Tahun 2007 tentang Penataan Ruang [The spatial planning act]. Jakarta: Pemerintah Republik Indonesia.

Republic of Indonesia (2008) Undang-undang No. 18 Tahun 2007 tentang Pengeloaan Persampahan [Waste management act]. Jakarta: Pemerintah Republik Indonesia.

Republic of Indonesia (2012) Peraturan Pemerintah No. 81 Tahun 2012 tentang Pengelolaan Sampah Rumah Tangga Dan Sampah Sejenis Sampah Rumah Tangga [The management of household waste and similar types of household waste]. Jakarta: Pemerintah Republik Indonesia.

Republic of Indonesia (2017). Presidential Decree Number 59, 2017. On Implementation of Sustainable Development Goals and Targets. Jakarta: Pemerintah Republik Indonesia.

Rutz, W. (1987). *Cities and towns in Indonesia.* Berlin: Gebruder Born-traeger.

Setiawan, B. (2017). Towards the New Urban Agenda of safe cities: Urban crimes in four Indonesian cities. *IOP Conf. Series: Earth and Environmental Science*, 99. DOI: 10.1088/1755-1315/99/1/012037.

Setiawan, B. (2018). The roles of planning education in the decentralization and democratization era: Lessons from Indonesia. In A. I. Frank & C. Silver (Eds), *Urban planning education: Beginnings, global movement, and future prospects* (pp. 219–231). New York and Dordrecht: Springer International.

Sullivan, J. (1992). *Local government and community in Java: An urban case study.* Singapore: Oxford University Press.

TPRP & Rekan (2017). Tugas Rumah Indonesia dalam Implementasi New Urban Agenda-Habitat III. Retrieved from https://tprprekanblog.wordpress.com/2017/03/04/tugas-rumah-indonesia-dalam-implementasi-new-urban-agenda-habitat-iii/ (accessed 7 April 2018).

UN (2014). World urbanization prospects: The 2014 revision. CD-ROM Edition. New York: United Nations, Department of Economic and Social Affairs, Population Division.

UN (2017a). The New Urban Agenda. New York: United Nations Conference on Housing and Sustainable Urban Development (Habitat III). Retrieved from http://habitat3.org/wp-content/uploads/NUA-English.pdf.

UN (2017b). Global indicator framework for the Sustainable Development Goals and targets of the 2030 Agenda for Sustainable Development. New York: United Nations. Retrieved from https://unstats.un.org/sdgs/indicators/Global%20Indicator%20Framework_A.RES.71.313%20Annex.pdf (accessed 7 April 2018).

UN Habitat (2016). Linkages between the New Urban Agenda and the 2030 Agenda for Sustainable Development. Retrieved from https://unhabitat.org/wp-content/uploads/2016/10/Linkages_Between_NUA_and_Agenda_2030_2016.pdf (accessed 18 August 2018).

Utami, C. F., & Ika, P. P. (2015). Factors determining the success of metropolitan institution: An experience from BKSP. Proceeding of the 5th International Conference of Jabodetabek Study Forum: Sustainable megacities: Vulnerability, diversity, and livability. Bogor, Indonesia, 17–18 March.

Wahyono, H. (2015). Perencanaan Kolaboratif antar Daerah di Perbatasan Kota-Kota Joglosemar [Collaborative planning on the border regions of Joglosemar, Central Java]. (Unpublished doctoral dissertation.) Yogyakarta: Universitas Gadjah Mada.

CHAPTER 9

A Vision in Which Every Family Has Basic Shelter

Noor Hasharina Hassan and Gabriel Y. V. Yong

1 Introduction: The Sustainable Development Goals

The Sustainable Development Goals[1] (SDGs) are a set of ambitious targets that seek to carry on and complete the work initiated by the Millennium Development Goals (MDGs), which tried to tackle the indignity of poverty. The SDGs were conceived at the United Nations Conference on Sustainable Development in Rio de Janeiro in 2012, 20 years after the first global conference on environment and development (known as the Earth Summit) in the same city, and were adopted by UN member states in September 2015. The SDGs also incorporate concerns raised in the Paris Climate Conference (COP21) and the Sendai Framework for Disaster Risk Reduction, both agreed and signed in 2015, to address the risks of climate change and natural disasters using a set of common standards and achievable targets. The 17 SDGs are meant to be viewed as interconnected targets, linking natural resource management to fostering peace and inclusive societies, reducing inequalities, and promoting economic growth.

1.1 *SDG 11: Sustainable Cities and Communities*

SDG 11 is concerned with rapid urbanization in developing nations, which has given rise to pockets of extreme poverty concentrated in many urban spaces, as governments struggle to accommodate the rising population in towns and cities. The scope of SDG 11 is broad; specific targets to be achieved by 2030 include access to adequate, safe, and affordable housing for all; provision of basic services and amenities such as access to public transport, electricity, water, and sewage systems; the upgrading of slums; inclusive and sustainable urbanization including integrated and sustainable human settlement planning and management; protecting and safeguarding cultural and natural heritage; safeguarding and improving the social, environmental, and economic conditions of vulnerable groups, including the poor; encouraging the provision of inclusive and accessible green and public spaces; supporting positive economic, social and environmental links between different nodes and regions through

1 See: http://www.undp.org/content/undp/en/home/sustainable-development-goals.html.

 | DOI:10.1163/9789004391949_010

improved planning and management; and assisting countries to improve their resource efficiency, mitigate against and adapt to climate change, be resilient to disasters according to the Sendai Framework for Disaster Risk Reduction, and achieve holistic disaster risk management.[2] This chapter will examine Goal 11 in the context of Brunei Darussalam. It will focus on two of these wide-ranging targets: the provision of adequate, safe, and affordable housing for all (including the upgrading of slums), and access to basic services. After tracing the evolution of the country's housing policies, the chapter ends with a critical discussion on whether the lofty vision of SDG 11 could truly be achieved in Brunei.

2 Historical Context

Brunei Darussalam came into being in the Sungai Brunei estuary, where large water settlement complexes had existed at various locations for 1,500 years or so before settling in the present location of Kampong Ayer in the latter part of the 19th century. The 20th century saw a migration onto land and a dispersal of the population, mainly to the Brunei-Muara District, where the capital city Bandar Seri Begawan is located, and to the coastal areas of the country. There was also in-migration from other parts of the country and from other countries to settle in Brunei Darussalam, either as permanent residents or as temporary expatriate or migrant workers. Foreign nationals have been part of the population since Brunei was a kingdom because it was a centre of trade, particularly between the 10th and 16th centuries (Druce, 2016). Today, foreign nationals who are temporary residents in the country make up just over 15% of the population (JPKE, 2016).

Since 1984, Brunei has been a nation state with a constitutional monarchy. It is still headed by His Majesty the Sultan as the Prime Minister. The government system shifted from comprising a Chief Minister and a State Secretary during the transition period from British Protectorate to independence (1959–84), to a ministerial system in which ministers have their own portfolios. Due to the country's small geographical size and population (just under half a million; see JPKE, 2016), Brunei does not have an extensive multi-level governance structure similar to other Southeast Asian countries such as Malaysia, for example, which has a federal government in which some powers are held by the central government and others are devolved to local governments to manage their

2 See: http://www.undp.org/content/undp/en/home/sustainable-development-goals/goal-11-sustainable-cities-and-communities.html.

own states or provinces. Rather, Brunei has a unitary system in which the central government plays a pivotal role with the Prime Minister's Office, under His Majesty the Sultan of Brunei, at its heart (Commonwealth Local Government Forum, 2018). There are a number of local administrative areas (based on the four districts in Brunei) which have their own District Offices: however, these offices have minimal or no autonomy. All major social, economic, and political decision and policy making, including welfare policies such as housing and education, is centralized and managed through the different central government ministries or departments. Of particular relevance to this chapter are the national housing schemes, which are managed by the Ministry of Development through the Housing Development Department.

This chapter focuses on changes in the housing situation during the transition from the settlements largely contained in Kampong Ayer to the present-day nation state of Brunei Darussalam. In particular, it examines the provision of housing by the government and discusses the extent to which this provision aligns with SDG 11 on basic shelter. It begins with a brief description of housing in pre-modern Brunei, followed by the situation and programs instituted by the British when Brunei was a British Protectorate (1888–1984), and the housing schemes introduced by the government of Brunei Darussalam following its Independence as a nation state in 1984. Table 9.1 (below) summarizes the history of Brunei's housing program since WWII.

2.1 *Pre-modern Kampong Ayer*

In 1386, it was estimated that there were 25,000 dwellings in the settlement complex of Kampong Ayer, as well as a large palace belonging to the first Muslim Sultan of Brunei, Sultan Mohammad Shah (Asbol bin Haji Mail & Tassim bin Haji Abu Bakar, 2016). In 1521, an Italian scholar and explorer, Antonio Pigafetta, who sailed around the world with Ferdinand Magellen, recorded Kampong Ayer as the largest offshore settlement and dubbed it the "Venice of the East", a major centre of trade for crafts and jungle products (Druce, 2016). According to Nicholl (1975, p. 11) the Spanish Governor in Manila, Francisco de Sande, described the city in 1578 as "very large and rich, ... [it] was built over a very broad and deep river and had the appearance of another Venice". The buildings were made of wood and their quality "excellent", many having stone works and gilding. The Sultan's palaces were "huge" in size and the city contained a "very sumptuous mosque, a very large and interesting building, quite covered with half relief" (ibid.).

Kampong Ayer was the centre of administration for much of the region during the reign of Sultan Bolkiah (1485–1524). It was a powerful, large kingdom and the gateway for international trade. It was also a centre from which Islam was propagated in the region, leading to the Castilla War,[3] with Spain declaring war on Brunei during the reign of Sultan Saiful Rijal (1533–81). The war had a serious impact on Kampong Ayer, which declined further as a maritime power and trading centre during the Civil War (1660–73) and reached its lowest ebb with the arrival of the British.

The Castilla War destroyed large parts of the settlement and the magnificent building-scape (Abdul Latif Ibrahim, 2012). The Sultan and most of the state dignitaries moved inland to Kota Batu, Meragang and Jerudong. The population declined sharply as a consequence of the war — and, quite probably, cholera and dysentery (ibid.). In 1601, Admiral Van Noort estimated the population of Kampong Ayer to be just 2,000–3,000, in 300 houses (Nicholl, 1975; see also Abdul Latif Ibrahim, 2012). The Civil War between Sultan Hakkul Abdul Mubin (who usurped the throne) and Sultan Muhyiddin (nephew of murdered Sultan Muhammad Ali) saw the relocation of the administrative centre to Pulau Chermin, an island at the mouth of Sungai Brunei estuary. In response, Sultan Muhyiddin moved his palace upriver to Kuala Tumasek, to avoid being attacked. As a result, a new village was established at Kuala Tumasek around the palace.[4]

At this point, Kampong Ayer was described as comprising "humble homes of wood or mats that were perched on stilts measuring between 6 to 12 feet" that were connected via wooden walkways (Abdul Latif Ibrahim, 2012). An 1840 account estimated the number of houses to be 15,000 (ibid.). It was divided into several villages or wards, segregated, in general, by type of occupation. The royal family resided on land at Kota Batu, overlooking Kampong Ayer. Brunei, and especially Kampong Ayer, underwent a long period of decline (1660–1904), with impoverished nobles living in poor conditions by the 20th century (LeBlanc, 2017). In 1888, Brunei became a British Protectorate, and accepted a British Resident under a treaty in 1906.

2.2 *British Protectorate Period: Dispersion/Resettlement*

In the first decade of colonial rule, the first British Resident, Malcolm McArthur, administered much of Brunei's development from the island of Labuan. The population of Kampong Ayer was estimated to be about 10,000 in 11 villages (Chevallier, 1911; see also Abdul Latif Ibrahim, 2012). McArthur initiated a program to improve the standard of living of the people in Kampong Ayer,

3 See: http://www.worldlibrary.org/articles/eng/Castille_War.

4 In historical studies, this is regarded as part of Kampong Ayer.

which included a resettlement scheme, because he regarded living conditions to be unsanitary. He was also convinced that Brunei would not develop unless the majority of its population migrated onto land (Fatin Nor Mazdina, 2010), where they could engage in food production on farms. When his program encountered resistance from the residents of Kampong Ayer, McArthur made the resettlement scheme mandatory. In 1911, he issued an order to discourage the building of any new houses in Kampong Ayer. The British Residency thus initiated a systematic transition from living in the Brunei estuary to living on land, with a concomitant shift in the basis of the economy from largely fishing and boat-making to farming and civil service.

According to Asbol Haji Mail & Tassim Haji Abu Bakar (2016, p. 113), Kampong Ayer lacked basic utilities, such as piped water and electricity, during the British Residency period. The inhabitants obtained water from several sources on land at Subok using row boats once or twice a week. Kerosene lamps, candles, and torches were used for lighting, and mangrove wood used for cooking. Human waste was discharged directly into the water, contributing to poor sanitary conditions. Furthermore, the close proximity of the houses, clustered together as a result of organic growth accompanying expansion of families, increased the risk of fire. The royal family and elites lived at Kampong Sultan Lama in Sungai Kedayan (Kedayan River), a major river that drains into Sungai Brunei, at the present location of Kampong Ayer. The administrative centre was moved onto land in 1908, founding Brunei Town (Asbol Haji Mail & Tassim Haji Abu Bakar, 2016).

2.3 *Discovery of Oil and Transition to Living on Land*

The discoveries of oil in 1929 and gas in 1963 were major factors in the development of Brunei. Brunei became one of the richest country in Southeast Asia, with comprehensive social welfare programs provided by its National Development Plans (NDP). In 1952, as part of the first NDP, the first Kampong Ayer residents from Kampong Bendahara Lama were resettled on land at Bunut. According to the Housing Development Department (HDD), each of the 28 families was granted a piece of land 2 acres in size (HDD, n.d.). They were provided with a financial incentive of BN$100 to destroy their old houses in Kampong Ayer, BN$100 to move onto land, BN$100 to take care of the land provided and another BN$100 to bear the cost of moving. In addition, each family was given BN$2 a day for 18 months while they waited for the plants and crops on their land to grow. This first program was followed by resettlement in Anggerek Desa, Delima Satu, Orang Kaya Besar Imas, Pancha Delima, Burong Pingai, Jaya Bukit, and Jaya Setia in the 1960s. Kampong Ayer residents involved in the program were either landless, identified to be in need of welfare assistance, displaced by government development projects, or victims of fire in Kampong Ayer (DTCP, 1997).

The Kampong Ayer Resettlement Department was established in 1970 to facilitate migration of Kampong Ayer residents onto land. The third NDP in 1975 included a National Resettlement Programme to continue the process of moving the population from the Brunei estuary onto land. The aim of the program was to improve the standard of living by providing people with comfortable houses with basic amenities and community infrastructure, such as schools, mosques, clinics, sports fields, and so on. It also aimed to reduce overcrowding and the close proximity of houses, which were characteristic of Kampong Ayer, and to halt any further deterioration (Fatin Nor Mazdina, 2010). However, the resettlement scheme did not abandon Kampong Ayer: conditions there were improved by the provision of clean, potable water, electricity, schools, and health clinics, and walkways connecting the houses were upgraded. Notwithstanding these improvements, the resettlement program did result in deterioration of the water village complex, as abandoned homes became dilapidated within the large, densely packed settlement.

During the British Residency period, houses in Kampong Ayer were made of wood, with wooden panels, nipah roofs, floors made of nibung or palm tree, and built on wooden poles or stilts. House style varied according to social classes. In the 1950s, the design and construction were modified, as modern building materials such as zinc roofing and wooden boards became available in Brunei (Fatin Nor Mazdina, 2010; Jones, 1997). Modernization brought amenities such as electricity and clean water supply to Kampong Ayer. Health services, emergency services, police, schools, and mosques also became available. Wooden stilts were shored up using concrete and plastic wraps; eventually, concrete pillars were produced and sold as house stilts. Human waste, however, was still discharged directly into the estuary.

Although Kampong Ayer had been a major centre of trade in the past, and specialized occupations could still be found there in the first half of the 20th century, the majority of residents were fishermen who depended on the water-based economy during this period. In the 1970s, it was still common to see traditional fishing equipment laid out in the porches of Kampong Ayer houses. However, expansion of the public sector following Brunei's independence in 1984 saw a shift in employment. It also brought about greater awareness of declining living conditions in Kampong Ayer due to overcrowding, poor sanitation, pollution, and fire risks. Indeed, large fire outbreaks destroyed Kampong Peramu in 1970, and the same fate befell Kg Sultan Lama, Kg Pemancha Lama, Kg Peramu, Kg Setia, Kg Setia Pahlawan Lama, Kg Saba Darat, Kg Saba, Kg Burong Pingai, and Sg Kebun in the 1990s.

3 Post-independence: Modern Brunei and Housing Development

Brunei gained full independence from Great Britain in 1984. The Resettlement Department was renamed the Housing Development Department (HDD). Through the National Housing Program (NHP), its mission is to provide safe, quality, and affordable housing for Bruneians, unlike earlier resettlement projects, which provided land and some financial aid. Two schemes have been initiated under the program: 1) the National Housing Scheme (NHS), which is an extension of the resettlement scheme, and is open to all citizens, including Kampong Ayer residents; and 2) the Landless Citizens' Scheme or Skim Tanah Kurnia Rakyat Jati (STKRJ) that is only open to local indigenous groups of Brunei who do not own land. In the early 1980s, three new areas were identified for resettlement, namely Mata Mata, Serasa and Lambak Kanan (see Figure 9.1, below). This was followed by the development of areas in Rimba and Mentiri. Housing provisions were allocated to government employees in Division II and III in the form of detached stilt houses.[5] During this time, the government also embarked on the development of a public rented housing scheme to deal with the shortages of the NHP for government employees. This further decreased the population of Kampong Ayer in the post-independence period.

Due to the growing demand for, and shortages of, houses and land for housing development, the HDD started building terraced houses in 2010 to condense development into smaller spaces. In 2012, it announced that it was no longer environmentally and economically feasible to provide everybody with landed properties. The government started to consider promoting vertical homes, in flats and condominiums. This stemmed from lessons learned from Singapore's experience with high-density, vertical residential development, which included recreational amenities to serve the social needs of the vertical community. A paradigm shift to high-density housing is considered an alternative strategy for sustainable development, and it is expected to cut waiting times for applicants as well as reduce repayment rates (more on this below).

3.1 *Minor Resurgence of Kampong Ayer*

By the 1970s, Kampong Ayer had deteriorated into a dilapidated, polluted water settlement complex with a population of 19,577 (14.4% of the country's population), according to the 1971 Census (JPKE, 2016). A large proportion of its inhabitants worked in civil service positions on land. After fires had destroyed

5 Building on stilts is a traditional house design; it meets cultural practices (large gatherings where guests are seated below the house), and acts as protection against floods which occur occasionally in certain areas of Brunei.

TABLE 9.1 History of national housing development in Brunei

Year of migration	Original village in Kampong Ayer	Relocation scheme & area
1952	Kg Pengiran Bendahara Lama	Bunut Berakas with provision of land for house and planting of rubber. Land size was 2 acres for each of the 28 families involved.
1953	Kg Burong Pingai	Burong Pingai Berakas, Anggerek Desa, Delima Satu, Orang Kaya Besar Imas and Panca Delima with provision of land (0.5 to 2 acres per family) for house and paddy farming. About 250 families were involved.
1958	Unstated	Jaya Bakti and Jaya Setia. Provisions for Jaya Setia included two parcels of land: 0.5 for a house and 4.5 acres to farm rubber trees. 54 families involved. For Jaya Bakti, each family was provided with 0.75 acres to build a house and to plant fruit trees. 37 families were involved. Government provided financial assistance to build houses for both villages.
1970	Establishment of the Resettlement Department to relocate people from Kampong Ayer to land.	
1976	Government conducted a nationwide survey to identify the demand for the Housing Scheme, including background and cultural information. The results helped identify citizens without housing or land to convert to property owners at a subsidized rate with a repayment period of 20–30 years. The survey identified type of housing design best suited to Bruneian cultural practices and the optimum population size for a resettlement area to be provided with amenities and facilities.	
1978	Government identified new resettlement areas to be spread across all four districts, rather than Brunei-Muara district only (the main administrative district in which Kampong Ayer is located).	
1980	National Housing Development Scheme (NHS) application opened to Bruneians; the first interviews of applicants were conducted in 1982.	
1983	Pg Bendahara Lama and Kg Sumbiling	Perpindahan Mata Mata and Perpindahan Serasa. Provision of houses to mainly Bruneian families affected by fire and government development scheme.
1984	Housing Development Department (HDD) was established.	
1986	First group of applicants received keys to their houses under the new housing scheme.	
2010–2011	Government introduced the idea of high-density living including vertical living to cut down cost of building/construction (economies of scale and making it affordable for monthly repayments) and reduce waiting time for applicants (from 18 years to 5–7 years). A terrace house under the NHS could cost about BND 43,000 but BND 200,000 on the open market. Government subsidizes up to 50% of housing costs. The Ministry of Development identified only 5% of Brunei Darussalam as developable land, leading to constraints for housing development and residential property construction in the context of a growing population. Bruneians applying for NHS will have to adapt to high-density living including vertical homes. The government views the need to build vertically as cost- and resource-effective, and necessary to meet the growing demands for affordable housing of Bruneians.	

SOURCE: HDD (N.D.) AND AUTHOR'S FIELDWORK RESEARCH

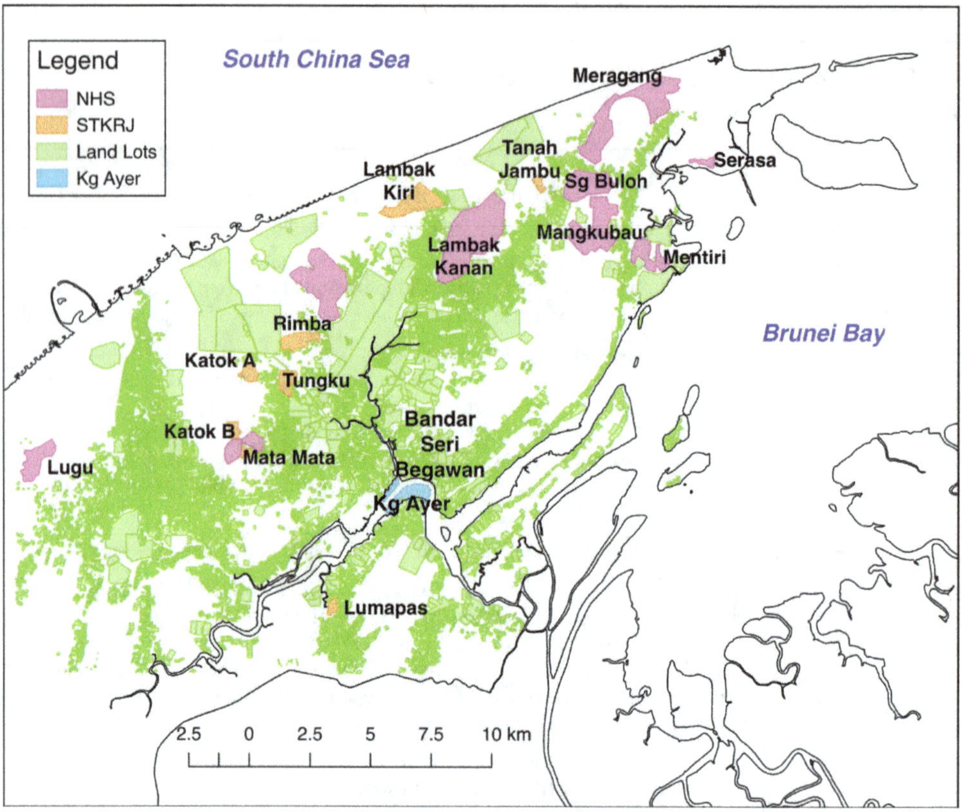

FIGURE 9.1 Dispersal of population from Kampong Ayer to land
Notes: Map showing distribution of national housing estates (NHS and STKRJ) and private and institutional land lots in Brunei-Muara District and part of Tutong District. In the 1900s, most of Brunei's population was concentrated in Kampong Ayer. Today, most of it is scattered through the land with 70% in the Brunei-Muara District

more than 1,000 houses, leaving the inhabitants homeless, the government issued an order on 15 March 1982 that prohibited renovations or extensions that would bring the houses in this densely packed area into ever-closer proximity (Fatin Nor Mazdina, 2010). It also set up programs to reduce fire risks (e.g., checking electrical wiring) and trained residents to respond to fire outbreaks. In spite of active resettlement via the NHP, improvements in the infrastructure in Kampong Ayer resulted in an increase in the number of villages to 40 with a population of 24,745 people in 1991.[6] Our own ethnographic research

6 When resettlement began, in 1906, there were only 11 villages and about 10,000 people in Kampong Ayer.

indicates that this increase was due in part to high fertility rates and, in some cases, in-migration of relatives from the region. However, improvement in the overall development of Brunei, coupled with better job prospects and better housing provision by the government, led to a steady decline of the population of Kampong Ayer in the 1990s. By 2001, it had dropped to 11,662 — a 47% decline from 1991.

In 1992, the Sultan Haji Hassanal Bolkiah Foundation (Yayasan Brunei) was established as a non-governmental agency that offers housing for the needy in Brunei. The Yayasan Brunei scheme was built on reclaimed land in an area occupied by Kampong Sultan Lama, a village in Kampong Ayer which was destroyed by fire in 1981.[7] It has two main projects: (1) housing support for the less fortunate (poor and disabled) and those living in unsafe accommodation; and (2) the construction of housing in Kampong Ayer under two schemes, Kampong Bolkiah 'A' and Kampong Bolkiah 'B', for residents of Kampong Ayer affected by the 1993 fire. Under the latter project, 226 houses were built in Kampong Bolkiah 'A' and 250 in Kampong Bolkiah 'B'. Two types of houses were constructed for these two schemes with a repayment period of 20 years: 26 were type 'A' houses, with two bedrooms, and the remainder were type B houses with four bedrooms.[8] Unlike the government's housing scheme, the Foundation provides and builds in Kampong Ayer. The first phase saw three villages (Kampong Sungei Bunga, Kampong Bolkiah A, and Kampong Bolkiah B) erected for victims of natural disasters such as fire. The houses built in these areas are at a safe distant from one another, organized in a regular grid structure, and fitted with the necessary facilities and infrastructure for a safe, hygienic, and comfortable standard of living. Crucially, they are the only villages in Kampong Ayer fitted with a (vacuum-type) sewerage system.

However, the new-style water village proved to be unpopular. *The Brunei Times* (2016) reported that there was growing concern over the increasing number of abandoned houses in Kampong Ayer, including 28 houses under this scheme. The main reason cited was a preference for housing under the NHS, such as in Lugu or Tanah Jambu. Another reason was the high travel cost and time to get to/from work, school, shopping in supermarkets, and other activities on land. According to neighbours, the abandoned houses were cannibalized by other residents in Kampong Ayer — a common occurrence, in which materials from abandoned houses are taken by other residents to shore up their own properties. There were also concerns that the abandoned houses could attract anti-social activities, such as drug abuse and prostitution, which occur in small pockets of Kampong Ayer where houses have been abandoned

7 See: http://bruneiresources.blogspot.com/2016/01/the-origin-of-kampung-sultan-lama.html.
8 See: http://www.yshhb.org.bn/11-info/role-and-contribution/11-welfare.

or rented out illegally (see Haji Ismail Hj Duraman, 2008).[9] The problem of the abandoned houses was an unintended consequence of the policy that prohibits recipients of housing from renting out or selling their old homes in Kampong Ayer.

In 2012, a rejuvenation project — part of a Kampong Ayer beautification strategy — saw the development of a set of modern homes in the area formerly occupied by Kampong Lorong Sikuna, which was destroyed by fire in 1995. It is a pilot project to upgrade Kampong Ayer initiated by the reigning monarch, His Majesty Sultan Haji Hassanal Bolkiah. These houses are said to conserve the cultural characteristics of houses in Kampong Ayer, but are more comfortable and safer; they are also more resilient, environmental friendly, and offer amenities similar to houses on land (Information Department, n.d.). Similar to the houses built by the Yayasan Brunei schemes, this housing project is divided into two types: type A is a single storey house with three bedrooms and a total area of 130 m^2 costing BN$65,000; type B is a double storey house with four bedrooms and a total area of 198 m^2 at a cost of BN$90,000. The materials used are said to be of high quality, including flame-retardant wooden walls; the houses are connected by roofed bridges; sewerage systems including treatment plants and pumping stations form part of the project. An electrical substation building, water piping systems, and fire protection systems are provided. The inhabitants of this housing are given permanent grants to occupy the property (ibid.). There have also been a number of private initiatives in and around the area which aim to promote tourism and the conservation of Kampong Ayer as part of the culture heritage of Brunei.

4 Discussion: Do Brunei's Development Plans Align with SDG 11?

Brunei's housing program started out as part of the British Residency program for the country at the beginning of the last century. By the time of its independence, Brunei had a national housing program to ensure that every family has a home. It should be pointed out that this vision refers only to citizens of Brunei Darussalam. This is relevant because there is a significant proportion of foreign nationals (30% in the 2001 census, 15% in 2016, according to JPKE, 2016) residing in Brunei temporarily, and even illegally.[10] There is also a sizeable group of non-citizens (officially 'stateless') who are granted permanent

9 Although *The Brunei Times* (ibid.) reported that the residents still felt safe in their own homes as the crime rate, according to them, is low.

10 This is a widely-acknowledged fact but there are no published data available.

residency, but are not allowed to own land. The government has, however, allotted parcels of land under Temporary Ownership Lease (TOL) conditions, renewable annually, to accommodate this particular group. To date, the HDD is still entertaining applications for housing under the NHS. Although the process has been plagued by delays, due to inadequate construction capacity to meet demand, the government continues, slowly but surely, to strive to achieve this vision of ensuring that every family has a home. However, to what extent is this vision driven by SGD 11? Can it be fully realized?

Through SDG 11, countries are to ensure that everybody has access to adequate, safe, and affordable housing and basic services, and to upgrade slums, by 2030. Brunei continues to work towards this goal especially through the affordable national or public housing scheme for Bruneians. Typically, Bruneians are dependent on private cars for mobility and the resettlements and post-independence affordable public housing sites are mainly located in suburban areas. The government provides infrastructure including improved roads to establish connectivity between housing estates and urban centres and services. Moreover, the 2012 rejuvenation and housing project in Kampong Ayer is seen as a means to protect and safeguard Brunei's cultural and natural heritage.

Consideration of the economic, environmental, and social sustainability mentioned in SDG 11 is thus evident in the recent Kampong Ayer project, in the resettlement program introduced during the British Residency, and in current housing schemes. Initially, the central government addressed issues related to protecting and improving the health and environmental conditions of Bruneians living in Kampong Ayer; at the same time, it ensured that people were able to improve and sustain their livelihoods through provision of land parcels. Regarding the call of SDG 11 for universal access to safe, inclusive, and accessible green and public spaces, in particular for women and children, older persons and persons with disabilities, the government provides such public facilities and social spaces in most of its public housing. Beyond this, Brunei produces short-term National Development Plans every five years as a way of meeting its long-term national development plan. In 2004, a national reflection by the council of ministers led by the Monarch and Prime Minister, His Majesty Sultan Haji Hassanal Bolkiah, resulted in the adoption in 2007 of a long-term strategic plan, entitled Wawasan 2035 (Vision 2035).[11] This includes an Infrastructure Development Strategy which aims to provide adequate social infrastructure, particularly in housing provision, transport and communication, health care and education, that reduces the income gap. This strategy also aims to adopt suitable legal and regulatory frameworks to

11 See: www.bruneiembassy.org/brunei-vision-2035.htm.

promote investments in social and industrial infrastructure, including privatization and Public–Private Partnerships (PPP)[12] in line with international best practice.

Brunei's housing program began almost a century before the SDGs were conceived. This means that Brunei's development is not a consequence of the SDGs; rather, it can be said that most of Brunei's development plans, including Vision 2035 on housing, align with the SDGs. Although Brunei participated in the Earth Summit at Rio in 1992, and the follow-up Summit on Sustainable Development 20 years later, it is unclear whether SDG 11 has significantly influenced the NHP. It might just have given it an added push to hasten the process to meet the housing needs of its people; it might be a coincidence that Vision 2035 aligns in many ways with the SDGs, including a similar target date (2035 vs 2030 for the SDGs). Of particular relevance to this chapter is the goal of Vision 2035 to provide the citizens of Brunei with the "best home" and "high standard of living" (Wawasan 2035, p. 12). Here, it does seem probable that the SDG process at least influenced the goals set in Vision 2035, because the country's leaders participated in the various international, regional, and inter-governmental programs on various sustainable development themes. Moreover, there is no single entity within the government that is responsible for Brunei meeting the SDGs, including for housing; sustainability is a government-wide effort and responsibility. The achievements, as well as shortfalls, of Brunei's NHP and policies could therefore be considered as efforts to realize SDG 11.

To date, Brunei's housing program has provided over 28,000 families with homes that are safe, durable, and equipped with modern amenities. Recipients have, in general, expressed gratitude to His Majesty and the government for what many refer to as their "dream home". This is also the case with the stateless permanent residents who were allowed to build homes on TOL land. However, while the NHP has been quite successful in meeting SDG 11 on the provision of adequate housing, a number of issues remain, including the matter of space.

4.1 *Coping with Rising Population/Demand*

Since the end of WWII, there has been a rapid increase in Brunei's population (Hairuni Haji Mohamed Ali Maricar, 1996). The population grew at an annual rate of 5.7% between 1947 and 1960 (Franz, 1980). It then declined, dropping

12 See: http://depd.gov.bn/DEPD%20Documents%20Library/NDP/PPP/PPP%20Guidelines.pdf.

to 2.8% in 1978; by then, the population had doubled from that of 1964 (ibid.). Franz (ibid.) notes that families in new housing areas are generally smaller, and suggests that the country's housing policy might have had some influence on family size and birth rates. House construction followed European designs, which cannot accommodate large families or households in the way that a traditional Malay house can. Flats, in particular, cannot accommodate traditional extended families. In-migration in response to labour shortages in the mid-1960s, late 1960s and mid-1970s further added to Brunei's population growth. These migration surges coincided, respectively, with the opening of Brunei airport, Muara deep water port and Brunei LNG plant, and infrastructure projects.

One implication of a rapidly rising population is a heightened demand for housing. The HDD has struggled to cope with housing demands of Bruneians. Between 1980 and 2015, application for houses grew nearly tenfold from 8,565 to 81,066 under the NHS (see Figure 9.2); an additional 7,613 applied under the STKRJ scheme. The high demand is due in part to the attractive low costs of repayments (BN$100–300 per month over 20–30 years depending on house type). The government also subsidizes up to 50% of the construction cost of the house (Md Zakaria Sarudin, 2011). The ongoing land shortage, rising

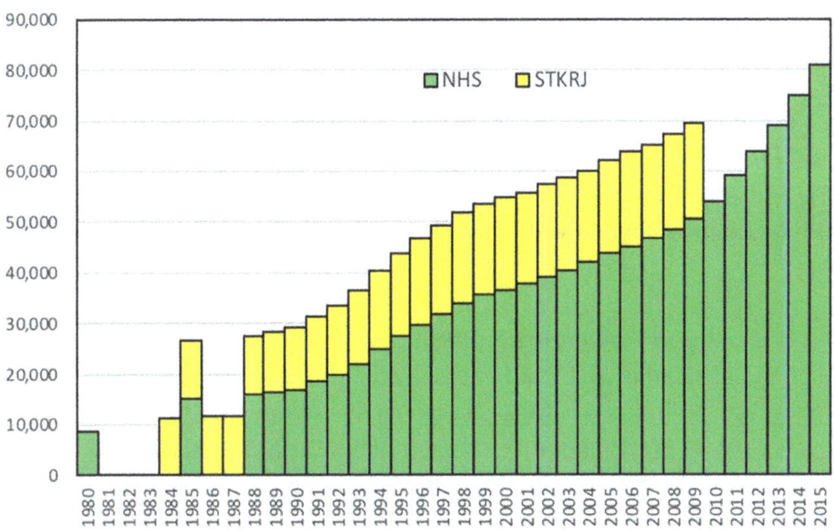

FIGURE 9.2 Total applications for NHS since 1980

NOTES: Y AXIS: CUMULATIVE FREQUENCY; X AXIS: YEAR

SOURCE: HDD, OPEN DATA INITIATIVE, THE E GOVERNMENT NATIONAL CENTRE (EGNC): HTTPS://WWW.DATA.GOV.BN/LISTS/DATASET/MDISPLAY.ASPX?ID=804

construction costs, and overall inflation also contributed to the high demand for the housing schemes.

To date, the total number of houses delivered is 22,562 under NHS (13,322 in Brunei-Muara District; 5,793 in Belait; 2,809 in Tutong; and 602 in Temburong) and 6,152 under STKRJ (5,158 in Brunei-Muara District; 792 in Belait; 170 in Tutong; and 32 in Temburong). In total, 18,480 houses were awarded to recipients in the Brunei-Muara District, 6,585 in Belait, 2,979 in Tutong, and 634 in Temburong. Only 27.8% of applicants under the NHS received houses. In the STKRJ, 32.2% applicants were successful.[13] The data show that the government's vision of providing each family with their own home is far from being realized; with the population increasing by approximately 1.4% per year (JPKE, 2016), this is a concern.

The HDD acknowledges that long delivery times (12–20 years) are a major impediment to achieving its NHS goals. This has resulted in shorter repayment periods, making repayment amounts high, given that 70% of applicants earn less than BN$1,200 per month (Md Zakaria Sarudin, 2011). In the period 2009–10, defaults in payment amounted to about BN$20 million (ibid.). The HDD cites scarcity of available land for housing development and growing demand for housing as major challenges and identified the high cost of subsidizing house construction and the lack of construction capacity as major factors (ibid.). The latter is mainly due to the reluctance of the Labour Department to grant permits for foreign labourers and professionals to carry out the work.[14] Nevertheless, the HDD hopes to reduce the waiting time in high-demand districts such as Brunei-Muara to less than five years by 2021, by building at least 10,000 new homes during each five-year NDP period (Md Zakaria Sarudin, 2011). The HDD is committed to providing affordable public housing with multiple schemes that consider social, environmental, and economic needs (ibid.).

4.2 *Housing for Temporary Residents*

The period of rapid modernization associated with the rise in oil price (1960s to 1980s) saw an influx of foreign nationals due to demand for professionals and labourers to implement development plans. Expatriate workers in Brunei are provided with housing befitting their status. Although low-wage migrant workers are supposed to be provided with adequate housing while they reside

13 HDD, Open Data Initiative: https://www.data.gov.bn/Lists/dataset/mdisplay.aspx?ID=804.

14 A limited foreign labour quota is imposed which results in inadequate capacity in the construction industry (see: http://energy.gov.bn/Lists/LatestHeadlines/DispForm.aspx?ID=1483 accessed 16 May 2018).

in Brunei, there have been cases where employers do not comply with their contractual responsibility and force workers to live in overcrowded dwellings and generally unsafe conditions. These may be houses owned by employers, some of which are in a dilapidated condition, or upper units of shop blocks, which are relatively cheap to rent. In general, these dwellings have piped water and electricity, although the electrical wiring tends to be in poor condition due to age, lack of maintenance or poor installation, often by unqualified contractors. As a member of the International Labour Organization (ILO), Brunei has laws to safeguard the welfare of migrant workers. Although some employers that have been found to have violated these laws have been prosecuted, others have evaded the enforcers. Consequently, there are migrant workers who continue to live in unsafe, overcrowded, and sometimes unsanitary conditions. There are no statistics available, as to date, there has been no systematic study of this issue.

4.3 *Slums*

Some locals, including the 'stateless' who were born in Brunei, who are not part of families that own land or houses, and cannot afford to rent, set up slum-like dwellings in various locations on the outskirts of emerging towns. There are also foreign nationals who live and work in the country illegally, either without a work permit or under a work pass offered by a local enterprise, for which they pay a substantial amount annually but work on a freelance basis. These are the most vulnerable groups in the country, with no safeguards. They tend to live in neglected old dwellings, including in Kampong Ayer, in more marginal areas of towns. The government's response to illegal workers is to conduct frequent raids of migrant workers' quarters following investigations.

Many slums or slum-like dwellings have been demolished. The residents are warned and told to seek other accommodation. Home owners of dilapidated old houses in some areas of Kampong Ayer have been resettled into the national housing projects to make way for a river corridor beautification project and the construction of bridges.

4.4 *Vertical Housing*

In a Legislative Council dialogue of 17 March 2016, the Minister of Home Affairs announced that the government is moving toward high-rise housing, with 2,500 apartment units planned for Lumapas, 420 units in Lambak Kanan and another vertical housing site under consideration in Lugu.[15] This

15 See: https://btarchive.org/news/national/2016/03/17/national-housing-scheme-only-bruneians.

supports HDD's plan to expedite delivery of homes, factoring in sustainable land utilization, construction costs, subsidies and owner's ability to pay (Md Zakaria Sarudin, 2011). The idea of living in flats is generally not met with enthusiasm. A survey by Noor Hasharina Hassan et al. (2011) found that the sampled population had a low preference for flats of the design typically found in Brunei. The most popular building type was the eco-building, followed by low-cost, open concept, smart buildings, and finally, modern condominiums. The study focused on cultural-architecture preferences and created a cultural typology based on the concept of proxemics. It found that current vertical designs (modern box-type flats) do not meet the proxemic needs of the people of Brunei. This is rooted in how Bruneians' lifestyle and housing evolved from Kampong Ayer. These findings could have implications for the success of the vertical housing program.

5 Concluding Remarks

For much of its early history, the population of Brunei resided largely in the Sungai Brunei estuary. While it was an impressive city and centre of trade during the 15th century, Kampong Ayer had deteriorated into a fishing settlement by the beginning of the 20th century, with poor housing and living conditions. Resettlement schemes initiated by the British Resident in 1906 began a systematic migration of the populace to land, which improved both living conditions and economic sustainability with the introduction of farming. The discovery of oil and natural gas provided Brunei with the means to modernize. The decades following WWII saw rapid development and population growth, both naturally and through in-migration. The transition to more local governance in 1959 saw the introduction of a NHP that continued the shift from living on water to living on land. Housing and living conditions were improved significantly. This included conditions in Kampong Ayer, where electricity, potable water, and durable building materials become available. By 2016, the majority of the country's population resided on land, with 70% living in the Brunei-Muara District. Only 5% remains in Kampong Ayer.

With a series of five-year National Development Plans, a long-term vision (Wawasan or Vision 2035) and the Housing Development Department to plan and implement housing programs, coupled with wealth from a robust oil industry, has Brunei met its own housing goal or SDG 11 on housing provision? The data indicate that it is progressing towards both its own goals and the SDG, but is yet to fully attain them. This assessment is based on the following.

1) The slow delivery of houses under the NHS that cannot keep pace with population growth or demand; over 65% (a third) of applicants are still waiting for houses, with waiting time averaging 15 years.
2) The NHS is facing financial pressures due to the large and continually rising number of applicants and the government's policy of subsidizing between 30% and 60% of the construction cost.[16] It is further hampered by declining availability of suitable land to accommodate new housing estates. The government is now hoping to address this problem through high-density living.
3) Although private developers are meeting some of the demand, the cost of private housing is generally high and unaffordable for the majority of the population. According to the HDD, even the low repayment costs of the subsidized houses provided under NHS are hard to meet for many low-income recipients, and defaulting on payments is a common problem.
4) Living conditions in Kampong Ayer have improved greatly. However, most homes do not have sewerage infrastructure. Raw sewage is discharged directly into the estuary contributing to water pollution and health risks.

Nevertheless, the HDD is committed to achieving its goal of providing every family with good living standards and affordable homes. Vertical housing is an option, as it would reduce waiting times, lower construction and repayment costs, and greatly reduce the physical footprint in a situation where land is scarce. The HDD would have to revise the subsidy and repayment arrangements to make the scheme and repayment more sustainable.

Living conditions for private properties, especially quarters for migrant workers and illegal residents, can be overcrowded and somewhat unsafe. This problem is being addressed by strengthening enforcement of labour regulation and potentially by some revision of policies. Generally, housing — including schemes outside the NHS — is relatively good with basic provisions incorporated. The size of traditional households in Brunei is large: in housing terms, this may be misconceived as overcrowding. However, this household structure and size are common in Brunei, with extended families sharing accommodation since the time when most Bruneians lived in Kampong Ayer.

The government is relentless in its efforts to minimize poverty and improve living conditions for its population through affordable public housing provision with flexible schemes that match Bruneians' income levels. The government and local agencies constantly find means for sustainable development

16 As revealed in the 13th Brunei Legislative Council Meeting in March 2017. See: https://borneobulletin.com.bn/subsidies-stay-prudence-advised (accessed 16 May 2018).

of housing, particularly affordable public housing, to allow more Bruneians to become home owners. These plans also take account of the country's cultural heritage and local practices. These strategies existed before the creation of the SDGs; nevertheless, it is evident that the Brunei government's plans on housing development align with the aims of SDG 11.

References

Abdul Latif Bin Haji Ibrahim (2012). *Kampong Ayer: Living memory*. Bandar Seri Begawan: Pentagram Design Sdn Bhd.

Asbol bin Haji Mail & Tassim bin Haji Abu Bakar (2016). To live on water: Lifestyle of the Kampong Ayer community during the British Residency Period, 1906–1941. In K. G. Ooi (Ed.), *Brunei: History, Islam, society and contemporary issues* (pp. 100–122). New York: Routledge.

Brunei Times (2016, June 14) Abandoned houses a rising concern among Kg Ayer residents. *The Brunei Times*. Retrieved from http://btarchive.org/news/national/2016/06/14/abandoned-houses-rising-concern-among-kg-ayer-residents (accessed 28 January 2018).

Chevallier, H. (1911). *Annual report. State of Brunei*. Kuching: Government Print Office.

Commonwealth Local Government Forum (2018). The local government system of Brunei Darussalam. http://www.clgf.org.uk/default/assets/File/Country_profiles/Brunei_Darussalam.pdf (accessed 23 January 2018).

Druce, S. (2016). The 'birth' of Brunei: Early polities of the northwest coast of Borneo and the origins of Brunei, tenth to mid-fourteenth centuries. In K. G. Ooi (Ed.), *Brunei: History, Islam, society and contemporary issues* (pp. 21–44). New York: Routledge.

DTCP (1997). *Town and Country Planning: ... Past ... Present & Future....* Brunei Darussalam Department of Town and Country Planning. Bandar Seri Begawan: Avesta.

Fatin Nor Mazdina Hj Hamdani (2010). Sejarah Kampong Ayer. *Pusaka: Berita Penyelidikan dan Kegiatan Pusat Sejarah Brunei*, 19, 14–25.

Franz, J. C. (1980). *The Sultanate of Brunei: Oil wealth and problems of development*. (M. Schmitz & A. Sharp, Trans.) Nuremberg: Wirtschafts und Sozialgeographisches Institut der Friedrich-Alexander Universitaet.

Hairuni Haji Mohamed Ali Maricar (1996). Population dynamics of Kampong Ayer. Paper presented at International Symposium on Kampong Ayer, Universiti Brunei Darussalam, Bandar Seri Begawan, 6–9 September.

Haji Ismail Haji Duraman (2008). Kampong Ayer Negara Brunei Darussalam: Warisan dalam Perkembangan. In Siti Norkhalbi Haji Wahsalfelah & Mohd Shahrol Amira Abdullah (Eds), *Kampong Ayer: Warisan Hidup, Cabaran dan Kesinambungan* (pp. 157–169). Bandar Seri Begawan: Universiti Brunei Darussalam.

Hassan, Noor Hasharina, Gabriel Y. V. Yong, Izni Azrein Noor Azalie, Khairunnisa Ibrahim, & Hairuni Ali Maricar (2011). Cultural consideration in vertical living in Brunei Darussalam. Paper presented at the 44th EAROPH Regional Planning Conference on Managing urban growth: Challenges for small cities, Bandar Seri Begawan, Brunei Darussalam, 8–10 November.

HDD (n.d). History. Brunei Darussalam Housing Development Department. Retrieved from http://www.mod.gov.bn/housing/bm/SitePages/Sejarah%20Perpindahan%20Tahun%201950an.aspx.

Information Department (n.d.). *Brunei Today*. Bandar Seri Begawan: Brunei Darussalam Information Department.

Jones, A. (1997). Urban conservation issues in Brunei Darussalam: The case of Brunei's water villages. *Planning Perspectives*, 12(4), 457–475.

JPKE (2012). *Tenth National Development Plan (2012–2017): Brunei Darussalam*. Bandar Seri Begawan: Department of Economic Planning and Development, Prime Minister's Office.

JPKE (2016). Population. Brunei Darussalam, Jabatan Perancangan dan Kemajuan Ekonomi. Retrieved from http://www.depd.gov.bn/SitePages/Population.aspx (accessed 28 January 2018).

LeBlanc, R. (2017). A Deleuze theory of urban morphology. Brunei water city. PhD Thesis. Victoria University, Wellington.

Md Zakaria Sarudin (2011). Brunei Darussalam National Housing Program: Future homes for the nation. Paper presented at the 44th EAROPH Regional Planning Conference on Managing urban growth: Challenges for small cities, Bandar Seri Begawan, Brunei Darussalam, 8–10 November.

Nicholl, R. (Ed.) (1975). *European sources for history of the Sultanate of Brunei in the sixteenth century*. Brunei: Muzium Brunei.

Wawasan 2035 (2007). Vision 2035. Brunei Darussalam Long-term Development Plan. Bandar Seri Begawan: Prime Minister's Office.

CHAPTER 10

The Missing Link: Sustainable Mobility for Sustainable Cities and Communities

Wendy Tan

1 Introduction

As Southeast Asian cities experience explosive economic growth and urbanization, various spatial claims and conflicts arise. One particular conflict is the fundamental mismatch between land use and transportation polices representative of cities and metropolitan regions in Southeast Asia. Here, the growing population and their increased car usage due to improving economic circumstances have led to a path of "automobile dependence" (Barter, 2004) with increasing mobility demand from all socio-economic levels, coupled with the urge to emulate misconstrued "Western" or "American" urban development ideals (Shatkin, 2008; World Bank, 2015). These cities have uncritically facilitated development and mobility in the form of modern skyscrapers and soaring freeways, thereby producing the familiar images of traffic congestion and haphazard urban development (Buijs, Tan, & Tunas, 2010).

Mobility, understood as the ability to move oneself, is a fundamental human right (UN, 1948). However, the impacts of fulfilling this right often clash with sustainable development ideals. For example, Bangkok, Thailand and Jakarta, Indonesia rank as the second and third most congested cities in the world (INRIX, 2017; TomTom, 2017). In addition to the negative effects of increased fossil fuel consumption leading to emissions and toxic fumes, and diminishing road safety (Bakker et al., 2017), traffic congestion also means that significant numbers of people spend a disproportionate amount of their time in relation to their income opportunities waiting in traffic, trying to get to work, to school or to other social needs. This has serious consequences for sustainable development in terms of increasing social inequalities and exclusion in these cities (UN Habitat, 2013). When certain parts of a city are gridlocked, basic access to healthcare, education, and other facilities can no longer be guaranteed. This contributes to the growing discussions on transport-related social exclusion (TRSE) within the fields of land use and transport planning (Kenyon, Lyons, & Rafferty, 2002; Lucas, van Wee, & Maat, 2016; Scheurer, Curtis, & McLeod, 2017).

 | DOI:10.1163/9789004391949_011

The conflict inherent in balancing social, economic, and environmental objectives makes the discussion of sustainable mobility both complicated and complex. This complexity is especially visible in the developmental state narrative in which the majority of Association of Southeast Asian Nations (ASEAN) member states find themselves, where the urgency and speed of economic growth sometimes overshadow environmental and social concerns (Johnson, 1999). It is therefore relevant to consider the complex role that land use and transportation planning have in achieving sustainable development futures. ASEAN members are currently at various stages in their pursuit of sustainable urban development, having advanced significantly since the ASEAN Vision 2025 was agreed upon (UN ESCAP, 2017). However, substantial structural challenges remain, notably those of access to healthcare and education, and the gap between income classes despite a growing 'middle' class (Development Initiatives, 2017).

Contrasting this initial progress with the 17 Sustainable Development Goals (SDGs) agreed by the United Nations in the 2030 Agenda for Sustainable Development (UN, 2015), one observes that there is much room for improvement. The SDGs, as measured by a variety of indicators, are meant to guide national policies to achieve an inclusive, sustainable, and prosperous future. The goals themselves, however, were formulated rather vaguely to achieve broad consensus and adoption by various nation states and across various levels and scales. Amongst the various goals and indicators, only SDG 11 (UN, 2015, 2017) carries an explicit reference to transport and sustainable mobility in terms of "safe, affordable, accessible and sustainable transport systems for all". However, the non-specificity of its indicators (measured as the "proportion of the population that has convenient access to public transport", Indicator 11.2.1) might prove challenging for implementation in practice.

To understand how SDG 11 ("Sustainable cities and communities") relates to sustainable mobility in the Southeast Asian context, this chapter will discuss the Goal through the lens of transportation and mobility planning to highlight the missing links and dilemmas. First, the fundamental concepts of sustainable mobility, its common strategies and its relation to social inequalities are discussed. Then, SDG 11 and its indicators are analysed along with the four elements of sustainable mobility. Next, by comparing active mobility projects and land use transportation integration policies in the cases of Singapore and Jakarta in relation to a similar fictitious journey in both cities, the policy context and implementation challenges of achieving SDG 11 are highlighted. The discussion ends with a reflection on potential roles for governments, markets, and civil society to reduce mobility inequality and exclusions.

2 Theory

Transport has been termed the "maker and breaker of cities" (Clark, 1958). In a context of rapid urban population growth, cities are bursting at the seams to accommodate residents and retain an acceptable quality of life. Congestion, pollution, and excessive commuting times are just some of the most direct and observable impacts experienced by those living in cities. Within the ASEAN region, population increases and unsustainable development strategies focused on growth and expansion have already led to measurable degradation in water and air quality, and displacement of local populace and businesses (Firman & Dharmapatni, 1994). Achieving a sustainable future therefore requires a better understanding of Southeast Asian city-regions through the lens of sustainable mobility instead of the traditional transport planning approaches (Buijs et al., 2010). In this section, the differences between sustainable mobility and conventional transportation planning are discussed and two commonly used strategies explained. The social aspects of transport planning and sustainable mobility are then reflected upon from the perspectives of transport-related social exclusion and the rights-based approach to mobility.

2.1 *Sustainable Mobility vs. Transportation Planning*

There has been a fundamental shift within the field of mobility and transportation planning in the last decades (Hull, 2008). No longer is the "predict and provide" narrative preached but rather a "predict and prevent" or "people and places" approach is adopted (Owens, 1995; see also Bakker et al., 2017; Banister, 2008). The technical-rationality approach to transportation planning, in which the emphasis is on increasing infrastructure efficiency to achieve a pre-determined goal in a transportation system, is outdated (Wilson, 2001). The vicious cycle of building more infrastructure (predominantly highways) for private car use and expanding traffic capacity has been proven to lead to more congestion, the need for more roads, and more private car use — with all the negative impacts that this entails (Ladd, 2012).

This has led to a paradigm shift in the field of transportation planning and its key debates, captured in the seminal text by Banister (2008, p. 73):

> It has often been said that transport planning is at a crisis point and that it underestimates the key challenges facing urban planners.... Local public transport, cycle and walking have become less attractive, and this in turn has resulted in the greater use of the car. Car dependence and the increased decentralisation of cities are difficult processes to reverse — this is the transport-led future. Sustainable mobility provides an alternative

paradigm within which to investigate the complexity of cities, and to strengthen the links between land use and transport.

With this fundamental change, the new focus in transport planning is to reduce travel in terms of trips and trip lengths, to encourage travellers to choose more sustainable modes of transport (that is, a modal shift from private car use to walking, cycling, or public transportation), and to increase efficiency of the transportation system. However, this can only be achieved if there is public and political acceptance. To make these changes palatable or appealing, individuals will have to be shown the potential gains and the concrete possibility of implementing sustainable mobility — and the risks of maintaining the status quo (Banister, 2008, 2011).

To attain sustainable mobility, a combination of the following four key elements must be present in any transport policy seeking modal shift (Banister, 2008):

1) adapting and investing in technology to increase efficiency (e.g., alternative fuels or information systems);
2) reflecting the real costs and externalities of transport (e.g., road pricing or additional fuel tax);
3) integrating land use with transport (e.g., proximity measures, compact developments);
4) behavioural change and acceptance (e.g., targeted personal information, campaigns promoting active modes such as walking or cycling).

2.2 *Active Modes and Land Use Transport Integration*

Within sustainable mobility discussions, two popular strategies which combine the above elements and which are adopted by cities are (1) active mobility policies to promote cycling and/or walking, and (2) land use transport integration (LUTI) policies in the form of transit-oriented development (Kenworthy, 2007).

The first seeks to change behaviours by providing incentives for modal shift towards more non-car based transportation such as walking and cycling, eventually encouraging more public transportation usage. The underlying motivations have always been possible emissions reductions from these modes, as well as the potential health and well-being advantages (Pucher & Dijkstra, 2011). For example, for longer journeys, if there were more connected and continuous cycle and walking paths, then an individual would be able to conduct their first-mile and last-mile journeys by those modes instead of having to rely on a private car to reach other transit options, or having to drive the entire trip. For shorter, daily trips for groceries or errands, for example, the strategy would then focus on reducing the time, cost or effort of walking or cycling in comparison to taking the car.

The second strategy combines compact, high-density developments around public transportation nodes to encourage transit use and increase the efficiency of current land use and transportation systems (Loo & du Verle, 2016). The foundation of this concept is found in the land use–transport feedback loop which contends that there is a mutual and cyclical relationship between land use patterns and changes in the transportation systems (Wegener & Fürst, 1999). Transit-oriented development (TOD) refers to mixed-used residential and commercial development that is oriented towards and in close proximity to a multi-modal transportation node (Mees, 2014; Tan, Bertolini, & Janssen-Jansen, 2014). These compact and dense settlements, where individuals can walk to take public transportation or can engage in retail, leisure or work in the direct vicinity through walking instead of driving long distances, are deemed to be sustainable for the environment but also to encourage economic and social activities.

2.3 *Mobility Inequalities and Exclusion*

Discussing sustainable mobility purely from an infrastructural or functional perspective, however, does not present a full picture. Its predominant contribution is in the alleviation of inequalities and social exclusion. This was a fundamental flaw in traditional transport planning, which was more preoccupied with the system and capacity improvements than with the individuals using the system, and their differentiated needs. The social aspects of mobility can be discussed through two lenses: transport-related social exclusion; and the rights-based approach to mobility.

TRSE focuses on how lack of access to transportation limits the participation of individuals in society. This is measured by access to employment, education, healthcare, and services (Preston & Rajé, 2007). Social exclusion, including that caused by lack of transportation access, is not unique to Southeast Asia. What is important is how pervasive and influential it can be in contributing to poverty and in preventing the social mobility implied by better education and employment opportunities. Access to basic services such as hospitals or schools and uneven spatial distribution of key facilities are made worse by the excessive congestion common to the region (Rama, Béteille, Li, Mitra, & Newman, 2015).

The rights-based approach is concerned with who might be excluded and marginalized due to their income levels or even ethnicity determining their spatial location, usually at the outskirts of cities where residences are cheaper (Coggin, 2017). This is compounded by the fact that most of these individuals, usually women or elderly already spatially isolated from main commercial activities, are less likely to be able to afford a private car to get to those activities (Thynell, 2017). As such, the lack of access to transportation systems leads to compounding social and economic inequalities. For example, if a disproportionate amount of an individual's working income is required simply to get to and from the work location, this restricts the ability to pay for food, healthcare, or education.

When discussing transport and mobility issues, it is thus imperative not to focus only on the functional and ignore the social aspects of sustainability. For sustainable cities and communities, "the accessibility planning (in its broadest sense) of public transport which is necessary to meet the travel needs of socially excluded people must be highly integrated with socially responsible land use, housing, health, education and welfare policies and programmes" (Lucas, 2012, p. 112).

3 Method

To understand how sustainable mobility can help in achieving sustainable cities and communities in Southeast Asia, this chapter first analyses SDG 11 and its indicators according to the four elements of sustainable mobility. As described above, these four elements are: adapting and investing in technology to increase efficiency; reflecting the real costs and externalities of transport; integrating land use with transport; and behavioural change and acceptance. To investigate how the combinations of these four elements could be implemented in practice, two cities in Southeast Asia are then compared for their recent sustainable mobility policies in active mobility projects (cycling and walking) and LUTI policies, illustrating the relationship between sustainable mobility and sustainable cities and communities. Comparing the city of Singapore and the city of Jakarta, the units of analysis are recent projects and policies showing the context for implementation. Here, Singapore is chosen as an exemplary case due to its constant innovations for sustainable mobility and transportation, such as the electronic road pricing system. Jakarta, the second most congested city in the world, is more representative of the sustainable mobility capacity of the other cities in the region. Each case will be described in general and in the context of its achievements in sustainable mobility in the last decade. The specific policies or projects of each city will be listed and analysed along with the four elements of sustainable mobility and how SDG 11 might or might not apply.

The recent projects and policies are selected based on the following criteria:

- status as the latest iteration of national or local plans and policies or projects within the last decade;
- clear objectives to encourage modal shift away from private car use through either active mobility (cycling and walking) or LUTI (TOD) policies, and/or other elements of sustainable mobility;
- clarity in governance structures and implementation strategy embedded in existing formal and informal institutions (that is, priority given to formalized strategies and plans).

To provide a complete context for comparison, a fictitious journey from a commuter town or suburb to the economic centre of each city will be constructed for the modes of private car, urban rail, bus, and walking. Each mode will be measured in terms of the distance, time, and cost involved in making such a journey: this is to ensure some internal validity given that the socio-political contexts of the two cities are vastly different. Distance, time, and costs are the key factors that affect travel behaviour and mode choice (Annema, 2013). Assessing the comparison between the two cities in terms of these factors, and in light of the approach to projects and policies, the perspective of sustainable mobility in relation to mobility inequalities and exclusion is then discussed.

4 Findings

4.1 *Sustainable Development Goal 11*

Much like the Millennium Development Goals (MDGs) which preceded them, transport is not a specific focus of the SDGs although it is implicitly present in various goals relating to health, energy, and equality. The SDG in which transport is most prominently mentioned is SDG 11 ("Make cities and human settlements inclusive, safe, resilient and sustainable") and specifically Target 11.2: "By 2030, provide access to safe, affordable, accessible and sustainable transport systems for all, improving road safety, notably by expanding public transport, with special attention to the needs of those in vulnerable situations, women, children, persons with disabilities and older persons" (UN, 2015; see also IAEG, 2016). The only other mention is found in SDG 9, Target 9.1.[1] It is ironic that in a document that avoids the term "transport", so many of the indicators are focused on measuring the effects of transportation systems in the forms of outputs and emissions (CO_2 or particulate matter). This does not match

1 Target 9.1 reads: "Develop quality, reliable, sustainable and resilient infrastructure, including regional and trans-border infrastructure, to support economic development and human well-being, with a focus on affordable and equitable access for all".

up to the state-of-the-art in transport planning discourse, namely sustainable mobility.

Target 11.2 in itself ticks none of the elements of sustainable mobility directly, except for a general increase in the efficiency of the public transportation system. When looked at in the light of transport-related social exclusion and rights-based approaches, it does call for special attention for the vulnerable groups of elderly, women, children, and disabled. However, how this can be implemented is unaddressed; Target 11.2 is to be measured by Indicator 11.2.1 as the "proportion of population that has convenient access to public transport, by sex, age and persons with disabilities" (UN, 2015). This is simultaneously specific and vague. Having a large proportion of the population able to access public transport "conveniently" says nothing about their eventual adoption of that mode for their travels, whether public transport will become their primary day-to-day form of transport, or whether it will help increase their access to employment, education or healthcare. The spatial element is also neglected. Social and spatial inequalities are de facto conditions in Southeast Asian cities. If the proportion with increased and convenient access to public transportation grows, but mostly consists of those of a higher social or income level (e.g., opening of new stations or increased bus routes in wealthy areas) then this indicator does not contribute much to improving sustainable cities or communities.

Target 11.3 provides more hope as it names inclusiveness and capacity for participation of civil society as key factors. The target reads: "By 2030, enhance inclusive and sustainable urbanization and capacity for participatory, integrated and sustainable human settlement planning and management in all countries". The two indicators accompanying the target are: 11.3.1 "Ratio of land consumption rate to population growth rate"; and 11.3.2 "Proportion of cities with a direct participation structure of civil society in urban planning and management that operate regularly and democratically". This is echoed partially (in terms of access to open space) in Target 11.7, although the emphasis there is on personal safety free from physical and sexual harassment: "By 2030, provide universal access to safe, inclusive and accessible, green and public spaces, in particular for women and children, older persons and persons with disabilities". Target 11.3 refers indirectly to integration of land use and transport as its indicator is weighted by "land consumption rate". Again, though, the lack of targeted measures is a cause for concern. Each member state is free to interpret these goals to their own national standards. There will, of course, be vast differences between the criteria that are considered important when discussing universal access in Thailand or in Norway.

By their very nature, these goals are meant to be general enough to enable adoption and adaptation by member states and are meant for monitoring purposes. However, one could have hoped for more ambition and more alignment with state-of-the-art concerns in transport and mobility planning for such future-oriented goals. The danger here lies in the context-less and governance-free goals that are proposed. This presupposes an ability to progress from current levels, and assumes that there is an acceptable baseline from which to progress. In reality, transport and mobility planning are crucial sectors that require combined expertise from multiple sectors and disciplines, as seen from the cases of Singapore and Jakarta below.

4.2 *Singapore*

Singapore is a dense and highly urbanized city with nearly 6 million inhabitants spread over about 722 km^2. Within the island state, there is a polycentric settlement patterns of satellite towns in each cardinal direction functioning as stand-alone hubs for employment, commerce, education, and healthcare (Buijs et al., 2010). However, most commuter journeys are still oriented towards the existing and new central business districts (CBD and Downtown Core) in the south of the island. The biggest challenges facing Singapore are lack of land and increasing population growth; these have informed decades of policies to curb urban sprawl while preserving nature areas for drinkable water supply (Barter, 2008). This section examines two projects — the increase of walking and cycling comfort under the Walk2Ride scheme and the National Cycling Plan, and the improvement of integrated transport hubs — which form part of the Land Transport Master Plan 2013.

4.2.1 Walk2Ride and National Cycling Plan

The implementation agency, Land Transport Authority (LTA), has been steadily developing sheltered walkways within 200 m of all public transport nodes and connecting all nodes to educational or healthcare facilities in order to shelter users from the tropical climate. With the new Walk2Ride programme, the sheltered walkway network now also connects to more services such as "public amenities, offices and residential developments within 400m of the stations; bus interchanges and LRT stations to developments within a 200m radius; selected bus stops with very high usage to trip-generating hubs within a 200m radius" (LTA, 2013a, pp. 24–25). The expansion of sheltered walkways is significant for the expected increase in the user group of the elderly and the less mobile, and appropriately offers more resting areas and signage at key points to guide pedestrians towards transit hubs. In addition, seven satellite towns (Tampines, Yishun, Pasir Ris, Sembawang, Taman Jurong, Changi-Simei and Bedok) identified as growth areas, as well as the new downtown area (Marina Bay), are being improved with dedicated cycling networks to connect these residential and commercial hubs, and better cycling facilities including a bike-sharing scheme under the National Cycling Plan (LTA, 2013a).

4.2.2 People-Centred Land Transport System

Singapore has adopted a comprehensive approach to transportation and land use planning since the 1970s (Lam & Toan, 2006). There is also a strong formal tradition of masterplans for the long term (i.e., 20 years), reviewed every five years (Yuen, 2011). Since the 1991 Constellation Plan, the national government, acting through its ministries for transport and national development and their various statutory boards, has enacted a transit-oriented development strategy to concentrate satellite towns and developments along transit corridors as integrated transport hubs (Yang & Leng, 2009). These hubs are part of the mass rapid transit network in Singapore developing from an east–west and north–south corridor to a fully-fledged wheel-and-spoke-network with five lines and 117 stations, with more satellite towns covered by either heavy urban rail (MRT) or light rail (LRT). The network will be expanded in the coming years to include four more lines. The transit services and hubs are run by the SMRT Corporation Ltd., a multi-modal land transport provider that was initially privatized but was recently delisted and is now owned by the LTA. In the latest iteration of the Transport Masterplan with a clear set of targets and guidelines (LTA, 2013a), the following objectives related to sustainable mobility are found:

1) improve bus and rail service efficiency through technical and information systems upgrading and expansion of rail network;
2) increase safety and accessibility for elderly and the less mobile;
3) reduce private vehicle reliance by restricting purchase and satellite-based road pricing system;
4) facilitate more walking and cycling through network extensions, sheltered walkways or bike-sharing schemes.

These objectives and measures are an extension of the authorities' decades-long plans to discourage private vehicle use, encourage shared mobility and integrate land use and transport developments (Diao, 2018). There is also a traceable rise in the concern for social and individual aspects (safety, accessibility, and comfort) — an increasingly "people-centred" approach — since the 2008 Transport Masterplan (LTA, 2008).

4.2.3 Singapore: Legacy and Innovation

In terms of sustainable mobility, Singapore, with its direct governance structure (i.e., a national plan supported by parallel ministries and implemented in coordination) fulfils all four elements of sustainable mobility with its previous and current master plans (Diao, 2018; Yang & Leng, 2009). Aside from efficiency increases for services through technology, there is a substantial effort to deter private vehicle ownership by making the real costs of car ownership explicit through the need to purchase a licence to own a car, and through dynamic road pricing (Han, 2010). The current master plan also coordinates the walking and cycling strategy to increase active mobility as a first- and last-mile option for its users. There has been a large-scale effort to gain citizen feedback through surveys and panels (LTA, 2008, 2013b). Cycling has yet to be mainstreamed in Singapore but at least 22% of all trips in 2011 were made using pedestrian modes (Palliyani & Lee, 2017). There has been a steady increase in public transportation mode share from 65% in 2008 to 71% in 2012 for those within close proximity of transit hubs (LTA, 2013b). The case of Singapore demonstrates that SDG Target 11.2 could work if the basic networks and infrastructure quality levels are already present. The proportion of population with "convenient access" could indeed be increased (Indicator 11.2.1) through TOD policies in tandem with complementary walking and cycling programmes. The majority of trips taken by public transport involve those who live close to it, thereby signifying high acceptance. In addition, the sheltered walkway and cycling facilities in Singapore are meant to cater to a vulnerable population by providing access, comfort, and safety (Indicator 11.3.2).

4.3 *Jakarta*

As the central node of the Jabodetabek metropolitan region, Jakarta city is the cultural, commercial, financial, and administrative centre and experiences large commuter flows from the surrounding eight cities (Hasibuan, Soemardi, Koestoer, & Moersidik, 2014). As one of Southeast Asia's largest megacities, it has more than 10 million inhabitants spread over 662 km². Within Jakarta city, the residential areas are on the outskirts (including the low-income housing developed in North Jakarta for relocated slum residents), while the employment and retail districts are in the middle (Feminin, Wiranegara, & Supriatna, 2018). After decades of rapid urbanization and motorization of traffic, bringing negative health and societal impacts, there has been some interest from local government and communities in adopting more sustainable transportation models and networks. This section focuses on the increase of cycling as an alternative mode of travel, backed by civil society, and the development of TOD areas on the city's north–south corridor.

4.3.1 Cycling Communities and Car-Free Days

The first cycling lane was opened in 2011, between Ayodia Park and Blok M in South Jakarta, made possible by generous donations equivalent to US$5,600 from the cycling community (Komite Sepeda Indonesia) (Rukmana, 2011). At the time of writing, the city boasts 26 km of dedicated — but fragmented — cycling infrastructure and the provincial government claims that more will be added to connect other parts of the city as part of the TOD projects planned for the north–south corridor (Indonesia Expat, 2018). Apart from these infrastructural projects, cycling does not represent a priority for the local authorities. In fact, most of the resources and capacities come from civil society, in the form of cycling enthusiasts such as the citizen-led group, Bike2Work (Pearl, 2015). Active mobility is in its early days in Jakarta, despite the car-free days which have been held every Sunday for the past decade. The issue is not just public acceptance of cycling as an option, but the very real dangers of traffic where a majority of the users are in motorized vehicles and have little understanding of non-motorized traffic (Wihanesta & Titis Bawono, 2017). In addition, walking is seen as challenging or dangerous due to the discontinued sidewalks, bad quality of the pavements, and the use of pedestrian spaces for illegal or informal businesses (Zulkifli, Hino, Pattinaja, & Tjahjani, 2009).

4.3.2 TOD on the North–South Corridor

Against a backdrop of democratization and decentralization of spatial planning for the metropolitan region (Rukmana, 2015), mass rapid transit was declared a national project in 2005 (although plans for it had existed since 1985). This activated the central government and the provincial government (DKI Jakarta) to start transit-oriented developments with loans provided by the Japanese government (MRT Jakarta, 2018a). Phase 1 of the 16 km north–south corridor started in 2013 and is currently under construction by PT MRT Jakarta — a private company owned mainly by DKI Jakarta that operates, constructs, and maintains urban rail infrastructure and facilities (ibid.). In addition, it develops and manages property and businesses around the stations. Plans for TOD were prompted by increasing pressures resulting from congestion caused by private vehicles, and the sprawling urban land use (MRT Jakarta, 2018b). Of the eight Phase 1 stations designated as TOD, PT MRT Jakarta is developing an integrated master plan for the stations Lebak Bulus, Fatmawati, Cipete area (with Cipete Station, Haji Nawi Station, Blok A Station), Blok M area (with Sisingamangaraja Station), and Dukuh Atas. Some of these station areas are being advertised to attract public–private investment partnerships (JIC, 2017). Eight principles guide the design of the station areas, namely the increase of mixed land use functions; high-density developments; quality of connections to different modes; quality of life in terms of safety; social equality for marginalized groups previously lacking accessibility; environmental sustainability; infrastructure resilience from disasters and climate change; and development of local economy and employment. In terms of sustainable mobility, PT MRT Jakarta's objectives for these areas are (MRT Jakarta, 2018b):

1) reduce private vehicle use and corresponding road congestion and air pollution;
2) support walking, and a healthy and active lifestyle;
3) increase access to employment and economic opportunities;
4) increase transit passenger numbers and ticket sale profits;
5) increase mode choice for the urban area.

4.3.3 Jakarta: Car-Oriented Late Bloomer

In terms of sustainable mobility, Jakarta is hitting most of the four elements with its TOD policy objectives. From adoption of new technologies — heavy and light rail and electronic ticketing — to the integration of station area development with transport system improvements, the local and regional government are investing in behavioural change and acceptance. The true test will be in the implementation process and adoption by its citizens beyond the rhetoric.

However, ongoing fuel subsidies and the lethargic governance response to sustainable transportation are indications of the car-oriented socio-political environment in Jakarta (Susilo, Joewono, Santosa, & Parikesit, 2007). With regard to the city's non-existent cycling policies, the good news is that there is an active civil society that is engaged and willing to motivate authorities to take a leap. Although now considered only as a leisure pursuit, the mainstreaming of cycling could, with the help of car-free days, start to create a mindset change in the public. The case of Jakarta thus shows that SDG 11 is too general when it comes to implementation. Even if the percentage of population (as per Indicator 11.2.1) could be increased through TOD policies, these individuals would still have to be able to walk or even cycle to those station locations. Given the large population, the rate of land consumption is already out of proportion (Indicator 11.3.1) and would be impossible to reverse without drastic de-urbanization. There are also other factors to consider, such as inhabitants who are currently disenfranchised in informal settlements and those who have been relocated to lower-income housing areas being even further disadvantaged in terms of access to employment and services (Feminin et al., 2018).

5 Comparing Singapore and Jakarta

Singapore and Jakarta might be incomparable in many respects; however, to give an impression of the transportation choices available to inhabitants, it is possible to compare the distance, time, and cost required for an individual to journey from her/his home in a commuter suburb to the central business districts of each city. This comparison is intended to give an insight into the issues relevant for sustainable mobility and eventual implementation of the SDGs. Taking a closer look at both cities, the northern suburb of Woodlands is identified for Singapore and North Jakarta for Jakarta. Both suburbs are located 17 km north of the central commercial area of their respective cities.

Table 10.1 shows that comparing a similar trip from the commuter town to the city centre 17 km away, in Singapore and Jakarta, yields the following results:

1) Distance: distances by car and on foot are similar for the two cities. However, for Jakarta, there are twice as many transfers needed to complete the journey by public transportation and a significant increase (of 300%) in walking distance for a door-to-door trip by public transportation.
2) Time: the car journey in Jakarta takes almost twice as long as in Singapore, while public transport (rail) requires 144% more time than in Singapore and public transport (bus) requires 46% more time. For walking, the journey in Singapore will take 20% more time than in Jakarta.
3) Cost: all modes are significantly more expensive in Singapore than in Jakarta, although in Singapore, public transport rates are differentiated depending on the user group (e.g., elderly or students).

Taking the above comparison at face value, it seems that it takes more time in Jakarta to complete our hypothetical journey, but it is significantly less expensive than a similar journey in Singapore. However, if we factor in the difference in average net monthly salary between Singapore (US$2,891) and Jakarta (US$350) (Numbeo, 2018), an inhabitant in Jakarta actually spends 4.4% of their daily net income to commute back and forth to work, while an inhabitant in Singapore spends only 2.8% (or 3.7% without discounts). This is a significant difference, especially considering that the Jakarta journey also takes more time and offers much less comfort (more transfers and greater walking distances). With a low to median income, our Jakarta traveller sacrifices not only a portion of their income, but also time which could theoretically be used for earning income, to get to and from work. This comparative exercise thus throws into sharp relief the issues of social and economic inequality as presented through mobility choices.

There are a number of contextual limitations and qualifiers to take into account. The reader should bear in mind that the two cities differ vastly in terms of economic welfare and governance structures. In addition, the cost of owning a car in Singapore is prohibitive, estimated at SG$18,000 (approximately US$13,000) per year (Wong, 2017). Acquiring ownership entails a process of

TABLE 10.1 Distance, time and costs comparison between Singapore and Jakarta for private car, urban rail, bus, and walking

	Singapore	Jakarta
Locations	Woodlands to downtown core	Marunda, North Jakarta to Central Jakarta
Distance (geographical)	17 km	17 km
Distance		
Car	27 km	22.8 km
Urban Rail	16 stops on urban rail, with 850 m walking	24 stops on urban rail, with 2,300 m walking
Bus	43 stops on bus, with 750 m walking	22 stops on bus, with 3,300 m walking
Walking	25 km	21.2 km
Time		
Car	32 mins	47 mins
Urban Rail	56 mins (incl. waiting) 38 mins, no transfer, with 11 mins walking	137 mins (incl. waiting) 79 mins, 2 transfers, with 28 mins walking
Bus	96 mins (incl. waiting) 74 mins, no transfer, with 9 mins walking	141 mins (incl. waiting) 68 mins, 2 transfers, with 41 mins walking
Walking	317 mins	263 mins
Costs		
Car	Car purchase, road tax, fuel, insurance and parking varies. Toll (ERP): SG$2.00/US$1.45	Car purchase, road tax, fuel, insurance and parking varies
Urban Rail	Standard ticket: SG$2.50/US$1.82 Adults: SG$1.89/US$1.37 Children/students: SG$0.58/US$0.42 Elderly/disabled: SG$0.87/US$0.63	RP11,500/US$0.78 One fare (3 hours) RP5,000/US$0.34
Bus	Standard ticket: SG$2.50/US$1.82 Adults: SG$1.89/US$1.37 Children/students: SG$0.58/US$0.42 Elderly/disabled: SG$0.87/US$0.63	Off-peak: RP2,000/US$0.14 Peak: RP3,500/US$0.24

Notes and sources: distances and transit options data from Google Maps: Woodlands to Downtown Core https://goo.gl/maps/2rBSMMDRF5k (last accessed 3 June 2018); Marunda to Central Jakarta https://goo.gl/maps/A8yygQf9Gu72 (last accessed 3 June 2018); costs from SBS Transit, transit fares overview https://www.sbstransit.com.sg/transport/trpt_fares_overview.aspx (last accessed 3 June 2018).

bidding for certificates of entitlement (COEs), which are limited and regulated by the authorities: private vehicle ownership is therefore only available to a small elite. These indirect taxes make it possible for the Singapore government to subsidize the costs of public transport. In direct contrast, the cost of car ownership is much lower in Jakarta, fuel is subsidized, and there is an accessible domestic market for private vehicles.

5.1 *Policies and Acceptance*

Both cities have some form of active mobility project and LUTI policy. The key differences are found in (1) how long these policies and projects have been adopted as part of a long-term strategy or plan, and (2) the stakeholders involved in decision making and implementation.

In Singapore, the comprehensive land use and transport master plan is the current iteration of a set of policies and strategies going back to the late 1960s. Singapore has had decades to refine and improve upon those plans and has consistently done so with sustainable transportation and mobility in mind (Diao, 2018; Yang & Leng, 2009; Yuen, 2011). The latest master plan shows not only network expansion (for rail and bus) but also investments to improve rider comfort and the quality of the pedestrian network. The plans encompass more than just transportation system improvement but also look at safety and inclusion of vulnerable groups. There is a linear and hierarchical governance structure — a single authority coordinating across multiple ministries and agencies but implementing on one level — which makes implementation appear effortless. SMRT, as the transport provider and at times developer, has a direct line to the ministries and implementing agency of LTA (LTA, 2018a, 2018b). However, what seems to be missing in the Singapore case is the voice and actions of civil society and non-governmental actors.

In Jakarta, the slow adoption of sustainable mobility policies in the form of transit-oriented development has been attributed to lack of political will and urgency, and pushback from car-oriented citizens unfamiliar with these types of policies (Susilo et al., 2007). The mass rapid transit (urban rail) development was expected to benefit the socio-economically marginalized groups, particularly women who were previously subject to safety and harassment issues. However, the new commuter lines have achieved limited success in this regard, as low-income residents in North Jakarta who have been relocated from slums are dissatisfied with their level of accessibility (Feminin et al., 2018). The most crucial barrier to sustainable mobility in Jakarta is not the lack of policies but rather the lack of policies that target the most basic network — sidewalks and pavements — required to provide access to all by increasing safety from motorized traffic, and connectivity for pedestrians. Compared with Singapore, there is also a much wider variety of stakeholders involved in Jakarta. Here, foreign governments play a significant role in supporting the TOD plans financially and in knowledge transfer (Turner, 2011), while the cycling network is mostly promoted and funded by civil society.

6 Discussions

Returning to the discussion of SDG 11, it becomes clear that the indicators related to transport are, on their own, insufficient to achieve sustainable cities and communities. Diving into the intricacies of mobility and transport planning, as witnessed by the cases of Singapore and Jakarta, shows that there are many more factors to consider in order to achieve sustainable transportation or development. The comparison of the fictitious journeys showcases the need to consider social, economic, and political contexts. Factors such as political will, governance structures, and public acceptance also play an important role. Singapore has been able to implement sustainable mobility due to a consistent strategy and streamlined governance structure. The differences between conventional transport planning and sustainable mobility can offer insights here, as they show that cross-sectoral coordination is required to move away from the "predict and provide" narrative to consider the socio-economic and cultural outcomes and impacts of mobility policies. The land use and transportation masterplan in Singapore shows that it is not sufficient to just consider quantities (i.e., proportion of population with access to public transportation); rather, we need to look at the potential for accessing those hubs, under what conditions access is granted (comfort, safety, or time and cost in relation to income levels), and to whom. Sustainable cities and communities are closer to being achieved when the marginalized and vulnerable are considered and prioritized in policies and implementation. Likewise, when civil society stakeholders are active, as in Jakarta, democratic rule and governance benefit even where policy capacity and implementation are lacking.

Making cities and human settlements inclusive, safe, resilient, and sustainable is an ambitious goal. It will require a change in the status quo and innovative measures. Innovations in mobility planning are made possible through coordinated change in governance and cultural shifts in transportation behaviour (Homrighausen & Tan, 2016). Political will and public acceptance can contribute significantly to the shift from a business-as-usual approach to a sustainable mobility approach when it comes to the transportation system. There are roles here for the government, private parties, and civil society. The case of

Singapore depicts the efficiency of a stable government with consistent policy goals, and demonstrates the need for a transparent governance structure that allows for community feedback and participation as shown in the latest master plans. The case of Jakarta highlights the potential for private investment opportunities with the realization of TOD projects and corresponding urban development. Both cases show that civil society has a role to play in reducing mobility inequalities, either by reflecting current imbalances (e.g., Singapore's masterplan feedback panels) or by stepping up and navigating policy change when political will is lacking (e.g., Jakarta's cycling community). The efficacy of SDG 11 could be improved if it were to move beyond a pure measuring instrument to affect and sustain systemic change by adopting the sustainable mobility approach to address mobility inequalities and social exclusion.

References

Annema, J. A. (2013). Transport resistance factors: Time, money and effort. In D. Banister, B. P. van Wee, & J. A. Annema (Eds), *The Transport System and Transport Policy* (pp. 101–121). Cheltenham: Edward Elgar Publishing.

Bakker, S., Contreras, K. D., Kappiantari, H., Tuan, N. A., Guillen, M. D., Gunthawong, G., … van Maarseveen, M. (2017). Low-carbon transport policy in four ASEAN countries: Developments in Indonesia, the Philippines, Thailand and Vietnam. *Sustainability*, 9(7), 1217.

Banister, D. (2008). The sustainable mobility paradigm. *Transport Policy*, 15(2), 73–80.

Banister, D. (2011). Cities, mobility and climate change. *Journal of Transport Geography*, 19(6), 1538–1546.

Barter, P. (2004). Transport, urban structure and 'lock-in' in the Kuala Lumpur Metropolitan Area. *International Development Planning Review*, 26(1), 1–24.

Barter, P. A. (2008). Singapore's urban transport: Sustainability by design or necessity? In T. C. Wong (Ed.), *Spatial planning for a sustainable Singapore* (pp. 95–112). Dordrecht: Springer Netherlands.

Buijs, S., Tan, W. G. Z., & Tunas, D. (2010). Governance of megacities. In W. G. Z. Tan et al. (Eds), *Megacities* (pp. 275–285). Rotterdam: 010 Publishers.

Clark, C. (1958). Transport: Maker and breaker of cities. *The Town Planning Review*, 28(4), 237.

Coggin, T. (2017). A right to transport? Moving towards a rights-based approach to mobility in the city. *South African Journal on Human Rights*, 31(2), 294–314.

Development Initiatives (2017). *Financing the Sustainable Development Goals in ASEAN: Strengthening integrated national financing frameworks to deliver the 2030*

Agenda. Report commissioned by UNDP's Regional Bureau for Asia and the Pacific. Retrieved from http://asean.org/storage/2012/05/Report-on-Financing-SDGs-in-ASEAN1.pdf.

Diao, M. (2018). Towards sustainable urban transport in Singapore: Policy instruments and mobility trends. *Transport Policy* [online].

Feminin, T. A., Wiranegara, H. W., & Supriatna, Y. (2018). Accessibility of low-income family flats in North Jakarta city. *IOP Conference Series: Earth and Environmental Science*, 106, 012030.

Firman, T., & Dharmapatni, I. A. I. (1994). The challenges to sustainable development in Jakarta metropolitan region. *Habitat International*, 18(3), 79–94.

Han, S. S. (2010). Managing motorization in sustainable transport planning: The Singapore experience. *Journal of Transport Geography*, 18(2), 314–321.

Hasibuan, H. S., Soemardi, T. P., Koestoer, R., & Moersidik, S. (2014). The role of transit oriented development in constructing urban environment sustainability: The case of Jabodetabek, Indonesia. *Procedia Environmental Sciences*, 20, 622–631.

Homrighausen, J. R., & Tan, W. G. Z. (2016). Institutional innovations for sustainable mobility: Comparing Groningen (NL) and Phoenix (US). *Transportation Research Procedia*, 19.151–163.

Hull, A. (2008). Policy integration: What will it take to achieve more sustainable transport solutions in cities? *Transport Policy*, 15(2), 94–103.

Indonesia Expat (2018). Jakarta plans to build more bicycle lanes. Indonesia Expat. Retrieved from http://indonesiaexpat.biz/news/jakarta-plans-build-bicycle-lanes/ (last accessed 3 June 2018).

IAEG (2016). Global indicator framework. Report of the Inter-Agency and Expert Group on Sustainable Development Goal Indicators (E/CN.3/2016/2/Rev.1), Annex IV. Geneva: UN Inter-Agency and Expert Group.

INRIX (2017). Interactive ranking & city dashboards: Hours spent in congestion. Retrieved from http://inrix.com/scorecard/ (last accessed 3 June 2018).

JIC (2017). Jakarta investment project digest 2017–2018. Jakarta: Jakarta Investment Centre. Retrieved from https://www.bkpm.go.id/images/uploads/berita_acara_iipc_fileupload/Jakarta_Investment_Project_Digest.pdf.

Johnson, C. (1999). The developmental state: Odyssey of a concept. In M. Woo-Cummings (Ed.), *The developmental state* (pp. 32–92). Ithaca, NY: Cornell University Press.

Kenworthy, J. (2007). Urban planning and transport paradigm shifts for cities of the post-petroleum age. *Journal of Urban Technology*, 14(2), 47–70.

Kenyon, S., Lyons, G., & Rafferty, J. (2002). Transport and social exclusion: Investigating the possibility of promoting inclusion through virtual mobility. *Journal of Transport Geography*, 10(3), 207–219.

Ladd, B. (2012). 'You can't build your way out of congestion' — Or can you? *disP — the Planning Review*, 48(3), 16–23.

Lam, S. H., & Toan, T. D. (2006). Land transport policy and public transport in Singapore. *Transportation*, 33(2), 171–188.

LTA (2008). What transport must do: Connect newsletter April 2008. Land Transport Authority. Retrieved from https://www.lta.gov.sg/content/dam/ltaweb/corp/PublicationsResearch/files/ReportNewsletter/Connect/Connect_April_FA.pdf (last accessed 3 June 2018).

LTA (2013a). Land transport master plan 2013. Singapore: Land Transport Authority. Retrieved from https://www.lta.gov.sg/content/dam/ltaweb/corp/PublicationsResearch/files/ReportNewsletter/LTMP2013Report.pdf (last accessed 3 June 2018).

LTA (2013b). Household interview travel survey 2012: Public transport mode share rises to 63%. Singapore: Land Transport Authority. Retrieved from https://www.lta.gov.sg/apps/news/page.aspx?c=2&id=1b6b1e1e-f727-43bb-8688-f589056ad1c4 (last accessed 3 June 2018).

Loo, B. P. Y., & du Verle, F. (2016). Transit-oriented development in future cities: Towards a two-level sustainable mobility strategy. *International Journal of Urban Sciences*, 21(Sup1), 54–67.

Lucas, K. (2012). Transport and social exclusion: Where are we now? *Transport Policy*, 20, 105–113.

Lucas, K., van Wee, B. P., & Maat, K. (2016). A method to evaluate equitable accessibility: Combining ethical theories and accessibility-based approaches. *Transportation*, 43(3), 473–490.

Mees, P. (2014). TOD and multi-modal public transport. *Planning Practice and Research*, 29(5), 461–470.

MRT Jakarta (2018a). Sejarah. MRT Jakarta. Retrieved from https://www.jakartamrt.co.id/mrt-jakarta/sejarah-mrt-jakarta/ (last accessed 3 June 2018).

MRT Jakarta (2018b). Kawasan Berorientasi Transit (TOD). MRT Jakarta. Retrieved from https://www.jakartamrt.co.id/konektivitas/transit-oriented-development-tod/ (last accessed 3 June 2018).

Numbeo (2018). Cost of living comparison between Jakarta and Singapore. Retrieved from https://www.numbeo.com/cost-of-living/compare_cities.jsp?country1=Indonesia&city1=Jakarta&country2=Singapore&city2=Singapore&displayCurrency=USD (last accessed 3 June 2018).

Owens, S. (1995). From 'predict and provide' to 'predict and prevent'? Pricing and planning in transport policy, *Transport Policy*, 2(1), 43–49.

Palliyani, S., & Lee, D.H. (2017). Sustainable transport policy: An evaluation of Singapore's past, present and future. *Journal of Infrastructure, Policy and Development*, 1(1), 112–17.

Pearl, H. (2015). Cycling the city: 'I have a dream that Jakarta should be like Copenhagen'. Retrieved from http://www.theguardian.com/cities/2015/sep/10/cycling-jakarta-dream-copenhagen-indonesia-bike (last accessed 3 June 2018).

Preston, J., & Rajé, F. (2007). Accessibility, mobility and transport-related social exclusion. *Journal of Transport Geography*, 15(3), 151–160.

Pucher, J., & Dijkstra, L. (2011). Promoting safe walking and cycling to improve public health: Lessons from the Netherlands and Germany. *American Journal of Public Health*, 93(9), 1509–1516.

Rama, M., Béteille, T., Li, Y., Mitra, P. K., & Newman, J. L. (2015). *Addressing inequality in South Asia*. Washington, DC: World Bank Group.

Rukmana, D. (2011). Indonesia's biggest city gets its first bicycle lane. Retrieved from http://thisbigcity.net/indonesias-biggest-city-gets-its-first-bicycle-lane/ (last accessed 3 June 2018).

Rukmana, D. (2015). The change and transformation of Indonesian spatial planning after Suharto's New Order Regime: The case of the Jakarta metropolitan area. *International Planning Studies*, 20(4), 350–370.

Scheurer, J., Curtis, C., & McLeod, S. (2017). Spatial accessibility of public transport in Australian cities. *Journal of Transport and Land Use*, 10(1), 911–930.

Shatkin, G., (2008). The city and the bottom line: Urban megaprojects and the privatization of planning in Southeast Asia. *Environment and Planning A*, 40(2), 383–401.

Susilo, Y. O., Joewono, T. B., Santosa, W., & Parikesit, D. (2007). A reflection of motorization and public transport in Jakarta Metropolitan Area: Lessons learned and future implications towards better transportation development in developing countries. *Proceedings of the Eastern Asia Society for Transportation Studies*, 6, 46–46.

Tan, W. G. Z., Bertolini, L., & Janssen-Jansen, L. B. (2014). Identifying and conceptualising context-specific barriers to transit-oriented development strategies: The case of the Netherlands. *Town Planning Review*, 85, 639–663.

Thynell, M. (2017). Roads to equal access: The role of transport in transforming mobility. *Transport and Communications Bulletin for Asia and the Pacific*, 87, 77–87.

TomTom (2017). TomTom traffic index: Measuring congestion worldwide. Retrieved from https://www.tomtom.com/en_gb/trafficindex/list?citySize=LARGE&continent=ALL&country=ALL (last accessed 3 June 2018).

Turner, J. (2011). Urban mass transit and social sustainability in Jakarta, Indonesia. Nairobi: UN Habitat.

UN (1948). Universal Declaration of Human Rights. New York: United Nations General Assembly. Retrieved from http://www.un.org/en/universal-declaration-human-rights/. (last accessed 3 June 2018).

UN (2015). Transforming our world: The 2030 Agenda for Sustainable Development. New York: United Nations. Retrieved from undocs.org/A/RES/70/1 (last accessed 3 June 2018).

UN (2017). Work of the statistical commission pertaining to the 2030 Agenda for Sustainable Development (A/RES/71/313), Annex (6 July 2017). New York: United Nations. Retrieved from undocs.org/A/RES/71/313 (last accessed 3 June 2018).

UN ESCAP (2017). Complementarities between the ASEAN Community Vision 2025 and the United Nations 2030 Agenda for Sustainable Development. Bangkok: United Nations. Retrieved from https://www.unescap.org/publications/complementarities-between-asean-vision-2025-and-2030-agenda (last accessed 3 June 2018).

UN HABITAT (2013). *Planning and Design for Sustainable Urban Mobility: Global Report on Human Settlements.* New York: Routledge.

Wegener, M., & Fürst, F. (1999). Land-use transport interaction: State of the art. TRANSLAND. Dortmund: European Commission.

Wihanesta, R., & Titis Bawono, E. (2017, September 12). Jokowi's bike: Where are the lanes? The Jakarta Post. Retrieved from http://www.thejakartapost.com/academia/2017/09/12/jokowis-bike-where-are-the-lanes.html (last accessed 3 June 2018).

Wilson, R. (2001). Assessing communicative rationality as a transportation planning paradigm. *Transportation*, 28(1), 1–31.

Wong, J. (2017). How much it really costs to own a car in Singapore. Channels News Asia. Retrieved from https://www.channelnewsasia.com/news/brandstudio/how-much-it-really-costs-to-own-a-car-in-singapore-9346730 (last accessed 3 June 2018).

World Bank (2015). *East Asia's changing urban landscape.* Washington, DC: World Bank.

Yang, P., & Leng, S. H. (2009). An Asian model of TOD: The planning integration in Singapore. In L. Bertolini, J. L. Renne, & C. Curtis (Eds), Transit-oriented development: Making it happen (pp. 91–108). New York: Taylor & Francis.

Yuen, B. (2011). Urban planning in Southeast Asia. Centenary paper: Urban planning in Southeast Asia: Perspective from Singapore. *The Town Planning Review*, 82(2), 145–167.

Zulkifli, M., Hino, Y., Pattinaja, A., & Tjahjani, I. (2009). Limitation of walking facilities in Jakarta: A visual analysis on actual condition and user's perception. *Proceedings of the Eastern Asia Society for Transportation Studies*, 7, 260–272.

PART 5

SDGs and the Environment, Clean Air and Water for All

CHAPTER 11

Transboundary Haze, ASEAN, and the SDGs: Normative and Structural Considerations

Helena Varkkey

1 Introduction

The term "haze" has been used in the Southeast Asian region to refer to "sufficient smoke, dust, moisture, and vapour suspended in air to impair visibility" (ASEAN Secretariat, 2008). Haze is a result of peat and forest fires, primarily in Indonesia and to a lesser extent Malaysia. Forest fires have been extensively recorded in Southeast Asia, particularly in Indonesia, since the 19th century (Eaton & Radojevic, 2001). While these fires have been naturally occurring during the dry seasons (around August to October) for generations, they have grown in severity and intensity over the years due to anthropogenic activities (Mayer, 2006). Disturbances to the landscape, either on a small scale by swidden farmers, or on a large scale due to commercial plantation activity, have caused the drying out of the land, making it more susceptible to fires. Fires are also sometimes intentionally set as a fast and cheap way to clear land (Colfer, 2002). Some fires get out of control, resulting in a belching of excessive smoke into the atmosphere which is then carried by the wind to the rest of the region. This creates "transboundary" haze, defined as such when "its density and extent is so great at the source that it remains at measurable levels after crossing into another country's airspace" (ASEAN Secretariat, 2008).

1.1 *Research Question*

Being a regional transboundary issue, most of the regional-level haze mitigation activities have been spearheaded by the Association of Southeast Asian Nations (ASEAN). ASEAN has recently adopted its 'Declaration on ASEAN Post-2015 Environmental Sustainability and Climate Change Agenda'. In this Declaration, ASEAN and its member states (AMS) affirm their commitment to move the sustainable development agenda forward through the achievement of internationally agreed development goals. Specifically, it declares ASEAN's commitment to "continue ... efforts to establish a balance among economic growth, social development and environmental sustainability as well as to strengthen ASEAN's commitments for the realization of the Post

 | DOI:10.1163/9789004391949_012

2015 Development Agenda and the attainment of the SDGs [Sustainable Development Goals]" (ASEAN Secretariat, 2015, p. 3). One of the issues that was given priority status in this Declaration was haze. The Declaration identified continued and improved cooperation over haze at the regional level as an important criterion in this agenda. Following the Declaration, AMS also adopted a 'Roadmap on ASEAN Cooperation towards Transboundary Haze Pollution Control with Means of Implementation', with a vision of a "transboundary haze-free ASEAN by 2020" (ASEAN Secretariat, 2016, p. 2; see also Naidu, 2016). Global goals such as the SDGs can sometimes play a role in actions at the regional levels of governance. However, this can be challenging, especially if there are compatibility issues between the international and regional levels. This chapter thus asks the question: to what extent can global goals such as the SDGs play an important role in galvanizing support and guiding actions at the ASEAN level?

1.2 *Significance of the Research*

Air pollution increases the risk of strokes, lung cancer, heart disease, and other serious health problems in populations all around the world. At the international level, key bodies of the United Nations have started to acknowledge the urgent need to address the challenges posed by air pollution to humans and the environment. In June 2014, The United Nations Environment Assembly under the United Nations Environment Programme adopted a resolution on air quality at its very first session. The resolution noted that "poor air quality is a growing challenge in the context of sustainable development, in particular relating to health in cities and urban areas" (Lode & Toussaint, 2016, p. 4).

Many low- and middle-income countries in particular suffer from bad air quality (Lode & Toussaint, 2016). The countries of Southeast Asia, which have experienced haze for decades, are a pertinent example of this trend. At its worst, the haze can travel to reach six Southeast Asian nations (Mayer, 2006). Indonesia, Malaysia, and Singapore, among the largest economies of the region, suffer the brunt of the haze almost every year (ibid.). A serious episode of haze in June and July 2013 caused the affected governments to declare repeated emergencies at the worst-hit areas, closing schools for days on end and restricting outdoor activities (Sun Daily, 2013). Since international goals for clean air are highly dependent on improved air quality at the regional level, it is important to study the compatibility between international goals and regional implementation efforts.

1.3 *Methodology*

This chapter adopts a qualitative approach to answer the research question. The data used during the study mainly consisted of public-domain documents like policy papers, data sets, and organizational reports. As the chapter concentrates on the theme of multi-level challenges for implementation, a qualitative comparison between UN-level documentation and ASEAN-level documentation related to air quality and haze issues was particularly important. A close reading of these documents, declarations, and reports was useful in discerning the past achievements, limitations, and foreseeable multi-level challenges of translating SDGs to implementable regional action in the case of haze. The chapter also makes use of the author's personal observations during high-level forums and other events to which she had access by virtue of her research interest in these areas.

2 Literature Review

Since the concept of sustainable development is relatively new, only recently gaining traction in governance mechanisms around the world, it is not surprising that many older regional organizations were not founded on the values of sustainable development. However, as regional projects are by nature fluid, this does not necessarily prevent them from moving towards broader sustainability objectives. Some regional institutions can be expected to undergo a gradual process called 'green integration'. This concept was introduced by Kojima and Bengtsson (2015) to refer to a regional process that facilitates and underpins a reformulation of the current development model — a rebalancing from a de facto strong focus on economic integration to broader sustainability objectives.

ASEAN is an example of an organization which, at its founding, paid very limited attention to notions of sustainability. Rather, the core developmental model of ASEAN as an organization has always been based on economic progress. This can be traced back to its founding documents, which consistently put economic growth before other regional objectives like social progress and cultural development (Elliott, 2003). This has resulted in an organizational focus on economic integration to boost economic growth, which is neither environmentally sustainable nor socially inclusive (Kojima & Bengtsson, 2015). This review of the literature explores whether ASEAN's normative and structural settings could accommodate a rebalancing of focus from economic integration to sustainable development.

2.1 *ASEAN Way Norms*

Engagement at the regional level among AMS and between AMS and ASEAN is broadly in accordance with the 'ASEAN Way' norms. These norms are spelt out in various ASEAN treaties, from its founding document, the 1967 Bangkok

Declaration, to the more recent 2007 ASEAN Charter (Koh, 2008). They are a set of behavioural and procedural norms that include the pursuit of consensus; the sanctity of sovereign rights and the related concept of non-interference; the principles of sensitivity and politeness; non-confrontational negotiation processes; behind-the-scenes discussions; an emphasis on informal and non-legalistic procedures; and flexibility (Kivimaki, 2001).

It is important to note that AMS do *not* blindly follow the ASEAN Way principles due to some deeply ingrained 'habit'. Rather, states pick and choose whether or not to adhere to the ASEAN Way principles, depending on whether it is in their interests to do so. While, for example, the European Union model of regionalism is characterized by the pooling of sovereignty, the ASEAN model is characterized by the maintenance of national sovereignty (Murray, 2010) and, by extension, national interest. This is the so-called 'myth' of the ASEAN Way (Nischalke, 2000). Indeed, there have been many instances where the ASEAN Way was deliberately ignored so that ASEAN states could pursue narrow understandings of their self-interests (Narine, 1998). Furthermore, no ASEAN member state had any serious reservations about the policy outcomes in most of these cases. Hence, an ASEAN member state can choose to *not* adhere to the ASEAN Way in order to preserve crucial interests, without suffering any consequences at the regional level (Nischalke, 2000).

The ASEAN norm of non-interference in particular is closely tied to the age-old adage that environmental protection is a Western agenda. In the early 1980s, many Asian countries were at the receiving end of continued criticism from Western environmentalists with regard to their vigorous forestry practices (Nossal & Stubbs, 1997). Mahathir Mohamad, Malaysia's Prime Minister at the time, fervently defended the region's position, arguing that "now the developed countries have sacrificed their own forests in the race for higher standards of living, they want to preserve other countries' rain forests — citing a global heritage — which could indirectly keep countries like Malaysia from achieving the same levels of development" (cited in Karim, 2008, p. 333). Mahathir also expressed similar views on the world stage at the United Nations Conference on the Environment and Development held in Rio de Janeiro in 1992 (Mohamed, 1992). In short, just as the ASEAN Way subscribes, he argued that the Western world should not have the right to 'interfere' in how Malaysia and other countries in the region manage their natural resources.

The ASEAN countries lauded Mahathir as a hero of the developing world for his brave statements, which lent legitimacy to the view that the developing countries should not be prevented from exploiting their own natural resources which are vital for their own development (Leigh, 1998). From that point, the protection of the environment within ASEAN was openly regarded as something that would threaten economic growth, development, and social cohesion in most of the AMS (D. M. Jones & Smith, 2002). While AMS recognized that they had common environmental problems, their material interests in addressing these problems arose from the importance of domestic economic progress and development. Thus, it was considered more important for these states to maintain the availability of and access to natural resources (Elliott, 2003) like timber and forest products to be used in the pursuit of development: there has been very little appetite for any reformulation of the current development with its focus on economic progress and development.

2.2 *ASEAN Institutional Structure*

The 'Cebu Declaration on the Acceleration of the Establishment of an ASEAN Community by 2015', signed by all AMS in 2012, formalized the pillar-based structure of the ASEAN Community (ASEAN Secretariat, 2012). Often stylized into a house-shaped structure with three pillars holding up the roof of the 'community', the pillars — namely the ASEAN Economic Community (AEC), ASEAN Political-Security Community (APSC), and ASEAN Socio-Cultural Community (ASCC) — represent the three integral focus areas of ASEAN community building (see Figure 11.1). Even though these pillars are generally pictured of equal size and thus seemingly of equal importance, the AEC is often regarded as the most important pillar and, indeed, a precondition for the realization of the other two pillars (VNA, 2015). Hence, many scholars would agree that in reality, community building revolves around the AEC (Olsen, Teoh, & Miyazawa, 2015).

While this sectoral, pillar-based structure is meant to improve focus on specific issues at the ASEAN level and thus increase the effectiveness of ASEAN as an organization in the development of the ASEAN community, the idea that issues of today can fit neatly (and in turn, be 'managed') in one of these three pillars is problematic when couched against current global understandings of sustainable development. The latter is, by definition, a cross-cutting concept, not confined to the realm of economy, politics, security, society, or culture. Hence, this cross-cutting nature of sustainable development seems unlikely to find a comfortable fit within ASEAN's current pillar-based structure (Olsen et al., 2015).

Taking these normative and structural considerations into account, it is questionable whether ASEAN would be a likely candidate to undergo the so-called process of green integration as described by Kojima and Bengtsson (2015). The ASEAN Way norms which uphold national interests and non-interference, coupled with the rigid pillar-based structure seem unlikely to support a rebalancing of focus from economic integration to broader sustainability objectives. Indeed, after more than four decades, there is little sign of

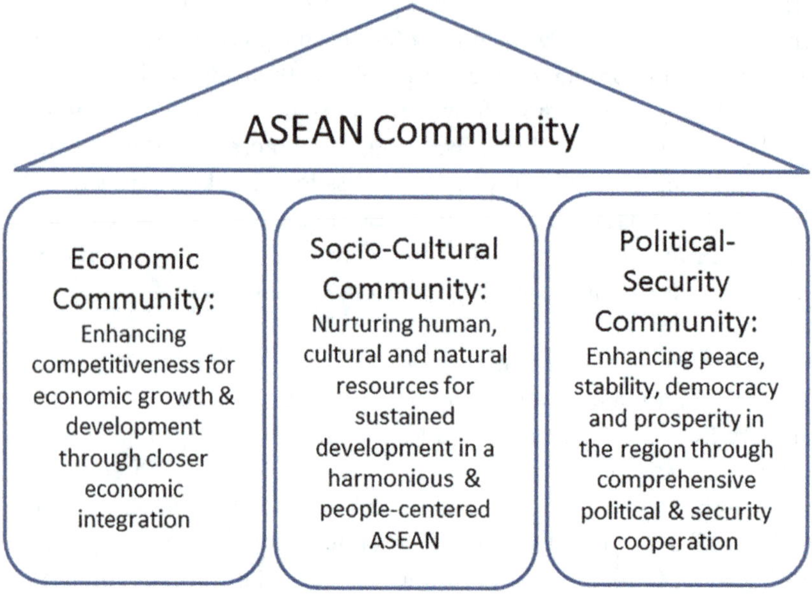

FIGURE 11.1 The three pillars of the ASEAN community

this happening. However, the adoption of the SDGs by ASEAN in 2015 fuelled renewed hope that these global goals would somehow inspire green integration in ASEAN. The following analytical section considers this within the context of the haze problem in the region.

3 Analysis

The literature review above examined some general regional-level normative and structural limitations that may stand in the way of green integration in ASEAN. Of course, global goals can play an important role in overcoming these limitations to facilitate a reformulation of the current development model towards sustainability objectives. However, as Kojima and Bengtsson (2015) point out, green integration is often more easily achieved when the issues involved are relatively uncontroversial. The issue of haze is extremely controversial.

The regional plantation sector, and palm oil specifically, is of vital economic importance for the main countries embroiled in the haze issue. Indonesia and Malaysia are currently the world's largest and second largest producers of palm oil respectively. In Indonesia, it contributes around 7% of GDP annually (Bank Indonesia, 2014; Das, 2014), and employs about 20 million people, both directly and indirectly (Simamora, 2011). In Malaysia, it steadily contributes

5–6% of the country's GDP (Clean Malaysia, 2015). Furthermore, Malaysia controls significant amounts of plantation land in Indonesia. Singapore is a major refiner and processor of palm oil derivatives, and also has plantation interests in both Indonesia and Malaysia. Many plantation companies in all these countries are politically well connected, and deeply entrenched in the regional patron–client business culture, allowing them to continue their operations — environmentally friendly or otherwise — almost with impunity (Varkkey, 2016a). Furthermore, environmental discussions surrounding haze management at the regional level remain carefully measured in order not to limit AMS' access to forests and their ability to exploit the land. Anti-palm oil sentiments, especially from the West, are often met head-on by Southeast Asian governments as direct threats to their national interests and sovereignty, and frequently couched in neo-colonialist arguments (Adnan, 2012).

3.1 *Air Quality, Haze and the SDGs*

The 2030 Agenda for Sustainable Development, brought into force on 1 January 2016, identified clean air as an integral element of the principle of sustainable development. Indeed, the SDGs detailed in the Agenda contain several targets and indicators specifically pertaining to air quality. Firstly, a target under Goal 3 ("Ensure Healthy Lives and Promote Well-Being for All at All Ages") aspires to "substantially reduce the number of deaths from hazardous chemicals and air, water and soil pollution and contamination, by 2030" (Target 3.9), as indicated by mortality rates attributed to household and ambient air pollution. Secondly, a target under Goal 11 ("Make Cities and Human Settlements Inclusive, Safe, Resilient and Sustainable") aspires to "reduce the adverse per capita environmental impact of cities, including by paying special attention to air quality and municipal and other waste management, by 2030" (Target 11.6), as indicated by annual mean levels of fine particulate matter in cities (Lode & Toussaint, 2016).

Many other SDGs are indirectly relevant to the issue. For example, at the recent UNESCO-led Expert Group Consultation on the Bioethical Perspectives on Haze in Kuala Lumpur (Malaysia) a representative from Malaysia's Prime Minister's Department identified another goal that speaks to the issue of air pollution. I attended this Expert Group Consultation, where haze was specifically discussed in the context of Goal 13 ("Take Urgent Action to Combat Climate Change and Its Impacts") (Razak, 2017). Additionally, a researcher from the Institute of Islamic Understanding Malaysia (IKIM), which is a member of UNESCO's International Bioethics Committee, identified Goal 12 ("Ensure Sustainable Consumption and Production Patterns") and Goal 15 ("Sustainably Manage Forests, Combat Desertification, Halt and Reverse Land Degradation, Halt Biodiversity Loss") as belonging to the list of SDGs which touch on this

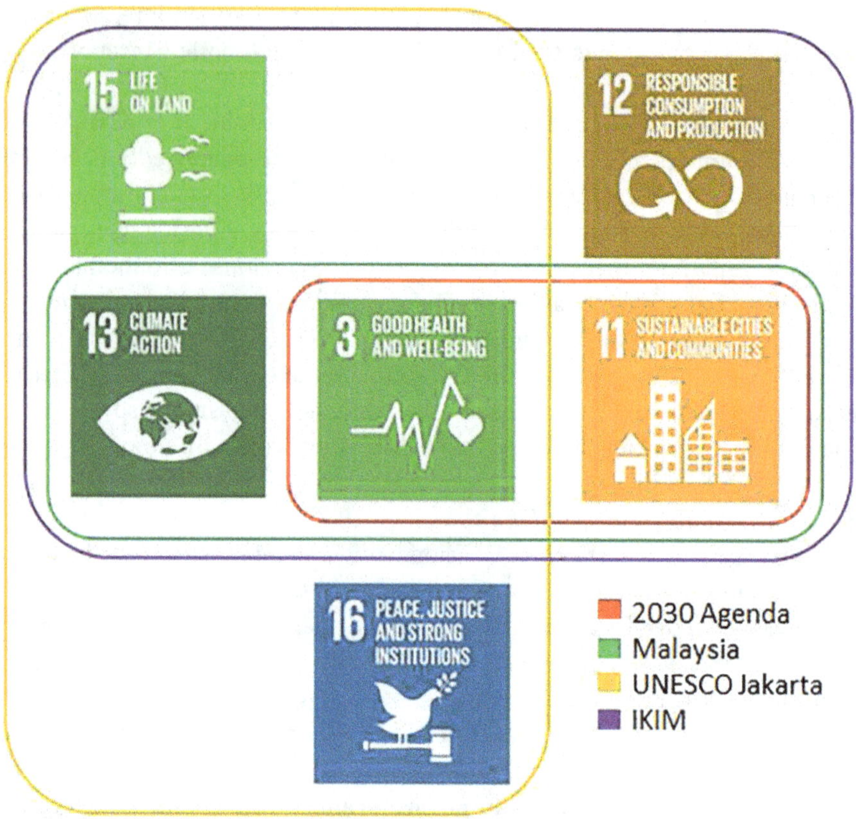

FIGURE 11.2 Organization-based Venn diagram of SDGs related to haze

issue (Shaik Mohd Salleh, 2017), while the UNESCO Jakarta representative added Goal 16 ("Promote Just, Peaceful and Inclusive Societies").

Each organization's views on SDGs related to air pollution and haze is illustrated in the Venn Diagram presented in Figure 11.2. The red outline indicates specific goals related to air pollution as spelled out in the 2030 Agenda. The other three groupings relate to haze in particular, with green identified by Malaysia, yellow by UNESCO Jakarta, and purple by IKIM. It is interesting to note that at the regional level — or, in other words, when haze as a type of air pollution is considered to be of particular interest to the region — the goals that can be associated with its reduction increase. This is because, while the specific targets identified at the global level by the United Nations General Assembly's Open Working Group (OWG) on SDGs focus on global problems and indicators, these goals can only be adopted and refined to local needs and priorities at the regional and national level. Furthermore, the link between a unique problem like haze and other goals only becomes apparent close to home.

As noted, the UNESCO Jakarta office also highlighted Goal 16 on Peace, Justice and Strong Institutions. The United Nations family is closely associated with regional haze activities. Together with the United Nations Economic and Social Commission for the Asia-Pacific (UN ESCAP), UNESCO Jakarta is part of the United Nations Development Group Asia-Pacific. ASEAN is a formal partner of UN ESCAP, and UNESCO Jakarta is located close to the ASEAN Secretariat; the two organizations have been engaging with each other over haze issues since at least 2006. This formal partnership was born during the ASEAN-UNESCO Concept Workshop on Human Security in South-East Asia which identified haze as a major human security issue in the region (Ong & Sané, 2006). It is telling that the UNESCO Jakarta office chose to highlight the importance of strong institutions in the efforts to mitigate haze: the significance of this is discussed below.

3.2 *Normative and Structural Considerations*

It is clear that the issue of haze is an extremely controversial one for all parties involved. This poses an additional challenge for green integration in ASEAN. At face value, haze mitigation would seem like a perfect platform to push green integration forward. Regional governance is critical for this issue area, since haze is clearly beyond the ability of one government to combat. It is also an issue which cuts across all three pillars of ASEAN. It is no surprise that efforts to address this issue over the past 30 years have overwhelmingly originated from the ASEAN level, as described below. Indeed, regional collaboration over haze quickly became a high-profile matter and was held up as one of the earliest and most promising examples of ASEAN cooperation over transboundary issues (Elliott, 2003). Since the haze could be seen as a result of the current development model (economic progress at the expense of the environment), it would not be a stretch to believe that efforts to mitigate this problem could help in reformulating the ASEAN's developmental orientation to one which is more in line with the sustainable development objectives.

Even though the SDGs do seem to complement sustainable development objectives at the ASEAN level, as indicated by various ASEAN declarations, there are several deep-rooted normative and structural issues within ASEAN which may reduce the effectiveness of the SDGs in working towards a haze-free ASEAN by 2020. The discussion below first analyses how the ASEAN norms have resulted in weak institutions, which are directly at odds with SDG 16 for Strong Institutions. Second, it highlights how the non-interference norm has resulted in resistance to ideas of the environment and development that originate from outside ASEAN, such as the SDGs. Third, it focuses on how ASEAN's sectoral, pillar-based structure does not fit well with broader global understandings of sustainable development. This has led to multi-level challenges in implementing haze mitigation activities.

The ASEAN Vision 2025 document, produced in 2015, emphasizes the complementarity of regional goals with the SDGs. Some of these are quite clear cut: both documents emphasize the people-centred nature of their goals. The 2030 Agenda sums up its bottom line as "no one will be left behind" while ASEAN's Vision 2025 pledges to create a rules-based, people-centred ASEAN. Both documents refer to unity or oneness, by including each person, each country, and each community on the journey towards sustainable development. This is clearly underlined by the Vision 2025's "One Vision, One Identity, One Community" (UN ESCAP, 2017). However, there are some normative differences between the two which may prove challenging when both organizations move from internationally agreed goals and targets to regional implementation arrangements. This section offers two norm-based arguments for why the SDGs may have limited potential for galvanizing support and guiding actions towards a haze-free ASEAN by 2020: weak institutionalism and aversion to interference.

3.2.1 Weak Institutionalism

The fact that the importance of strong institutions in mitigating haze has been highlighted by UN representatives and not by Southeast Asian-based organizations raises a flag on the compatibility and complementarity between the SDGs and haze mitigation. The UN OWG on SDGs defines "strong institutions" as "effective, accountable, and inclusive … function according to legitimate laws, not prone to arbitrariness and abuse of power, and capable of delivering public services to everyone" (UN, 2016, p. 2). ASEAN is generally not known to be a "strong institution" according to these broad criteria. Rather, the preferred 'ASEAN Way' of regional engagement has weakened the ASEAN's capacity to create and enforce strong institutions to serve the collective regional interests of mitigating haze. As a result, ASEAN haze institutions have not been effective or inclusive, and are prone to abuse by elites. Indeed, it can be argued that this is how AMS prefer it to be. The 'myth' of the ASEAN Way can also be used to explain the ineffectiveness of ASEAN haze mitigation efforts to date (Varkkey, 2016a).

The first ASEAN-level activity that specifically addressed haze was in 1992, with the Workshop on Transboundary Pollution and Haze in ASEAN Countries (ASEAN Secretariat, 1995). This was followed by several other soft-law initiatives

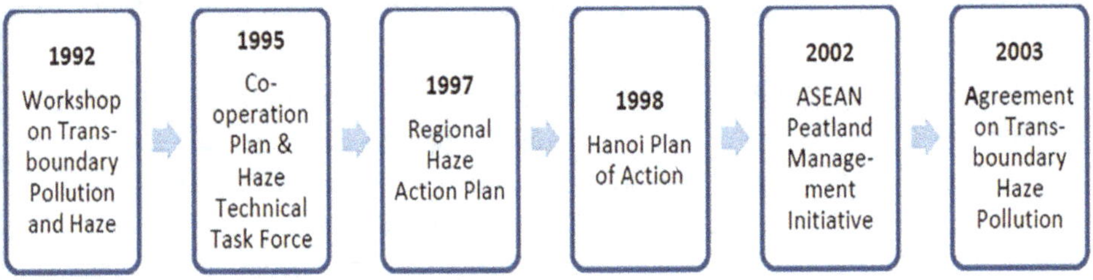

FIGURE 11.3 Timeline of major ASEAN haze mitigation efforts
SOURCE: VARKKEY (2016A)

(see Figure 11.3). In the literature, binding agreements, institutionalized systems, and the sectoral integration of policies are often identified as some of the most important factors that create strong institutions (Murray, 2010; Shelton, 2003). A close examination of the early ASEAN initiatives on haze reveal that these ASEAN Way-compliant initiatives rarely contained these important factors for success. For example, the Cooperation Plan and Haze Technical Task Force was found to be much less formal or binding than comparable documents in other regions, such as Europe, for curbing transboundary harm (Tay, 2008). The plan was largely a listing of general actions that governments *ought* to take to prevent and mitigate forest fires (Severino, 2006). The lack of explicit operational directives rendered it ineffective (Tay, 2008).

Likewise, the 1998 Regional Haze Action Plan (RHAP) continued to focus on national plans and capabilities instead of regional initiatives (Tay, 1998). Like the Cooperation Plan, the RHAP exemplified a "soft law" approach and was also not a legally binding agreement (Tay, 2008). It was left to the governments concerned to decide what was to be included in their national plans, with the freedom to bypass or equivocate on matters raised in the RHAP (D. S. Jones, 2006). There were no mechanisms under the plan for any member country to ensure that the other member countries fulfilled their obligations (Parliament of Singapore, 1998). As a result, most of the ASEAN initiatives on haze were useful in generating a massive amount of information on the issue (Severino, 1999) but not in terms of effective implementation of haze mitigation activities. The focus on national plans and the lack of legally binding agreements ensured that states were largely free to pick and choose the regional initiatives that best suited their narrow economic national interests.

The 1997–98 haze episode, which was the most severe the region had witnessed at the time, sparked renewed outcry from the public and civil society. This backlash prompted AMS to agree to establish a legally binding mechanism to address haze and appease civil society. In 2001 the ASEAN Agreement on Transboundary Haze (ATHP) was proposed, to provide legally binding support for the RHAP (Florano, 2003). The ATHP is notable for being one of the few legally binding ASEAN environmental agreements to come into force (ASEAN Secretariat, 2004). The agreement was ratified by all AMS in 2014, with Indonesia being the final member to accede (Soeriaatmadja, 2014). Even when the AMS agreed to adopt the legally binding ATHP, however, this was done while still closely adhering to the ASEAN Way. The result was a seriously watered-down document that lacked hard legal mechanisms and that continued to protect current national economic interests (Varkkey, 2016a).

Furthermore, Indonesia, the country which accounts for the highest amount of smoke haze, delayed ratifying the ATHP for more than a decade after the agreement came into force. Indonesia's prolonged snub of the ASEAN Way principle of consensus (after the other nine AMS ratified the agreement) did not incur any serious consequences for the country, even though the limited early success of the ATHP in curbing regional haze was often linked to Indonesia's non-ratification. This further reinforced the argument that states who do not comply with the ASEAN Way, will not suffer any ill-effects. Indeed, even after ratifying the ATHP, Indonesia did not make any overt efforts to comply with the areas within the Agreement that it had wilfully ignored before ratification.[1]

Patron–client networks (both local and cross-border) within the region's plantation sector have had a strong influence over the national interests of Indonesia, Malaysia, and Singapore. This is due to the mutually beneficial relationship between government officials (patrons) and well-connected businessmen (clients). Hence, elite-centric issues such as continued access to, and ability to exploit, natural resources took priority over citizens' health and safety. Because of this, the states involved *chose* to adhere to the ASEAN Way when dealing with the haze. By insisting on ASEAN Way principles like national sovereignty and self-determination (Narine, 1998), AMS have been able to shape collective mitigation initiatives at the ASEAN level in accordance with their most pertinent interests at the time — those of the elites.

For example, one important area which has yet to be institutionalized at the ASEAN level is the measurement of haze. SDG Targets 3.9 and 3.d, as prioritized by the Malaysian government in combating haze, highlight the importance of managing risks to prevent the adverse effects of bad air quality. Proper and standardized air quality measurement systems are important to accurately manage risks. While the three major countries affected by haze have air quality measurement stations set up in strategic urban and rural locations, there is no ASEAN standard for measuring the haze. While Singapore and Indonesia

1 See detailed discussion of Indonesia's post-ratification behaviour in Varkkey (2017).

currently use PM (particulate matter) 2.5 to measure air quality, Malaysia is still using PM10, citing yet-to-be-upgraded monitoring facilities. A measurement of PM2.5 is considered to be world standard and more accurate because it can measure the smaller and more dangerous particles (up to 2.5 microns as opposed to 10 microns) which can more easily enter the body through respiration. This divergence in measurement systems explains the sometimes drastic difference of air quality indicators in cities as close to each other as Singapore City and Johor Bahru across the causeway. It has been argued that, for Malaysia, a PM10 measurement usually results in a less severe reading; this gives a positive (good for tourism and business) but misleading reading which may present the problem as less severe than it really is (Wong, 2015), with potentially dire consequences, especially to health.

Hence, it is not surprising that the SDG Goal 16 for Strong Institutions, a high priority for the United Nations, is not on the radar of the regional organization in charge of haze mitigation, ASEAN. This highlights a fundamental non-compatibility between the two institutional ideologies which raises the question of whether, in such cases, the ideals of one can be used to inspire the other. It also highlights the fact that even though the SDGs are supposed to be the embodiment of universal values regarding sustainable development, some of these values — like the importance of strong institutions in this endeavour — may not, in fact, be so universal.

3.2.2 Aversion to Interference

Another aspect of compatibility and complementarity between the UN's 2030 Agenda and ASEAN's Vision 2025 is that the former focuses on the global community, while the latter focuses on the regional community. The SDGs are meant to be aspirational goals, which should then be operationalized at the regional level (UN ESCAP, 2017). However, problems may arise if the aspirations behind the goals are not shared between the global and regional level. Although the SDGs were unanimously agreed upon at the United Nations General Assembly, in reality, some regions may not share some of these aspirations. Letchumanan (2015), a former ASEAN Secretariat staff member turned academic, speaks from experience when he claims that there is a general feeling among the AMS that fires and haze are rather unrelated to global issues. This chapter started out by explaining how, at the global level, air pollution reduction is an integral element of sustainable development. However, if the AMS do not acknowledge haze as a global air pollution issue, these two levels of governance may be speaking at cross purposes.

The literature review section above highlighted the neo-colonialist sentiments rooted in the non-interference norm in ASEAN. Nowhere is this more obvious than in the ongoing scientific and political debate on whether peatlands should be used for commercial agricultural development (Kaat, 2010; Schreier-Uijl et al., 2013). While peatlands play an important role in the global climate balance, they are also potentially valuable new frontiers for agricultural development. Many of the peat scientists who argue that tropical peat is unsuitable for agricultural development happen to be of Western origin. A group of Malaysian peat researchers oppose these arguments on developmental grounds, arguing that peatlands can and should be developed for agriculture, especially palm oil (Melling, Chua, & Lim, n.d.). This group of researchers received significant funding from the state government to host a recent major international conference focusing on peat. Having won the bid, the researchers used this platform to avidly and openly push their agenda at the international level (Wijedasa et al., 2016). The same group of researchers are now being engaged by the Indonesian government to help restore and manage Indonesia's degraded peatlands (Sun Daily, 2016).

Such opposition to the Western position, coupled with the AMS' tendency to view external interference with suspicion, further calls into question the basic compatibility between the global SDGs (agreed upon in a forum which is traditionally considered to be controlled by 'the West') and regional goals, and indeed casts doubt on whether AMS actually see the region's fires and haze as related to global issues at all. Even though it is far-fetched to assume that AMS view the SDGs as 'interference', deeply ingrained sentiments such as these could inhibit the potential of the aspirational SDGs to act as a global driver in operationalizing and implementing effective solutions to this regional air pollution problem. In short, it is possible that ASEAN's aversion to interference could result in resistance to ideas of sustainable development that originate from outside ASEAN, including the SDGs, especially when they are at odds with regional understandings of natural resource use and its link to economic progress. Such 'right to development' views held by many Southeast Asian states are linked to another aspect of regional haze governance, the tension between the economy and the environment, as expounded in the next section.

3.2.3 Institutional Structure

At the global level, the SDGs were crafted according to the understanding that the three dimensions of sustainable development — economic, social, and environmental — are indivisible, and that issues within these dimensions should be approached in an integrated and balanced manner (UN, 2015). In practice, however, the AMS have agreed that the region's broad environmental agenda should be treated solely as an element under the socio-cultural pillar, arguably the least important pillar within the Community (Tay, Lee, & Yi, 2017).

The shortcomings of the pillar-based structure are highlighted by the issue of haze pollution. The fires and haze are human-made disasters, reflecting the underlying unsustainable systems of production in the region's plantation sector, making this very clearly an economic issue (Tay et al., 2017). At the same time, the severe impact on public health highlights the human security dimension of the haze. Hence, the haze is indeed a cross-cutting issue. However, despite its clear linkages to economics (and to a lesser extent, security), haze is only discussed among environment ministers under the socio-cultural pillar (see Figure 11.4). Environment ministries in the region are generally ill-equipped to enforce existing regulations or to design, implement, monitor, inspect, and enforce effective environmental and sustainable development policies. Furthermore, because environmental protection is still regarded as a niche area, these ministries often languish in the lower ranks of the government hierarchy with relatively little power (Anbumozhi, 2017).

The ASEAN Community does in fact have several mechanisms to coordinate work on cross-cutting issues parked under one pillar. The first is the ASEAN Coordinating Council (see Figure 11.5). The Council is represented by Foreign Ministers and meets at least twice a year. It is tasked with (a) coordinating the implementation of agreements and decisions in the ASEAN Summit; and (b) coordinating ASEAN AEC, APSC, and ASCC Community Councils to enhance policy coherence, efficiency, and cooperation among them. However, the Coordinating Council functions more as a series of high-level meetings rather than as a working-level forum within ASEAN at which national focal points can convene for in-depth discussions on how to coordinate cross-pillar actions. To date, these meetings have not gone far beyond preparing documentation and reports for the ASEAN Summit. For example, the environment ministers report to the Council, which will receive a separate report from the AEC ministers. At this stage, it is extremely challenging to try to address cross-cutting issues (Tay et al., 2017).

The second is the ASEAN Secretariat. It is important to have a transparent, accountable, and above all strong Secretariat that could potentially tip the balance towards sustainability objectives. However, despite having responsibility for inter-pillar coordination, the Secretariat is generally known to be understaffed and under-resourced, considering the scale of its operations and responsibilities (Anbumozhi, 2017). AMS have deliberately denied the Secretariat the resources and mandate necessary to ensure good governance at the regional level (Kim, 2011: 422). In reality, the Secretariat is subordinate to national secretariats (Beeson, 2007; Solingen, 1999), even though on paper this relationship is one of coordination (see Figure 11.5). Decision making in ASEAN is thus shaped according to the interests of the AMS, and the Secretariat (and

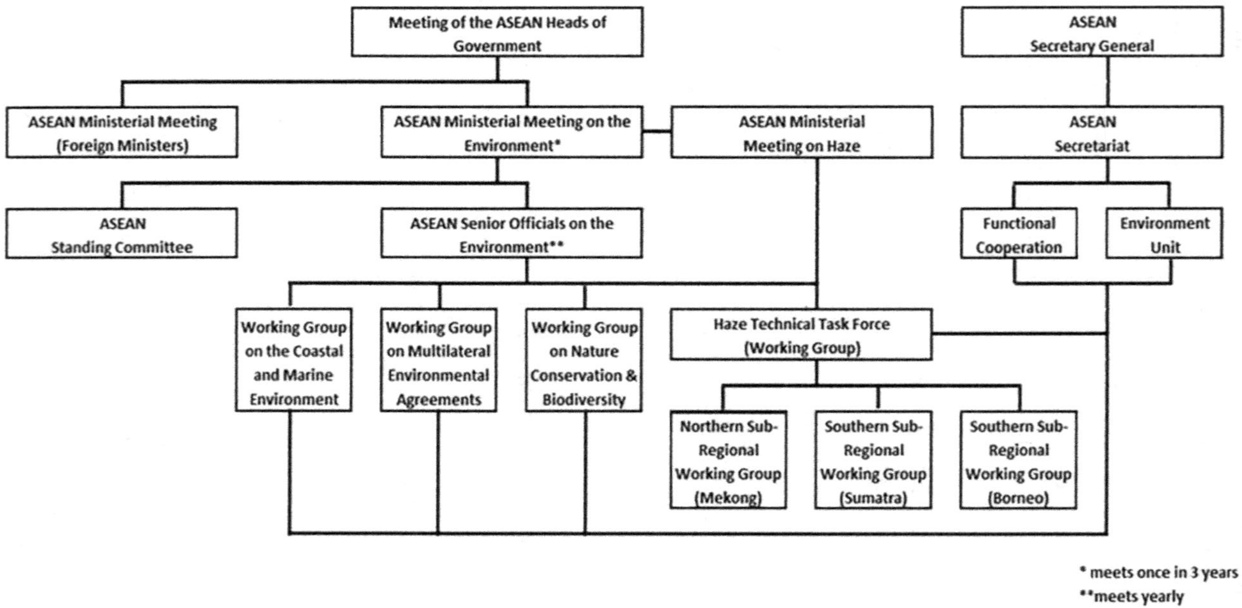

FIGURE 11.4 Organizational structure of ASEAN Environmental and Haze Cooperation
SOURCE: ASEAN SECRETARIAT, 2002 IN VARKKEY (2016A)

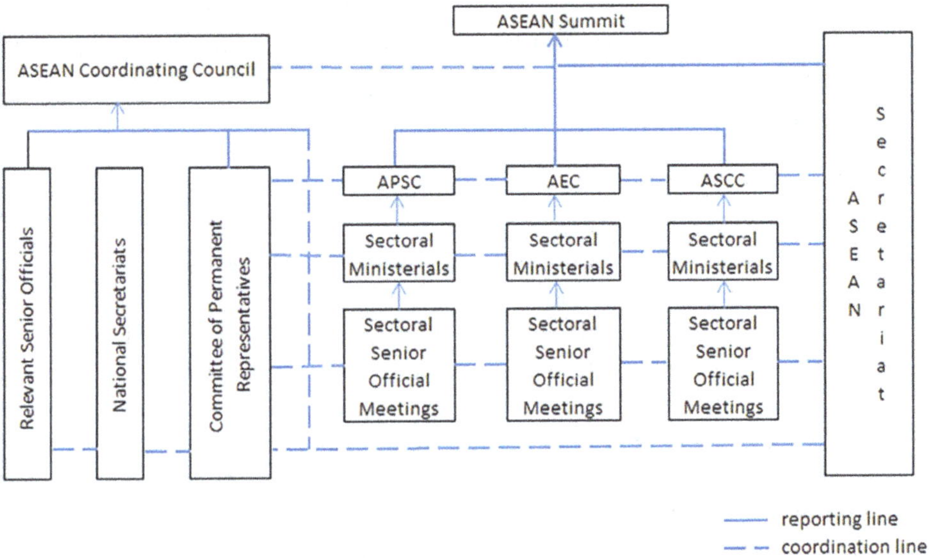

FIGURE 11.5 Coordination mechanism of ASEAN integration
SOURCE: ADAPTED FROM FRENCH (2016)

by extension the organization) has an inherently weak capacity to act in the regional interest (Varkkey, 2016a). A clear illustration of this is the Secretariat's very limited organizational capacity to determine the direction of negotiations of the ATHP, or to push for more stringent outcomes. Throughout this process, AMS sought to protect elite interests by maintaining their power of veto (by denying the Secretariat a strong mandate) over effective policy innovation on haze mitigation (Cotton, 1999). For example, the Secretariat currently also houses the ASEAN Haze Coordinating Centre. Indonesia's ratification of the ATHP was meant to kick-start the process of a separately staffed Centre in Riau, but this is yet to materialize and the Secretariat continues to shoulder this responsibility with its limited resources (Varkkey, 2017).

There is also a lack of formal mechanisms for civil society participation in ASEAN. Civil society, including businesses and other stakeholders, could play a useful role in consultation and implementation of regional objectives towards achieving the SDGs. However, ASEAN has long been considered an elitist organization, with a clear division between the elite leaders and the wider society. Indeed, for the first 30 years of ASEAN, there was a mutual distrust between civil society and ASEAN leadership. This is changing slowly; Article 13 in the recently adopted ASEAN Charter does include some language on people's participation in ASEAN. Platforms for civil society engagement with ASEAN have been established, such as the annual ASEAN Civil Society Conference and the

ASEAN People's Forum. However, these have not been operationalized in the ASEAN structure (Wahyuningrum, 2013), as illustrated above. In the context of haze, ASEAN has mostly been engaging with one civil society group, the Global Environment Centre (GEC). GEC, an NGO focusing on peatland conservation in the region, is recognized as an official partner of ASEAN for haze mitigation (ASEAN Secretariat, 2003). While there are many other civil society groups in the region focusing particularly on haze issues — including CERAH (Malaysia) and P. M. Haze and H.E.A.T (Singapore) — these groups have only limited access to the regional organization.

It can thus be argued that the sectoral, elitist, and pillar-based nature of the ASEAN structure does not fit well with broader global understandings of sustainable development as a cross-cutting issue. While this sectoral structure has served ASEAN well in promoting conventional socio-economic development in areas such as trade and education, it is not suitable for promoting sustainable development, which requires more cross-sectoral coordination and policy integration (Olsen et al., 2015). Taken together with the controversial nature of the problem, this makes haze management an unlikely candidate for green integration at the regional level.

4 Conclusion

ASEAN was a late adopter of the predecessor to the SDGs, the Millennium Development Goals (MDGs). There was an almost 10-year delay between the global adoption of the MDGs and the integration of the MDGs into ASEAN's work (Kojima & Bengtsson, 2015), with the Joint Declaration on the Attainment of the Millennium Development Goals in ASEAN in 2009. While ASEAN was quicker to integrate the SDGs into its agenda, evidence from ASEAN's MDG experience should be a reliable indicator of the importance placed by ASEAN on global goals as galvanizers of support and guiders of actions at the local level. While the haze problem can be framed through multiple SDG goals, there are several deep-rooted normative and structural issues within ASEAN which are likely to reduce the usefulness of the SDGs in achieving a transboundary haze-free ASEAN by 2020. It is telling that the Roadmap towards this vision of a haze-free region itself makes no mention of the SDGs in its 32-page document (ASEAN Secretariat, 2016). Transboundary haze is viewed by AMS as a uniquely regional problem with uniquely regional sensitivities. This explains the general wariness and reluctance of ASEAN and AMS to accept outside help, for example, from agencies like the United Nations through mechanisms such as REDD+, despite their own inability to resolve the problem at the regional level for more than three decades (Nathan, 1998). With this track record, it is unsurprising that global goals such as the SDGs seem to be gaining little traction in the actual operationalization of haze mitigation activities, apart from occasionally being acknowledged in documents.

Overall, this chapter cautions against over-emphasizing the importance of the SDGs in haze mitigation. The findings of this research indicate that solutions for haze should come from within ASEAN and AMS, not from outside. The likelihood of such an outcome is higher if the issue is not couched in 'us against them' terms, coloured by anti-Western sentiments. As long as ASEAN's non-interference norms and ideas of environmental exploitation for economic growth prevail, there seems to be limited opportunity for global goals such as the SDGs to play a serious role in galvanizing support and guiding actions at the regional levels of governance.

That being said, this does not mean that efforts to promote global, holistic ideas of sustainable development at the ASEAN level should stop. AMS' frames of thinking must continue to be encouraged to shift from seeing the haze as an environmental issue only, to one that is inextricably linked to economic and social concerns — in other words, to sustainable development (Tay et al., 2017). Universal goals such as the SDGs could have a positive motivational effect on ASEAN in the quest for sustainable development, but may be still ahead of their time in the region. This does not need to be permanent: countries like Singapore have shown evidence of shifting their priorities away from the strictly economic towards a more holistic concern for the well-being of their society in the context of haze (Varkkey, 2016b) and other countries could follow. However, such major paradigm shifts take time and effort from communities at all levels — national, regional, and global.

References

Adnan, H. (2012, January 25). Cries of EU biofuel discrimination grows louder. *The Star*.

Anbumozhi, V. (2017). Ensuring ASEAN's sustainable and resilient future. In M. Baird & L. Maramis (Eds), *Building ASEAN community: Political-security and socio-cultural reflections*, Vol. 4 (pp. 309–323). Jakarta: Economic Research Institute for ASEAN and East Asia.

ASEAN Secretariat (1995). ASEAN meeting on the management of transboundary pollution. Kuala Lumpur: ASEAN.

ASEAN Secretariat (2002). ASEAN agreement on transboundary haze pollution. Kuala Lumpur: ASEAN.

ASEAN Secretariat (2003). ASEAN peatland management initiative. Manila: ASEAN.

ASEAN Secretariat (2004). 4: Transnational issues. In *ASEAN Annual Report* (pp. 1–10). Jakarta: ASEAN Secretariat.

ASEAN Secretariat (2008). Information on fire and haze. HazeOnline, Jakarta. Retrieved from https://haze.asean.org/.

ASEAN Secretariat (2012). Cebu Declaration on the acceleration of the establishment of an ASEAN community by 2015. Retrieved from http://asean.org/cebu-declaration-on-th-acceleration-of-the-establishment-of-an-asean-community-by-2015/.

ASEAN Secretariat (2015). Declaration on ASEAN post-2015 environmental sustainability and climate change agenda. Kuala Lumpur: ASEAN.

ASEAN Secretariat (2016). Roadmap on ASEAN cooperation towards transboundary haze pollution control with means of implementation. Retrieved from http://asean.org/wp-content/uploads/2012/05/Haze-Roadmap.pdf.

Bank Indonesia (2014). National accounts. Retrieved from http://www.bi.go.id/sdds/series/NA/index_NA.asp.

Beeson, M. (2007). *Regionalism and globalization in East Asia: Politics, security and economic development.* New York: Palgrave.

Clean Malaysia (2015). Just how big is Malaysia's palm oil industry? Retrieved from: http://cleanmalaysia.com/2015/12/09/just-how-big-is-malaysias-palm-oil-industry/.

Colfer, C. J. P. (2002). Ten propositions to explain Kalimantan's fires. In C. J. Colfer & I. A. P. Resosudarmo (Eds), *Which way forward? People, forests and policymaking in Indonesia* (pp. 309–321). Singapore: Institute of Southeast Asian Studies.

Cotton, J. (1999). The "haze" over Southeast Asia: Challenging the ASEAN mode of regional engagement. *Pacific Affairs*, 72(3), 331–351.

Das, N. (2014). Creating demand for sustainable palm oil through tariff policies in India & Indonesia. Oxford: GIST Advisory.

Eaton, P., & Radojevic, R. (2001). *Forest fires and regional haze in Southeast Asia.* New York: Nova Science Publishers, Inc.

Elliott, L. (2003). ASEAN and environmental cooperation: Norms, interests and identity. *The Pacific Review*, 16(1), 29–52.

Florano, E. R. (2003). Assessment of the 'strengths' of the new ASEAN agreement on transboundary haze pollution. *International Review for Environmental Strategies*, 4(1), 127–147.

French, C. (2016, November 22). ASEAN: A community of opportunities. [Presentation] Retrieved from http://slideplayer.com/slide/9294130/28/images/15/.

Jones, D. M., & Smith, M. L. R. (2002). ASEAN's imitation community. *Orbis*, 46(1), 93–109.

Jones, D. S. (2006). ASEAN and transboundary haze pollution in Southeast Asia. *Asia Europa Journal*, 4(3), 431–446.

Kaat, A. (2010). Expansion of oil palm plantations on Malaysian peatlands based on faulty science. Retrieved from: https://malaysia.wetlands.org/News/tabid/509/ID/2488/ramsar.wetlands.org/.

Karim, M. S. (2008). Future of the Haze Agreement: Is the glass half empty or half full? *Environmental Policy and Law*, 38(6), 328–334.

Kim, M. (2011). Theorizing ASEAN integration. *Asian Perspectives*, 35, 407–435.

Kivimaki, T. (2001). The long peace of ASEAN. *Journal of Peace Studies*, 38(1), 5–25.

Koh, K. L. (2008). Regional and state level environmental governance. ASEAN's environment governance: An evaluation. Paper presented at the UNITAR/Yale Conference on Environmental Governance and Democracy, New Haven, CT, 10–11 May.

Kojima, S., & Bengtsson, M. (2015). Greening Asia's integration: An urgent challenge. In H. Hamanaka & H. Mori (Eds), *Greening Integration in Asia: How regional integration can benefit people and the environment.* (pp. 1–10). Kanagawa: Institute for Global Environmental Strategies (IGES).

Leigh, M. (1998). Political economy of logging in Sarawak, Malaysia. In P. W. Hirsch & C. Warren (Eds), *The Politics of Environment in Southeast Asia* (pp. 93–106). London: Routledge.

Letchumanan, R. (2015). Towards a transboundary haze-free ASEAN by 2020: Prevention and collaboration. *RSIS Commentaries, 243.*

Lode, B., & Toussaint, P. (2016). Clean air for all by 2030? IASS Policy Brief. Potsdam: Insitute for Advanced Sustainability Studies (IASS).

Mayer, J. (2006). Transboundary perspectives on managing Indonesia's fires. *The Journal of Environment & Development*, 15(2), 202–233.

Melling, L., Chua, K. H., & Lim, K. H. (n.d.). Managing peat soils under oil palm. Retrieved from http://tropicalpeat.sarawak.gov.my/modules/web/pages.php?mod=download&sub=download_show&id=16.

Mohamed, M. (1992). Statement by His Excellency Prime Minister Dr Mahathir Mohamed of Malaysia at the United Nations Conference on Environment and Development. Rio de Janiero: UNCED.

Murray, P. (2010). The European Union as an integration entrepreneur in East Asia: Yardstick or cautionary tale? Paper presented at the Australian Political Studies Association Conference, Melbourne, 27–29 September.

Naidu, S. (2016, August 11). ASEAN adopts roadmap for transboundary haze-free region by 2020. *Channel NewsAsia.*

Narine, S. (1998). ASEAN and the management of regional security. *Pacific Affairs*, 71(2), 195–214.

Nathan, D. (1998, April 6). Asean gets UN to help keep haze out. *The Straits Times.*

Nischalke, T. I. (2000). Insights from ASEAN's foreign policy co-operation: The 'ASEAN way', a real spirit or phantom? *Contemporary Southeast Asia*, 22(1), 89–112

Nossal, K. R., & Stubbs, R. (1997). Mahathir's Malaysia: An emerging middle power? In A. F. Cooper (Ed.), *Niche diplomacy: Middle powers after the cold war* (pp. 147–163). London: Macmillan.

Olsen, S., Teoh, S., & Miyazawa, I. (2015). ASEAN community and the Sustainable Development Goals: Positioning sustainability at the heart of regional integration.

In H. Hamanaka & H. Mori *Greening Asia's Integration: How regional integration can benefit people and the environment*, (pp. 59–78). Kanagawa: Institute for Global Environmental Strategies (IGES).

Ong, K. Y. & Sané, P (2006). Foreword. Paper presented at the ASEAN-UNESCO Concept Workshop on Human Security in South-East Asia, Jakarta, 25–27 October.

Parliament of Singapore (1998). ASEAN region haze action plan (fulfillment of obligations) (1998–01–15). Singapore: Parliament of Singapore.

Razak, M. (2017). Haze and Sustainable Development Goals (SDGs). Paper presented at the UNESCO-IKIM Expert Group Consultation (ECG) on Bioethical Perspective of Haze, Institute of Islamic Understanding Malaysia (IKIM).

Schreier-Uijl, A. P., Silvius, M., Parish, F., Lim, K. H., Rosediana, S., & Anshari, G. (2013). Environmental and social impacts of oil palm cultivation on tropical peat: A scientific review. Kuala Lumpur: Roundtable for Sustainable Palm Oil.

Severino, R. C. (1999). Fighting the haze: A regional and global responsibility. Address at the Final Regional Workshop of the Regional Technical Assistance Project on Strengthening ASEAN's Capacity to Prevent and Mitigate Transboundary Atmospheric Pollution, Jakarta, 21 June.

Severino, R. C. (2006). *Southeast Asia in search of an ASEAN community: Insights from the former ASEAN Secretary-General*. Singapore: ISEAS.

Shaik Mohd Salleh, S. M. S. (2017). The bioethical dimension of haze: Cultural and bioethical perspectives. Paper presented at the UNESCO-IKIM Expert Group Consultation (ECG) on Bioethical Perspective of Haze, Institute of Islamic Understanding Malaysia (IKIM), 18 September.

Shelton, D. (2003). *Commitment and compliance: The role of non-binding norms in the international legal system*. New York: Oxford University Press.

Simamora, A. P. (2011, March 26). SBY vows to protect palm oil interests. *Jakarta Post*.

Soeriaatmadja, W. (2014, September 16). Indonesia's parliament agrees to ratify Asean haze pact. *Straits Times*.

Solingen, E. (1999). ASEAN, quo vadis? Domestic coalitions and regional co-operation. *Contemporary Southeast Asia*, 21(1), 30–53

Sun Daily (2013, June 23). PM declares haze emergency in Muar and Ledang. *Sun Daily*.

Sun Daily (2016, April 2). Indonesia to test M'sian peatland compaction method to prevent forest fires. *Sun Daily*.

Tay, S. S. C. (1998). South East Asian forest fires: Haze over ASEAN and international environmental law. *Reciel*, 7(2), 202–208.

Tay, S. S. C. (2008). Blowing smoke: Regional cooperation, Indonesian democracy, and the haze. In D. K. Emmerson (Ed.), *Hard choices. Security, democracy and regionalism in Southeast Asia* (pp. 219–240). Singapore: ISEAS.

Tay, S., Lee, C., & Yi, L. (2017). ASEAN approaches to environmental protection and sustainable development: Cooperating across borders, sectors and pillars of regional community. In S. Tay & J. P. Tijaja *Global megatrends: Implication for the ASEAN economic community* (pp. 98–122). Jakarta: ASEAN Secretatiat.

UN (2015). Transforming our world: The 2030 Agenda for Sustainable Development. New York: United Nations. Retrieved from https://sustainabledevelopment.un.org/post2015/transformingourworld.

UN (2016). Peace, justice, and strong institutions: Why they matter. New York: United Nations. Retrieved from https://www.un.org/sustainabledevelopment/wp-content/uploads/2017/01/16-00055p_Why_it_Matters_Goal16_Peace_new_text_Oct26.pdf.

UN ESCAP (2017). Southeast Asia subregion challenges and priorities for SDG implementation. Bangkok: United Nations Economic and Social Commission for Asia and the Pacific. Retrieved from http://www.unescap.org/sites/default/files/Southeast%20Asia%20Subregion%20Challenges%20and%20Priorities%20for%20SDG%20Implementation%202017.pdf.

Varkkey, H. (2016a). *The haze problem in Southeast Asia: Palm oil and patronage.* London: Routledge.

Varkkey, H. (2016b). Transboundary haze and human security in Southeast Asia: National and regional perspectives. *Georgetown Journal of Asian Studies*, 3(2), 42–49.

Varkkey, H. (2017). "In 3 years we would have solved this": Jokowi, ASEAN and transboundary haze. *Jurnal Studi Pemerintahan: Journal of Government and Politics*, 8(3), 277–295.

VNA (2015, November 16). Three pillars of to-be-formed ASEAN community. *Vietnam News Agency.*

Wahyuningrum, Y. (2013, November 22). ASEAN & civil society advocacy. Retrieved from https://www.slideshare.net/wahyuningrum1/civil-society-engagement-in-asean-yuyun-wahyuningrum.

Wijedasa, L. S., Jauhianen, J., Kononen, M., Lampela, M., Vasander, H., LeBlanc, M.-C., … Varkkey, H. (2016). Denial of long-term issues with agriculture on tropical peatlands will have devastating consequences. *Global Change Biology* (online first), 23(3), 977–982.

Wong, C. (2015, September 27). Malaysia must use global standards to measure haze. Wong Chen, Opinion. *Malay Mail Online*.

CHAPTER 12

Emerging Spaces of Citizenship: Grassroots Communities and Water Governance in Indonesia

Maharani Hapsari

1 Introduction

Debates on governance in the context of the Sustainable Development Goals (SDGs) in Indonesia and in Southeast Asia more generally highlight the need to connect plural modes of development (Thanh, 2008). This includes discussions on strong democratic participation through formal and informal political means that link citizens and the state. There has been increasing scholarly and policy attention to *bottom-up* approaches. Participatory development, grassroots innovation, and community development are some of the terms that converge into a normative vision in which citizens assume substantive roles in everyday development policy making (Brecher and Costello, 1998; Christopherson, 2008; Cornwall and Brock, 2005; Seyfang and Smith, 2007). This is a response to the limits of the *top-down* approach in global poverty eradication. In the words of Escobar (1992), development emancipation only makes sense when individuals actively participate in transforming their everyday life and practices.

This chapter discusses the role of local actors, namely grassroots communities,[1] in shaping the politics of SDGs with regard to achieving Goal 6 (clean water and sanitation). It examines forms of citizenship exercised in the relations between grassroots communities and the state. It focuses on political strategies that allow the development agenda to be shaped in daily political practices, and explores challenges within the movements as they seek to transform state–society relations. Working on state–society relations as an arena of political transformation, the question which this chapter seeks to explore is: how have grassroots communities engaged in water governance and what forms of citizenship are invented in the interpretation of SDGs as a global development policy framework in the local context?

1 The term 'grassroots communities' refers to the "realm of individuals within the communities in which they live, of face-to-face relationships, of small-scale meetings where most people in the room know one another's name" (Read and Pekkanen, 2009, p. 1).

 | DOI:10.1163/9789004391949_013

The chapter draws on a case study to capture the interpretation of 'development from below', using primary data taken from personal communications, focus group discussions (FGDs) with grassroots community activists working on water issues in Daerah Istimewa Yogyakarta Province, notes from workshops, and field observation since January 2017. Secondary data are taken from national and local government documents, community organizations' websites, and mass media articles covering the activities of grassroots communities. The analysis assembles the views of grassroots community members on what makes water political, how relations between society and the state are restructured, and the areas of contestation around which grassroots actors organize themselves to achieve political objectives.

2 Interpreting SDGs at the Grassroots: Literature Review

Interpreting the SDGs demands a shift of focus from state-centric analysis to one that includes the views of more diverse actors in everyday policy making. As Falk argues, there are two major narratives of globalization that shape individuals and groups in today's world: *globalization-from-above* which relies on the preoccupation of leading state actors and agents of capital formation; and *globalization-from-below*, which is being shaped by transnational social forces influenced by such concerns as the environment, human rights, resistance to patriarchy, and community diversity (Falk, 1993, p. 39). Not only is the latter more inclusive in terms of actors, but understanding development from below is also inseparable from examining the political context that informs citizen participation. This may be associated with exclusionary practices that disadvantage citizens and with the presence of proper government facilitation (Makhoul, 1999; Nel, Hill, & Binns, 1997). Hickey and Mohan (2005) argue for a micro-politics of community practices that works in relation to good governance and the state. Different modes of engaging with development processes are seen as ways to create spaces of transformation, and this also requires recognizing the issues of power and difference (Cornwall, 2005). There is a need to examine and redefine the concept of development when dealing with everyday challenges of poverty eradication.

In order to capture the agency of citizens in responding to various development challenges, the notion of citizenship itself is important. Citizenship is associated with status and practice, which situate citizens in the arena of contestation over development policies. Stokke sees two major strands in citizenship studies: a cultural turn, which highlights cultural differences and group-differentiated rights within the liberal citizenship model; and a global turn, which envisions postnational, denationalized, and transnational forms of citizenship (Stokke, 2017, p. 24). Both cultural and global turns see citizenship not as given, but as subject to continuous contestation.

Instead of seeing status and practice as merely the logical consequences of rights assigned by the state, more effort must be directed towards understanding status and practice as the product of power struggles, in response to the prevailing political order of development policy making. Escobar advocates alternative visualizations of development, as an interest in culture and local knowledge, a critical stance with respect to established scientific discourse, and the defence and promotion of localized, pluralistic grassroots movements (Escobar, 1992, pp. 418–419). Questions converge around "who should participate in decision-making; at what geographical and political scales should governance institutions operate; and what is the appropriate role of market or non-market criteria in allocation of water" (Woodhouse & Muller, 2017, p. 226).

Modern citizenship embodies four interconnected dimensions: membership, legal status, rights, and participation (Stokke, 2017, p. 25). Citizens reproduce the boundaries of their belonging, in which they are able to see the scope of their political influence within a particular civic space. The reproduction process considers the local context of participation as key to democratic governance with a precondition of decentralizing modes of rights fulfilment and participation among communities at the grassroots and in their relations to the state (Jones & Gaventa, 2002). In constructing the citizenship project, there are links between the particularity that characterizes communitarianism and the social drive towards general society (Procacci, 2006).

Development policy making involves forces of inclusion/exclusion and the management of social differences. In this chapter, the participation of grassroots communities is examined critically, as it mostly takes place within unequal power relations. It is also subject to the exclusion of others, or making some segments in the grassroots more knowledgeable than others (Jones & Gaventa, 2002, p. 23). Harris, Stokke, and Törnquist (2013) suggest three crucial factors in the interplay of strategies and capacities of political actors. These include the location of political actors in the political terrain, and the politicization of issues, interests, ideas, and identities. The third factor is concerned with how and at what levels actors mobilize support for their policies (modes of political inclusion) (Harris, Stokke, & Törnquist, 2013, pp. 19–20).

There are various political aspects that can be explored in the formation of citizenship. Buytaert et al. (2014) focus on how citizens engage in active scientific knowledge-generation processes. This includes joint knowledge creation and exchange, strengthening legitimacy of generated knowledge, improving citizens' access to decision making, feedback loop mechanisms, and the fostering of long-term citizen involvement. There are also a number of more operational strategies through which spaces of citizenship are being created, such as participatory appraisal, participatory budgeting, community workshops and seminars, all of which enable the formation of a public sphere (Jones & Gaventa, 2002, p. 22). The scope of transformation is not limited to citizen–state engagement, but includes the power relations in participatory spaces in which citizens build collective action.

3 The Water Policy Landscape

Yogyakarta Special Province is facing rapid demographic change as a result of a massive inflow of population to the region. In 2016, the population of Yogyakarta was 3.7 million people, living in one city and four districts (Statistics Bureau of D. I. Yogyakarta Province, 2016). Yogyakarta is home to both permanent and temporary residents who all depend on good-quality water and sanitation. This demographic factor intertwines with a socio-economic factor brought about by globalization. The government currently acts as water provider to 140,663 households, mobilizing supply from ground water, springs, river, and reservoirs (Statistics Bureau of D. I. Yogyakarta Province, 2015). The rest of the population relies on private providers and community-based water providers to access water services. The tourism industry produces competing demands, with the need to accommodate around 2.5 million domestic and international tourists annually.[2] Ensuring an adequate quantity and quality of water, therefore, is a priority.

Current challenges to water provision within the political domain of Yogyakarta Special Province need to be situated within a broader context of development politics in Indonesia. The country's dominant development paradigm, as made explicit in Article 33 of the Constitution of the Republic of Indonesia, sees water as a resource controlled by the state to be continuously reproduced for the benefit of the people. When translated beyond the legal text, there are multiple dimensions to be considered in asserting the role of the people in policy making, from local to national level. Water as a basis of

2 The number of star-rated hotels in 2016 was 54 units, with 5,460 rooms. The number of non-starred hotels was 521 units, with 9,963 rooms (Statistics Bureau of D. I. Yogyakarta Province, 2016, pp. xvii–xviii).

the nation's livelihoods is organized across a number of vertical and horizontal policy-making processes, in which citizens are embedded.[3]

While lack of infrastructure remains the challenge for many, the distribution of access and control over water resources has become increasingly salient. Data from the National Development Planning Agency (BAPPENAS) show that 71.14% of Indonesians have access to drinking water, while 28.86% of the population (75.35 million people) are still deprived of such access. With regard to sanitation, data from the Ministry of Health show that as recently as 2015, some 62 million people in rural Indonesia (53% of the rural population) did not have proper access to sanitation (Ministry of Health, 2016). These figures cast light on aspects of welfare that require the government and society to cope with social inequality.

From a top-down perspective, government engagement with the SDGs is mostly regarded as an effort to benchmark national policy practices with international development objectives. National targets in the water and sanitation sector include management and conservation of rivers, dams, and other water reservoir facilities. In terms of resource allocation, the budget for water governance is mostly derived from the national government, followed by local government and the private sector (Kusmayadi, 2015). Given the extensive coverage of the SDGs, policies around water and sanitation co-exist and intersect with policies from different sectors.

Indonesia's national regulatory framework is historically built upon Water Law 11/1974, which allowed the government to utilize water resources for the purpose of supporting the national development agenda. The shift from centralized to decentralized politics since the end of the New Order era has brought mixed results across different subnational contexts, particularly with

3 Nationally, authority to regulate water resources is distributed across six ministries: Public Works, Forestry, Environment, BAPPENAS (National Development Planning Agency), Agriculture, and Home Affairs. The Directorate General of Water Resources under the Ministry of Public Works is responsible for large water infrastructure projects. At the provincial level, authority lies in the hands of the Office of Water Resources Management (mostly referred to as Dinas SDA) and the Office of Public Works (Dinas PU). At the district and municipality level, the distribution of tasks is organized on the basis of rural–urban territorialities. Districts tend to deal more with irrigation policy and forestry issues while the municipalities handle urban water problems and sanitation. Apart from these specific government agencies, there are also other government actors involved in water resources management. One important group of actors are the so-called River Basin Organizations (RBOs). These include: Balai Pengelolaan Sumber Daya Air (Balai PSDA), which functions as provincial authority for water resources management; PJT (Perum Jasa Tirta), a joint governance between state and enterprises/corporations under the supervision of the Ministry of State Owned Enterprises; BBWS (Balai Besar Wilayah Sungai), in charge of major river basins; and BWS (Balai Wilayah Sungai), which has responsibility for minor river basins. For a more detailed explanation, see Wieriks (2011, pp. 20–22).

regard to the reshaping of relations between citizens and the state in managing the allocation and redistribution of water resources. In 2004, the government enacted Water Law 7/2004, which regulated water conservation, the utilization of water for livelihoods, and control over water-related hazards. In 2015, political contestation caused the Constitutional Court to annul this Law and to reinstate Water Law 11/1974.[4] Article 7 of Water Law 7/2004 makes provision for water utilization rights and water business rights. The business rights are provided by national or local government, depending on the appropriate scope of authorization needed by the private sector. Many civil society elements claim that this provision is unconstitutional, as the provision of water business rights is likely to result in a diminishing of the government's role in meeting the rights of citizens.

In addition to the role of the state and the communities, the role of the private sector in water governance in Indonesia is worthy of note. There has been a mixed public response to the role of the private sector in water provision. In places without state-driven water infrastructure projects, the presence of the private sector is very often inevitable, but private water providers are also present in places where state infrastructure is already available. Privatization on a large scale may have profound implications for existing forms of water provision organized by the government and communities. A strong state position is needed to ensure that the role of the private sector in water management meets the state's social objective, which is to provide equal basic access for all citizens.

Waves of privatization have taken place in various regions in Indonesia since 1997 when the country entered into its democratization era. Problems around the negative impacts of the privatization of water service, however, have led to social protest and a citizen lawsuit in 2012 against two private companies operating in Jakarta. In October 2017, the Constitutional Court ordered the termination of water privatization and the restoration of public management (Zamzami, 2017). At the time of writing (mid-2018), a new Water Law is being drafted. Communities negatively affected by privatization have expressed a strong desire to return water services to the public domain, restoring to the state the responsibility for ensuring universal access to water (Marwa, 2017). In the face of such aspirations, a clear indication by the government of how the role of communities is being taken into account is notably absent.

4 The court's ruling emphasized the state's presence in water management. State control over water is manifest in policy making, supervision, regulation, management, and water affairs. Consumers should be freed from paying water management costs. If the water source is insufficient, the state is responsible for providing clean water through water distribution (Ardhianie, 2015).

Indonesian society has witnessed a normative shift in relations between state and citizens from an autocratic mode of governance to one that is characterized by a search for democratic governance. Table 12.1 shows the shifting water issues across major historical episodes in the post-independence era (Old Order, New Order, and the period of *Reformasi*). It is important to examine further the implications of this shift for the development of grassroots initiatives.

TABLE 12.1 Shifting modes of water governance in Indonesia

Period	Modes of governance	Water issue	Development approach
Old Order (1945–1967)	– Autocracy (personal leadership) – Top-down decision making – Concentration on national administration – Limited role of civil society	– Massive engineering projects – Legacy of neglect of water infrastructure during the colonial era	– Nationalization of water management, construction, and development of public works – Techno-centric approach with the support of anti-Western nationalistic policies
New Order (1968–1998)	– Autocracy (personal leadership) – Top-down decision making – Concentration on national administration – Limited role of civil society – Greater role for national private companies	– Over-population (encroachment of the river banks, illegal housing) – Poor operation and maintenance of existing infrastructure	– Water as part of growth-oriented developmental state – Massive role of foreign donors from the Western countries (bilateral, multilateral and non-lateral) – Significant role of Dutch government as the catalyst of water-related policies especially in the urban areas – Increasing role of the Asian Development Bank and Japanese investors in water infrastructure projects in remote and rural areas
Reformasi (1998–present)	– Democracy (participatory leadership) – Bottom-up decision making – Increasing role of provincial and district administration – More active role of civil society – Private–public partnerships (PPP)	– Tension between executive power (the Ministry of Housing, Infrastructure and Planning) separated from the Ministry of Public Works and the decentralized agencies – The role of private sector in water resource management	– Major role of World Bank in the drafting of the Water Law 7/2004 – Water project funding by government – Introduction of Integrated Water Resources Management (IWRM) – Sporadic engagement of civil society in policy making

SOURCE: ADOPTED FROM FULAZZAKY (2014) AND WIERIKS (2011)

4 Emerging Grassroots

Grassroots community initiatives in the water sector are a response to the rapid pace of demographic change, industrialization, and modernization, all of which produce unequal development across different segments of society. They are products of power relations around the asymmetric distribution of material resources and uneven exposure to modern social forces, such as capital, knowledge, and technology. In Yogyakarta, grassroots communities around water issues develop sporadically within the population living on the periphery of modernizing urban areas, mostly associated with socially and economically vulnerable social groups.

This study documents the experiences of three such grassroots communities. The first is Paguyuban Air Minum Masyarakat Yogyakarta (PAMMASKARTA) or the Association of Yogyakarta Drinking Water Providers, which was established by Governor Regulation 27/2015 on the Provision of Drinking Water. Membership covers Yogyakarta Municipality and four other districts (Sleman, Bantul, Gunung Kidul, and Kulonprogo).[5] This association of grassroots communities was initiated through a university-based community programme, and later institutionalized under the supervision of the Provincial Office of Public Works. The governor regulation recognizes the legal right of this community to use local water resources for public service purposes. It also facilitates and supports capacity building. Organizing ground water extraction, members of PAMMASKARTA are connected to each other through knowledge exchange, capacity building in community-based water provision, and mobilization of access to the community-based water extraction technology that sustains their activity. There are currently around 1,000 grassroots communities registered in the Provincial Office of Public Works within the regulatory scope of PAMMASKARTA. Usually, water provision is organized in the form of water cooperatives. Each unit of PAMMASKARTA applies a tariff for users, which is below the tariff charged by the regional state water company (Perusahaan Daerah Air Minum, or PDAM). Consumers have regular access to water services and the tariff is determined through a collective process involving the cooperative management and users.[6]

The second is Banyu Bening Community, a grassroots initiative developed around alternative provision of fresh drinking water through rainwater

5 The model encourages synergy between grassroots communities and their village counterparts in devising programmes to deliver water to village inhabitants and to surrounding villages.

6 Interview with Unggul Anggito Ardi, PAMMASKARTA consultant, 4 May 2018.

harvesting.[7] The community started operating in 2012. It works on the premise that natural water needs to be retained as long as possible before it reaches the ground, making it useful for various purposes for the benefit of human livelihoods. For this community, managing rainwater is an important part of preventing water-related disasters (drought and floods) when it is applied to water-scarce areas. During the rainy season, when implemented on a large scale, rainwater harvesting may also regulate the water volume and prevent rivers from overflowing.

In this community, activities include promoting tree planting upstream as part of a water conservation strategy. There has been concern that rapid industrialization upstream has led to the disruption of water sources that are supposed to supply the needs of the population downstream. Artisanal mining practices around the summit of Mount Merapi have been of particular concern, as they are perceived to have adverse effects on water catchment. Conservation efforts range from planting tree species to rain harvesting to building environmentally friendly water drainage. The community does not have any formal organizational structure. Networking develops on the basis of intensive personal and informal relations among activists across localities in Indonesia. The community provides free drinking water to local households within a small outreach, maintaining that the rainwater that they harvest is not for sale.

The third grassroots community is the Association of River Community in Yogyakarta, or Asosiasi Komunitas Sungai Yogyakarta (AKSY). AKSY was formed in 2016; it currently consists of 16 communities working along the riverbank from upstream of the Mount Merapi summit to downstream. The association facilitates coordination among river activists in Yogyakarta, working towards a sustainable river that is based on strengthening capacity in every river community. Activities cover conservation, settlement arrangement, and a river school programme. The association aims to promote activities not only at the local level, but also at national and regional levels.[8]

These three communities are different in terms of how they seek to transform existing development politics. While Banyu Bening and PAMMASKARTA orient their programmes towards direct daily provision of water for communities, AKSY works on the level of connecting grassroots communities through policy advocacy by making connections with government agencies from

7 This community was founded in 2012 by Kamaludin and Sri Wahyuningsih in Tempursari Village, Sardonohardjo Sub-District, Sleman, Yogyakarta.

8 Minutes of AKSY's General Meeting, Yogyakarta, 7 December 2017.

various sectors. This diversity has developed gradually as these communities find their way through the maze of policy networks operating at the provincial and district levels.

5 Politics at the Grassroots

Power relations are a feature of the daily struggles over access and control of water provision as Yogyakarta Special Province evolves into a globalizing industrial site. The development of commercial water infrastructure projects, when not governed by strong state regulation, has led to competition over water resources between the affected communities and commercial water providers. This takes place within a complex governance scheme of water provision that involves various actors. Contestation emerges around the organization of knowledge, issue framing, and the formation of formal and informal political channels through which grassroots communities articulate their demands.

5.1 *From Technocratic Knowledge to Social Knowledge*

In many countries, modern technology has been at the centre of state policy making in the water sector, and the same is true in Yogyakarta Special Province. Scientists and academics generally have a prominent role in the formulation of a water policy framework. The provision of technology to produce fresh water, especially for urban citizens, has been organized as a public project driven by state investment. Technical aspects have been dominant in determining power relations among actors.

Grassroots communities are engaging in contestation of knowledge on the basis of which their relations to water as a development resource are being organized. The current situation indicates that these communities are struggling to make institutional science (dominated by scientists) more relevant to addressing problems around water provision and management. At the grassroots, there are various strategies to mobilize collective knowledge in order to address water and sanitation problems. Communities have different ways of engaging with modern science; this can be attributed to different ways of knowing local water problems and the ways that these differences are appreciated in the decision-making process. The development of grassroots activities is built upon fluid knowledge relations between communities, the state, and institutional scientists in academia or think tanks. Modern science is part of the daily interaction and there are efforts to incorporate it within the broader social and political context.

For grassroots communities, having limited access to modern science is not necessarily a constraint when it comes to meeting collective needs. Institutional knowledge tends to be centralized in the hands of development experts; it is rarely inclusive and is prone to marginalizing stakeholders. However, this does not necessarily undermine community capacity to assemble local knowledge initiatives. The distributive consequences of access to modern knowledge are also seen as an opportunity for local innovation, as one activist highlighted: "We shall not be conquered by science. If we conquer science, we can do innovation" (Sri Wahyuningsih, Banyu Bening Community).

Banyu Bening community applies electrolysis technology to turn harvested rainwater into potable water. Most of the appliances needed are available on the local market, although there is also the flexibility to use much more complicated material that can be sourced from other regions. The rainwater harvesting installation has been modelled on a prototype developed by academics at a state university in Yogyakarta; the dissemination of the technology is 'ear to ear' as volunteers interact with government officials, academics, non-governmental organizations (NGOs), and other public actors.

PAMMASKARTA utilizes various forms of technical knowledge supported by industrial materials commonly used in government-led projects.[9] Many water provision communities are using standardized technology introduced by the District Office of Public Works as they build water infrastructure facilities for local communities. There are, however, some units that have developed technologies using simple local materials that are suitable for providing water to a small number of households, for example using bamboo pipeline installations. This illustrates the unequal access to industrial knowledge that persists in the relations between water users and regimes of modern knowledge. Some members of PAMMASKARTA have mobilized financial resources by seeking loans from local banks to develop infrastructure.[10] Knowledge dissemination works through regular informal meetings among the management personnel representing each cooperative.

In AKSY, acquiring collective knowledge that allows communities to map and identify local issues in a systematic manner remains a challenge. There is a clear need to strengthen the management of collective knowledge so that issue identification can be organized through sustainable, accountable, and transparent platforms. Intra-community organization of knowledge is useful

9 The range of technology in PAMMASKARTA varies from the use of bamboo pipeline systems to modern water infrastructure installations to deliver water from springs to households.

10 Interview with Unggul Anggito Ardi, PAMMASKARTA consultant, 4 May 2018.

as it highlights detailed problems at a micro level. However, more needs to be done to connect with the policy level, and the role of the state as the ultimate provider of access to water. There has been an effort to link local-based knowledge that enables these communities to identify each other and to consolidate their very diverse activities. This is mostly done through initiatives from external actors when community-level data are needed to support donor-based or government-led projects. The technical and social aspects of river policies need to be balanced in order to improve the quality of people's livelihoods.[11]

Due to their limited scope of work, grassroots communities rarely engage in circulating public information on a wide scale. Elitism and the top-down approach continue to dominate the way information is produced at the local level.[12] Mostly, these communities rely on personal association to spread information; recently they have begun to access modern communication and information technology. There are sporadic initiatives to assemble collective knowledge, mostly by collaborating with NGOs and academia, leading to participatory mapping. However, there are barriers to mobilizing data at the community level that would allow for the identification of local issues. The main challenge of building a collective knowledge platform lies in the need for intensive coordination to ensure regular information production within reliable social parameters. Other concerns include making knowledge accessible to the public and ensuring that it contributes to collective mobilization strategies.

Communities have relied on government-produced information to understand the local context while continuing to search for mechanisms to connect different stakeholders. So far, each community collects documents using different methods, depending on the type of issue that they deal with. Communities who have worked directly with the government generally tend to have better-organized data, although such data may only be part of an ad hoc programme. Others focus on dealing with short-term issues. Whether standardized data would be desirable as part of a collective strategy is still debated, given the prioritization of different issues across different communities. Meanwhile, even public data produced by the government are not necessarily accessible to a broad audience. This raises the issue of transparency: the public may not be sufficiently aware of government projects that will affect their access to water.

---

11 Interview with Arianto Nugroho, volunteer of Winongo Asri Communication Forum — Forum Komunikasi Winongo Asri (FKWA). FKWA was established in 2009.

12 This tendency is associated with the legacy of the New Order era in which communities did not participate substantially in producing public information, but were positioned as the objects of government policies.

The gap in information exchange reinforces the elitist nature of knowledge dissemination.

Grassroots communities are realizing that collaboration with academia, NGOs, and government bodies is becoming increasingly important. Collaborative projects with local universities form part of the search for more socially embedded solutions to local problems. For example, there have been efforts to develop closer links between communities and universities using collaborative community empowerment programmes (Kuliah Kerja Nyata — KKN). Universities send their students to work with communities for two-month periods, working on thematic programmes based on collaborative assessment. In a shorter programme, academia has also engaged in collaborative class projects across universities using participatory action research methods.

5.2 *Bringing Water to the Public Debate: Health and Community Empowerment*

In the context of Yogyakarta, and particularly of the three communities described above, health and community empowerment surface as two major issues that drive grassroots activism. Communities have different strategies to bring their issues to a broader public audience. Banyu Bening community stresses the importance of increasing citizen awareness of health issues that arise from water consumption. As noted in a discussion with a community representative: "The human body is a laboratory that knows best which substance is good and bad in sustaining human life.... People have options to consume different types of water with various consequences to health.... The quality of good water is important to disentangle the roots of health problems in the human body".[13]

In their encounters with government agencies and academia, there has been tension over the implementation of water standard criteria established by the Ministry of Health. For the community, unequal power relations do not allow the public to have proper information as regards the benefits of rainwater to human health. This problem, according to activists, arises mainly because of a lack of support from the government in monitoring water quality and including rainwater in the scheme of formal government drinking water provision.[14] Consequently, more efforts are needed on the part of grassroots actors. Likewise, there is an urgent need to work on democratic access to

13 Excerpts from discussion in Banyu Bening community with Kamaludin and Sri Wahyuningsih, 6 May 2018.

14 FGD with AKSY members, 26 August 2017.

knowledge that allows the public to understand more about different forms of drinking water.

Since it was established, Banyu Bening has rapidly developed a social network at the national level. It maintains intense communications with environmental activists, academics, and government officials through personal relations and its presence on social media platforms. The community is active in linking its campaign on rainwater with other issues, mainly conservation and riverbank planning. The ability of this community to be active across a range of issues benefits its campaign.

For AKSY and PAMMASKARTA, community empowerment has been the main focus. Both communities are oriented towards policy advocacy and are very much influenced by the idea of professionalization of their public activities as a precondition for communities to stand equal with the government in development policy-making processes.[15] In working towards equalization, however, activists are constrained by their intra-community issues. This requires them to devote extensive time and energy to a very particular agenda, mostly dealing with socio-economic vulnerabilities. Meanwhile, when it comes to addressing the overall development agenda, they need to consolidate inter-community issues as a basis for collective policy advocacy. In fact, some more organized communities prefer to work unilaterally to avoid complex negotiation processes with other communities.[16]

In the case of AKSY, and especially since the volcanic eruption of Mount Merapi in 2010, riverbank planning has become an important policy issue for local government. Grassroots communities have organized themselves to take part in resource redistribution under an agenda of development and disaster risk reduction. AKSY has organized around this idea by connecting activities in major riverbanks from upstream to downstream. In general, the government tends to follow community programmes that enable the community to push for solutions to problems they are facing in their daily life. This is against the backdrop of weak capacity within local government to integrate the needs of the communities within the broad framework of development planning. From 1990 to 2000, government was seen as the adversary; recently, however, there has been a shift towards more harmonious relations.

There is an increasing awareness among grassroots community members that citizens could take a more active role in the redistribution of developmental resources by engaging in local-based planning that involves community members from the beginning of a programme to its implementation. Focus

15 Interview with Arianto Nugroho, volunteer of FKWA, 4 May 2018.

16 FGD with AKSY members, 23 September 2017.

group discussions raised the concern that communities would only act as the "crew of the government". Although the role of the government is important in terms of establishing the legal standing of the communities, this should not obstruct the autonomous position of communities in articulating criticism when there is a failure to address public grievances.[17] As one representative of FKWA, which is a member of AKSY, explained:

> One of the weaknesses of grassroots communities is how to map comprehensively the problems and proposed policies across members of AKSY. We often conduct events involving public who have concerns about riverbank problems in Yogyakarta and this has been facilitated well by government. It did produce a short-term supply of material resources, but there is a need to locate this within the logic of long-term collective action that is normatively desirable for all members of AKSY. If communities are able to create a masterplan in which they can see their contribution in the longer term, this will help government to allocate state resources much more effectively to support local development.[18]

Within this context, unmanaged disparity among AKSY's member communities in documenting knowledge and in managing local issues has produced internal tension. This is demonstrated when they have to negotiate with the government to address grievances raised by citizens within their scope of work. Some established communities are able to develop strategic relations with the local government as a result of their experience; meanwhile, newly emergent communities face difficulties situating themselves within the existing grassroots networks due to unfamiliarity with the political situations that shape state–society relations.

Communities are also facing the issue of under-representation, with intergenerational gaps and elitism involving members associated with stronger decision-making power. Communities both upstream and downstream work and mobilize resources mainly outside the policy radar of the local government. Currently, most government programmes concentrate on issues in the midstream, which are represented by different communities. This often creates internal tension among community activists representing different political goals.[19]

There was an attempt by the secretariat of AKSY in 2012 to categorize its member communities based on their managerial capacity to organize programmes.

17 FGD with AKSY members, 23 September 2017.
18 Interview with Arianto Nugroho, volunteer of FKWA, 4 May 2018.
19 FGD with members of AKSY, 23 September 2017.

Ideally, the classification would correspond to the level of access to resources provided by the government. However, those who considered themselves in a marginal position, compared to relatively more organized groups, did not necessarily agree with such an approach. There was a concern that, if not properly done, this approach would simply continue to expose inequality among member communities rather than resolving it. Further debate is taking place to negotiate a more conducive strategy that may address under-representation.

PAMMASKARTA also shares the idea of community empowerment as a priority issue, with quite a different emphasis. This is particularly concerned with how member communities are able to manage existing government-led water projects, both in terms of ensuring that physical infrastructure works sustainably and of increasing social trust among water provision beneficiaries. When first initiated, water infrastructure development is financed by the government; once up and running, management is transferred into the hands of the local community. This assumes a certain capacity on the part of the local community, and also assumes that it has a sense of belonging to the programme. In reality, this is not always the case. So far, most PAMMASKARTA members rely on government-led programmes that are largely technical. Local Public Works Offices often deliver training to improve the capacity of cooperatives to maintain water infrastructure facilities that are provided by the government. However, the social dimension of managing water cooperatives is not well institutionalized in the government programmes. There is a lack of institutional knowledge to manage social tensions or to deal with management weaknesses. In Yogyakarta, many government water projects stop functioning because of a lack of citizen participation after the initiation of the project, or because of inadequate management capacity.

Institutional constraints are attributed to bad governance that disrupt social trust within the grassroots communities and create barriers to the provision of good-quality clean water for consumers. A critical aspect of this is the fact that water cooperative management is very often imbued with personal interest over securing economic livelihoods. This can lead to resource capture practices by some individuals. There are cases in which water tariffs are subject to electoral politics when members of the management intend to gain power through political candidacy. Not only does this subvert the accountability of management, it also reproduces bad governance.

The redistributive aspect of economic welfare in water provision services has often been neglected. Water service provision has wider socio-economic repercussions beyond the provision of infrastructure. It involves people who dedicate their labour to engage in the provision of a public good which is supposed to be provided by the government. People who run PAMMASKARTA, for example, generally receive limited economic remuneration, in line with the non-profit orientation of the community-based system. The link between water provision and decent livelihoods, moreover, is an area that needs to be addressed through proper redistributive regulation by the government.

5.3 *Exploring Formal and Informal Political Channels*

Quite intense communication networks connect grassroots communities, NGOs, and other stakeholders. Efforts to rescale local issues occur through both formal and informal channels. Formal channels involve the bureaucratic structure of communities from village level up to the national level; the informal channel is built upon close personal connections between the grassroots and elites in central or local government bodies, and is the channel most commonly used in Yogyakarta. Such forms of engagement provide a useful temporary resource but do not offer reliable long-term support for grassroots communities' activities.

In terms of formal engagement with the government, PAMMASKARTA is the most institutionalized of the collectives, whose activities are clearly defined within government regulations. On an everyday basis, however, the level of institutionalization of their public function depends on how each unit is able to work beyond the technical aspect of water provision. The relationship between a water cooperative and its local consumers is particularly important in sustaining a collaborative water management framework. The ability of PAMMASKARTA members to engage the government in a timely manner when the presence of the state is needed is crucial within the scope of the government's role as currently defined. The extensive range of local water infrastructure that needs to be maintained is not well served by being held in the hands of the provincial government.

In AKSY, there is more reliance on the informality of state–citizen communications, which helps to maintain long-term relationships. However, it remains important to locate these informal relations within a broader goal of transforming the system to serve the poor. Formal work with the government is organized on the basis of Memoranda of Understanding with agencies such as the Major River Basins Agency (Balai Besar Wilayah Sungai — BBWS), the Regional Development Planning Agency (BAPPEDA), and the District or Provincial Environmental Agency. This takes the form of, for example, the organization of community events and community assistance programmes, in developing riverbank planning, conservation, and livelihoods provision.[20]

In AKSY, there are two approaches to engagement with the state, which determine the mode of issue rescaling. In one approach, AKSY plays a role in

20 Interview with Arianto Nugroho, volunteer of FKWA, 4 May 2018.

mobilizing inter-community channels when one community needs help and support to raise an intra-community issue. The other approach focuses on utilizing available channels such as mass media, academia, and NGOs. In both cases, the role of AKSY is central to articulating specific grievances when the strategies of individual communities do not give those communities sufficient voice to be heard due to bureaucratic complexity.[21]

Banyu Bening community is the most informal among the three, working on the basis of individual networks. A government presence is not always considered to be a determining factor, especially when grassroots activism does not pursue a proactive approach in rights claiming. Greater emphasis is therefore given to how a community is able to find solutions to its everyday problems; the role of the state is perceived to be the protection and improvement of public services on behalf of citizens' interests, rather than the imposition of bureaucratic obstacles.[22] In a context in which the top-down culture represents a structural barrier, a large segment of the grassroots population lacks the political motivation to engage seriously with formal decision-making processes. Moreover, in the absence of strong local engagement, especially with the local government at the village, subdistrict and district levels, there is a disconnect between grassroots development practices and the sphere in which such initiatives are to be institutionalized and recognized through legal formal means.

When communities face cases that are not resolved at the local level, there are efforts to raise support from a broader set of stakeholders. As suggested above, the engagement of grassroots communities with more formalized and professionalized NGOs allows them to rescale issues and to make them more public. To date, there has been no systematic attempt to identify channels of communication and to organize engagement with the state that involves broad-based coalitions. The historical New Order era relations between the state and civil society (see Table 12.1) also seem to characterize much of the contemporary relationship between grassroots communities and the state. Avenues of interaction between grassroots communities and state officials are most often found in community events and public discussions. Events such as river-cleaning programmes are useful in keeping state officials informed about the environmental conditions facing communities. When these kinds of programmes work well, they allow more fluid interactions that can extend to wider grassroots initiatives.

---

21 FGD with members of AKSY, 23 September 2017.

22 Interview with Kamaludin and Sri Wahyuningsih, founders of Banyu Bening community, 26 April 2018.

On the whole, personal connections seem to dominate the approach of grassroots communities when dealing with the state. Not only do personal relations affect the mobilization of resources, benefiting only a limited number of individuals at the grassroots level; they also contribute to exploitation by political elites who use such connections to pursue their own vested interests. This potentially leads to an ineffective policy system. The government is still seen as reluctant to integrate grassroots communities into the policy cycle and its policy responses have not addressed the root causes of poverty associated with access to and control over water resources.[23]

Government regulations to protect existing grassroots activities are limited; they are mainly related to conservation, the utilization of water resources, and the management of water-related hazards which could lead to disasters. Through Law No. 6/2014 (the Village Law), grassroots communities are challenged to engage progressively with political processes at the local level, i.e., the village government. In this context, an issue to be addressed in the near future is the expansion of knowledge exchange on the mechanisms and processes of policy making.[24]

Policy advocacy is hindered by the failure to systematically explore and exploit the formal channels that are currently available. Activists often refer to the difficulty of aligning their claims and grievances with the existing policy task distribution. For their part, government officials claim that government agencies are not allowed to initiate any policy in the absence of a higher regulation that protects their authority in implementing any response to community needs. This results in slow policy responses to urgent needs at the local level, and continuous delays in the making of everyday policy.

6 Conclusions

This chapter has discussed struggles over decent access to water and forms of citizenship at the grassroots level that seek to transform water governance through a more inclusive development framework. Systemic conditions create challenges but also produce modalities by which local actors claim some participation in the development agenda. Water governance links the technocentric dimension of water with socio-political organization that may allow citizens to substantially contribute to development policy-making processes.

23 Interview with Arianto Nugroho, volunteer of FKWA, 4 May 2018.

24 FGD with members of AKSY, 23 September 2017.

Citizenship is concerned not only with ensuring that politics is open to substantive civic participation. It is also relevant for the building of coalitions among various stakeholders, allowing water and sanitation issues to be identified on the basis of wider social legitimacy. The process of building coalitions and alliances is crucial to the formation of a collective political agenda. It also allows multi-stakeholder engagement in extending policy options that may facilitate the fulfilment of community rights to water and sanitation, especially at the local level. Local knowledge organization must be strengthened in order to adequately address complex causes and consequences of development problems.

The link between local development practices by grassroots communities and the international policy framework to achieve the Sustainable Development Goals will need to be developed much more intensively. With the diverse characteristics of resource mobilization and issue identification, and the pluralistic composition of actors, policies that are oriented towards rescaling local practices will be problematic if they neglect the possibility of different articulations across the domain of grassroots community activism. Grassroots communities work within fluid networks, and develop asymmetric relations with state and private actors at different scales of interaction. This has consequences for the way that relations between society and the state are (re)formulated over time.

References

Ardhianie, N. (2015, March 3). What next after the water law annulled. *The Jakarta Post*. Retrieved from http://www.thejakartapost.com/news/2015/03/03/what-next-after-water-law-annulled.html.

Brecher, J., & Costello, T. (1998). *Global village or global pillage: Economic reconstruction from the bottom up*. Cambridge, MA: South End Press.

Buytaert, W., Zulkafli, Z., Grainger, S., Acosta, L., Alemie, T.C., Bastiaensen, J., … Zhumanova, M. (2014). Citizen science in hydrology and water resources: Opportunity for knowledge generation, ecosystem service management, and sustainable development. *Frontiers in Earth Science*, 2(26), 1–21.

Christopherson, A. (2008). Book review. Local and regional development by Andy Pike, Andrés Rodrigues-Pose and John Tomaney. *Economic Geography*, 84(2), 241–242.

Cornwall, A. (2005). Spaces for transformation? Reflections on issues of power and difference in participation in development. In S. Hickey & G. Mohan (Eds), *Participation: From tyranny to transformation? Exploring new approaches to participation in development* (pp. 75–91). London: Zed Books.

Cornwall, A., & Brock, K. (2005). What do buzzwords do for development policy? A critical look at "participation", "empowerment", and "poverty reduction". *Third World Quarterly*, 26(6), 1043–1060.

Escobar, A. (1992). Reflections on "development": Grassroots approaches and alternative politics in the third world. *Futures*, 24(5), 411–436.

Falk, R. (1993). The making of global citizenship. In J. Brecher, J. B. Childs, & J. Cuttler (Eds), *Global visions: Beyond the new world order* (pp. 39–52). Boston, MA: South End Press.

Fulazzaky, M.A. (2014). Challenges of integrated water resources management in Indonesia. *Water*, 6, 2000–2020.

Harris, J., Stokke, K., & Törnquist, O. (2013). *Politicizing democracy: The new local politics of democratization.* London: Palgrave Macmillan.

Hickey, S., & Mohan, G. (2005). *Participation: From tyranny to transformation? Exploring new approaches to participation in development.* London: Zed Books.

Jones, E., & Gaventa, J. (2002). Concept of citizenship: A review. IDS Development Bibliography 19. Brighton: Institute of Development Studies.

Kusmayadi, R. (2015). Water supply in Indonesia towards SDG. Paper presented at the International Forum at JWWA General Assembly and Conference 2015, Saitama, Japan. Retrieved from http://www.jwwa.or.jp/jigyou/kaigai_file/seminar_05/F-06_PERPAMSI(F).pdf.

Makhoul, J. (1999). Development from below: An exploratory study of the concept and process of development from lay people's point of view. (Doctoral dissertation.) University of Wollongong, Wollongong.

Marwa (2017). Changing water regime: Remunicipalisation in Indonesia under the global hegemony of privatisation. (Undergraduate thesis.) Department of International Relations, Universitas Gadjah Mada.

Ministry of Health (2016). Menuju 100% Akses Sanitasi Indonesia 2019. Jakarta: Ministry of Health. Retrieved from http://www.depkes.go.id/article/view/16060100003/menuju-100-akses-sanitasi-indonesia-2019.html.

Nel, E., Hill, T., & Binns, T. (1997). Development from below in the 'new' South Africa: The case of Hertzog, Eastern Cape. *The Geographical Journal*, 163(1), 57–64.

Procacci, G. (2006). Governmentality and citizenship. In K. Nash & A. Scott (Eds), *Blackwell companion to political sociology* (pp. 342–351). Oxford: Blackwell Publishing.

Read, B., & Pekkanen, R. (2009). Local organizations and urban governance in East and Southeast Asia. London: Routledge.

Seyfang, G., & Smith, A. (2007). Grassroots innovations for sustainable development: Towards a new research and policy agenda. *Environmental Politics*, 16(4), 584–603.

Statistics Bureau of D. I. Yogyakarta Province (2016). Tourism statistics. Yogyakarta: Badan Pusat Statistik Yogyakarta.

Statistics Bureau of D. I. Yogyakarta Province (2015). Water supply statistics of Daerah Istimewa Yogyakarta. Yogyakarta: Badan Pusat Statistik Yogyakarta.

Stokke, K. (2017). Politics of citizenship: Towards an analytical framework. In E. Hiariej & K. Stokke (Eds), *Politics of citizenship in Indonesia* (pp. 23–53). Jakarta: Yayasan Obor Indonesia.

Thanh, V. T. (2008). Narrowing the development gap in ASEAN: Approaches and policy recommendations. In H. Soesastro (Ed.), *Deepening economic integration: The ASEAN economic community and beyond* (pp. 147–155). ERIA Research Project Report 2007-1-2. Chiba: IDE-JETRO.

Wieriks, M. (2011). Water governance and policy networks in Indonesia: The challenges of a decade of water sector reformation. (MSc Thesis.) TU Delft, Delft.

Woodhouse, P., & Muller, P. (2017). Water governance: An historical perspective on current debates. *World Development*, 92, 225–241.

Zamzami, I. (2017). A new wave of water privatization in Indonesia. Amsterdam: The Transnational Institute.

PART 6

SDGs and the Economic Agenda for Inclusive Economic Growth and Decent Work for All

CHAPTER 13

Economic Reform and Sustainable Development in Vietnam

Tran Dinh Lam

1 Introduction

After Vietnam was partitioned in 1954, the North was subjected to a highly centralized planned economy in which all activities and decisions were controlled by the central government, a model which proved advantageous during wartime. After the reunification of the country in 1975, the Vietnam government believed that the command economy would bring prosperity; it decided to implement a policy that sought the cooperation of all Vietnamese citizens, aiming to transform Vietnam into an industrialized nation in 20 years. In order to pursue this strategy, the government proceeded to reform private industry and the commercial sector, confiscating all of the business premises, transportation means, and houses of the comprador bourgeoisie[1] (Thu vien quoc gia Vietnam, 2012) in Southern Vietnam, and forcing them to participate in the New Economic Zones Programme launched in 1978–1979. All institutions for the circulation of goods and banks for production and consumption based on the modern market economic model were abolished.

The centrally planned economy had many adverse consequences, including increased public discontent and slower economic growth. The economy grew at a rate of 0.4% per annum, much lower than the population growth rate, which led to a negative growth in GDP per capita. Vietnam became one of the five poorest countries in the world. The average rice output and rice mobilization rate for government spending dropped from 211 kg per capita in 1976 to 157 kg per capita in 1980. Economic conditions continued to worsen, and Vietnam had to import 450,000 tons of rice annually during the period 1975–1987 (Đặng Phong, 2016b, p. 380). By 1987, 2 million people were on the verge of starvation. Over a 20-year period, hundreds of thousands fled the country, many in small ships and boats, becoming known as the boat people. This wave of boat people peaked in 1978 and 1979 as 1 million fled Vietnam for Indonesia, Thailand, Philippines, Hong Kong, and Malaysia (Wikipedia, n.d.).

1 On 25 August 1976, the Vietnamese government announced its policies aiming to shift private industry and the commercial sector to socialism.

 | DOI:10.1163/9789004391949_014

In the mid-1970s, Vietnam experienced incursions from across the Cambodian border by Pol Pot's Khmer Rouge, supported by China, beginning with an attack on Phú Quốc Island in April 1975 and later assaults on the southwest border in 1977. The previous war had barely ended when a new one began. Vietnam launched many counter-attacks to help Cambodia fight the genocidal regime of Pol Pot, leading to new challenges for the economy. ASEAN countries changed policy; from cooperating with Vietnam, they now began to impose embargoes against Vietnam. Emphasizing regional security rather than justice, and not fully appreciating the situation, the world did not understand that Vietnam was helping Cambodia prevent further genocide. Vietnam's intervention isolated the country for years.

Because of adverse economic conditions, many local authorities in the South took the decision to reduce their dependence on the central government; this enabled them to purchase rice at market price instead of the stipulated government price, eliminate the use of coupons, and shift to a one-price policy. The spirit of these inventive acts was then adopted by the government; the coupon mechanism was completely eliminated in the second quarter of 1989. With this decision, all prices began to follow the market rule of equivalent exchange, the free market was promoted, and people were allowed to develop their own business strategy. The pay policy statement in Southern Vietnam also helped central government make amends for its inefficient policy management. For example, the local authority at An Giang Province developed a state treasury model which differed from that of the North and allowed it to become more active in state budget management; the central government later applied this model to the entire country (Đặng Phong, 2016a, p. 384).

In agriculture, farming was run in the form of collectivized production (Khoán việc). In this form of production, land was owned by the cooperative and the return to the farmer was determined by the number of hours of work, measured by a cooperative official placed on the farm. There were so many disadvantages to this approach, particularly lack of incentives and management capability, that it led to a decrease in production before it was terminated by Resolution No. 10 of the Politburo[2] (Phạm Thị Quý, 2015).

The agricultural planned economy controlled by the cooperative was ended in 1985 and land was returned to the farmer, under the Khoán hộ policy favoured by the people. This created a considerable incentive that led to an increase in agricultural production. People had enough rice to eat and, for the first time since reunification, Vietnam began exporting rice. In 1989, rice exports reached 1.4 million tons, making Vietnam the third largest rice exporting

2 A 'renewal resolution' in agriculture overcame the previous inappropriate policy which had discouraged production.

nation in the world, after Thailand and the USA (Đặng Phong, 2016b, p. 381). Since then, Vietnam has increased rice exports and stabilized annual export levels at 3 to 4 million tons.

Vietnam also changed its policy regarding foreign investment in 1987 (Công báo, 1998). However, as with the import/export market, the government remained the exclusive agent in signing contracts with foreign companies. Because trade was monopolized in this way by the central government, the economy faced many obstacles, including a shortage of raw materials. Leaders of the Ho Chi Minh City government did their best to overcome this shortage, seeking advice from economists in the South who specialized in market mechanisms. These experts in turn used their networks in Singapore and Hong Kong to acquire raw materials such as yarn, tobacco, fuel oil, etc. All of these raw materials were priced in dollars but were paid for in kind with items such as dried shredded squid, dried shrimp, and peanuts, thus bypassing official valuation. Goods were delivered to Buoy No. 0, offshore in international waters. In order to support the import/export business, city leaders promulgated an official dispatch for this activity:[3] businesses did not have to acquire a letter of credit from the bank or use the official import/export channel.

Socio-economic life had been revived step by step. Many Southern locals could sell agricultural products to Ho Chi Minh City and purchase industrial products, which were then sold to other people. In 1982, import and export performance in the South, particularly exports to capitalist countries, was almost five times higher than that of the central region (Đặng Phong, 2016a, Table 16.1). Table 13.1 presents export data for the period 1976–1985.

TABLE 13.1 Vietnam: total export turnover, 1976–1985

Year	1976	1977	1978	1979	1980	1981	1982	1983	1984	1985
Total export turnover regulated by central government (millions of rubles)	222.7	322.0	326.8	320.5	335.9	366.2	435.5	530.7	619.0	698.0
Exported to capitalist countries (millions of US$)	43.1	60.2	46.9	50.6	39.0	49.1	32.9	59.4	n.a.*	n.a

* n.a.: data not available
SOURCE: ĐẶNG PHONG (2016A, P. 448)

3 Dispatch No. 15 UB dated 4 August 1980 of The People's Committee of Ho Chi Minh City.

2 Poverty Alleviation for Ethnic Groups and Reduced Inequalities

The Vietnamese economy experienced many positive changes as a result of reforms which transformed the planned economy into a market one, as well as the abolition of the agricultural cooperative model. In 1993, the government allowed farmers to transfer their land use right — a pivotal decision that incentivized farmers to continue working and consequently alleviated poverty in the 1990s. Open policies boosted exports and created many job opportunities in the field of services provision, helping the economy develop in all areas. According to 2017 research by the World Bank, Vietnam is a model of poverty alleviation. Based on a poverty standard of US$1.03 per day, the percentage of poor people fell dramatically, from 77% in 1993 to 10.7% in 2012 (Eckardt, Demombynes, & Chandrasekharan Behr, 2016). This means that the poverty rate in Vietnam is lower than that of China, the Philippines, Pakistan, and India. However, estimates from the World Bank (ibid., p. 81) indicate that poverty alleviation among minority ethnic groups stagnated from 2012 to 2014 and that, by 2020, those who remain poor are most likely to be members of ethnic minorities.

Although the process of economic reform has brought about many changes for all Vietnamese citizens, the wealth gap between ethnic minorities and majorities (Kinh and Hoa) is still very large, to the extent that those belonging to the minority groups may fall into poverty again unless inclusive investments are made. In a field study in Gia Lai Province, Thanh Phan, Tran Dinh Lam, and Van Mon (2011) show that the only way to alleviate poverty and help the Bahnar people integrate into society is to improve their human capital capability. Once trained, Bahnar officials prove efficient in carrying out government policy, creating a trustworthy image of their people to other ethnic groups. Realizing the importance of education in development, the government promulgated the 'Educational development strategy 2001–2010' (Cơ sở dữ liệu quốc gia về văn bản pháp luật, 2001). During that period, 250 ethnic boarding schools and over 100 day schools were built.

Despite such measures, the World Bank estimates that the inequality between ethnic minority children and Kinh and Hoa children is still considerable and proving difficult to reduce (World Bank & Ministry of Planning and Investment, 2016). In addition, there are health and nutrition gaps between the children of ethnic minority groups and Kinh and Hoa children. Figure 13.1 and Figure 13.2 illustrate some of these discrepancies. Because of differences in how material life is viewed in the culture of different ethnic groups, the government needs to implement synchronous activities to raise children's academic level, reduce malnutrition, and increase awareness of the importance

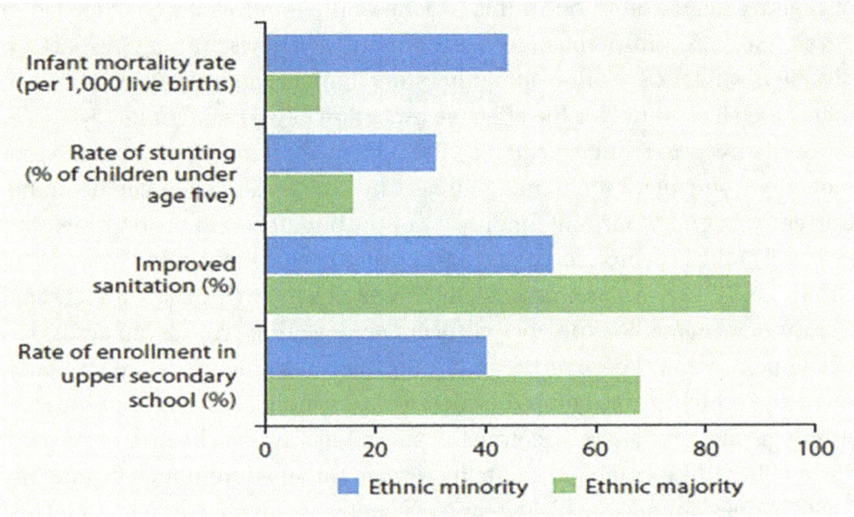

FIGURE 13.1 Indicators of inequality for children of ethnic minority groups
SOURCE: WORLD BANK & MINISTRY OF PLANNING AND INVESTMENT OF VIETNAM (2016, P. 40)

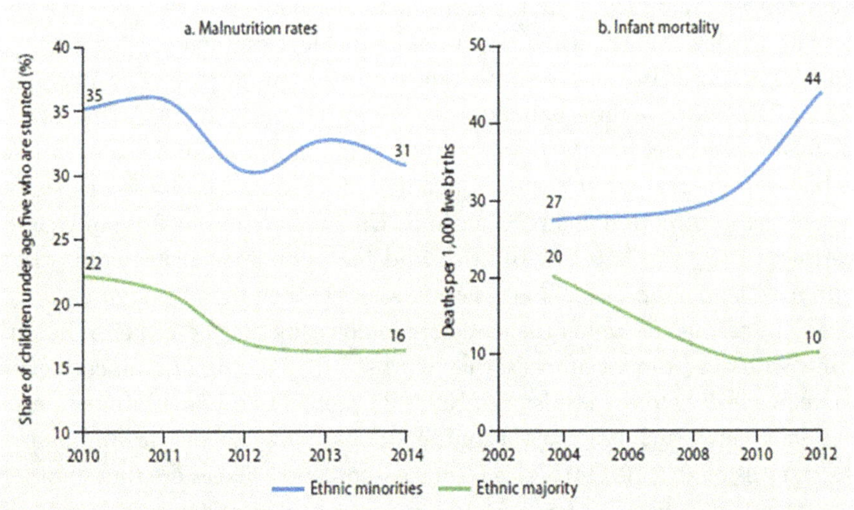

FIGURE 13.2 Child health and nutrition by ethnicity
SOURCE: WORLD BANK & MINISTRY OF PLANNING AND INVESTMENT OF VIETNAM (2016, P. 70)

of sanitary living conditions if it is to achieve the mindset changes needed to develop a more prosperous life. Furthermore, awareness raising, respect for diversity, avoidance of ethnic prejudices, and human capital training improvements are all essential for the effective execution of government policy.

Another way to reduce inequality and poverty for minority groups is to promote domestic migration to urban areas where high-salary jobs can be found. Currently, migrants account for only 1% of the population in the big cities, but in Binh Duong Province, migrants make up 11% of the population (World Bank Group, 2016). This is a promising sign for reducing the wealth gap in Vietnam.

However, inequalities do not only occur among ethnic groups but across the whole population. Those at the bottom of the social hierarchy are still poor, no matter which ethnic group they belong to, including the Kinh and the Hoa. Although the annual average growth rate in Vietnam has been around 5.5% since 1986, with income per capita increasing threefold from 1990 to 2014, rising GDP does not necessarily guarantee a happy life for all. For that, social justice is also needed. After many difficult years, the Vietnamese social model is now oriented towards ensuring the independence of all people and social justice, with the aim of eradicating discrimination between economic sectors, opportunity inequality, and the ever-increasing wealth gap. Vietnamese society has changed dramatically but the continuation of joint efforts by everyone, especially the law makers, will be essential for the development of a just society. Policies proposed by the government must always consider the basic question: who will benefit from this policy?

To give one example: the financial success of many present-day enterprises is based in the real estate sector, which does not require productive capacity, but rather information and relationships. The increase in this form of wealth can widen the gap between the rich and the poor. Specifically, enterprises purchase land from farmers at a very low price; they then obtaining a land use right certificate, which they are able to do easily thanks to their relationships with government officials; they then sell the land on at a much higher price. According to the Ministry of Natural Resources and Environment, legal actions against such practices accounted for 95% of 1,539 lawsuits in the first six months of 2017 (Tuổi trẻ, 2017). Transparent land policies are thus another important step towards a just society.

3 Economic Growth and Decent Work

A study by James Feyrer (2011) shows that commercial expansion and increased foreign direct investment (FDI) can create many job opportunities.

After a long period of implementing a closed-door policy and rejecting private enterprise, the Vietnamese economy was exhausted. Since recognizing this, the government has, over the past 30 years, been reforming the economy to promote growth, with the private sector playing a significant role. The private sector currently provides jobs for 20 million people — a crucial factor in improving income levels and social mobility. Vietnam now has over 650,000 registered private enterprises compared to 40,000 in 1990 and almost none in 1989. According to estimates by the World Bank Group (2016), the public sector has been responsible for many losses while the non-public sector — including private enterprises, family businesses, and the foreign investment sector — has accounted for about 70% of the country's growth. Private enterprises also attract investment from banks, leading to a growth in onward investment and job creation. Nowadays, local governments in the Mekong Delta have adapted to changes in the market and acknowledge the importance of small and medium enterprises. Cooperation among regions is encouraged to share the production of goods with one another and solve the problem of (un)employment.

Although there has been noteworthy economic development in Vietnam, it is only meaningful if it brings about long-term sustainability and more jobs. The government has introduced many policies promoting sustainable development, one of which is to encourage the growth of private enterprises — a major contributor to the national budget and to job creation. Economic development will be sustainable if local enterprises can attract labour; however, business activities will only be efficient if an effective management model is adopted. Around 70% of the population lives in the rural areas, accounting for over 43% of the labour force, but they contribute only 16–17% of GDP. Currently, farmers still face difficulty in selling their products and connecting to consumers, which means that produce is wasted. The most effective way to solve this problem is to sign contracts with partners who agree to buy the produce before the harvest, allowing farmers to plan their production. The Vietnamese President has held discussions with 300 farmers in which he advised them to consider to whom they will be selling their products before they plant their seeds; he also encouraged enterprises to follow market signals in order to continue creating more jobs (Hương Quỳnh, 2018).

4 Affordable and Clean Energy

Vietnam's economic transition has been remarkable. However, a dedicated and impartial team of officials is needed to move the country towards a more sustainable development path. Recently, the decentralization of power to local governments and an increase in personal responsibility has caused serious environmental incidents in areas near the Formosa steel plant, damaging the coast of four central provinces and raising a red flag regarding the quality of economic development (Ling Tong, 2017). According to the Minister of Natural Resources and Environment, the incident caused serious economic, social, and environmental impacts for the immediate and long term, directly affecting fishing, aquaculture, seafood processing, salt making, fishery logistics services, and tourism, and producing chaos, insecurity, discontent, and anxiety. A recent news item showed people's opposition to the Vinh Tan power plant's practice of dredging and dumping trash into the sea, causing environmental incidents. Due to the levels of environmental pollution surrounding the plant, authorities had to inform residents not to use contaminated groundwater (Huỳnh Hải, 2017).

In the wake of the Formosa case and other incidents, the Vietnamese government expressed its determination to pursue sustainable economic development; at the same time, the Prime Minister emphasized and warned domestic and foreign enterprises that Vietnam would not damage its environment for the sake of economic development (Tuổi trẻ, 2016). The government has decided to develop Vietnam's economy using clean energy. A study by Nguyen Khac Nhan (2017) shows that Vietnam would be able to produce clean energy on a large scale by 2050. The development of wind and solar energy is especially important because thermoelectric plants may interfere with the sustainable development of Vietnam's economy. Given advances in science and technology, wind power is getting cheaper. With a long coastline, Vietnam has extremely favourable wind conditions for energy development (Pham Van Ninh, 2003), giving it more potential to develop wind power than Laos and Cambodia (Vu Tu Anh & Dam Quang Minh, 2006). Mr Nguyễn Anh Tuấn, director of the Renewable Energy Centre of the Institute of Energy, Ministry of Industry and Trade, says that Vietnam has the potential to develop renewable energy (Quỳnh Nga, 2017b). In theory, Vietnam could produce up to 9.1 million MW of renewable energy, much higher than the total capacity of its current power system, which was 41,000 MW as of 2016. Specifically, biomass electricity could generate up to 15,000 MW; biogas energy over 177,200 MW; energy from waste material over 9,000 MW; wind energy up to 2.1 million MW; and solar energy up to 6.8 million MW. The World Bank (2010) has also studied conditions for generating wind energy in Southeast Asia, emphasizing the potential for developing small wind power plants in rural Vietnam. A study on the advantages of wind power in rural areas shows that 41% of the land area of Vietnam could be used for developing small wind power plants, compared with only 13% in Laos, 9% in Thailand, and 6% in Cambodia (Vietnam

Management Consultant, 2017). Lazard's survey indicates that the Levelized Cost of Energy (LCOE) of coal power is around 6–14.3 cent/kWh while the LCOE of solar power and wind power is around 3–6 cent/kWh.[4]

Since 2004, Vietnam has passed several relevant resolutions including Resolution No.41-NQ/TW on environmental protection, sustainable development, and modern technology, and Resolution No.55/2014/QH13 with regulations on renewable energy development, environmentally friendly production, and consumption. On 25 November 2015, the government released an important legal document outlining a renewable energy development strategy to 2030, with a broader outlook to 2050. On 17 November 2017, the Vietnamese government passed Resolution No.120 on sustainable development in the Mekong Delta with regard to protection of land and humans, renewable energy development relating to forests, and coastal protection. The government also showed some determination by passing Decision No.11-QĐ TTg-2017 to promote solar power projects (Bộ Công Thương, 2017).

In 2017, the Ministry of Industry and Trade also released Circular 16 under Decision No.11 on encouraging the development of solar power in Vietnam and purchasing power output from solar power plant projects. Electricity from wind power and solar power are sold at a price of 3–3.6 cents/kWh compared to electricity from coal power which is priced at 7.5–8 cents/kWh. Moreover, Vietnam is planning to develop renewable energy projects in the Central Highlands and the South, including a solar power plant project with a potential capacity of 300 MW which is being considered for Ninh Thuan Province. Currently, there are 35 communities in the Mekong Delta using solar energy as their main electricity source for daily activities. In Bac Lieu Province, the largest wind power plant went into operation in 2016 with a capacity of 99 MW. There are five wind power plants operating in Vietnam with a combined capacity of 160 MW. These sites have also become popular with tourists.

Despite the enormous potential, however, in practice, the development of renewable energy in Vietnam is still very limited. As of 2016, only 159 MW of wind power produced was connected to the national grid; electricity generated from solar power reached 6 MW, of which only 0.18 MW was connected to the grid; and the share of biogas power in electricity production was no more than 2.5 MW (Trung Chanh, 2017). At the same time, Electricity of Vietnam (EVN), the country's largest power company, has plans to build new thermal power plants across the region, in spite of strong opposition. This includes an additional 14 coal power plants to be built at Cửu Long Delta on EVN plan. Jim Yong Kim, President of the World Bank, has warned that this could be a disaster

4 See:https://www.lazard.com/perspective/levelized-cost-of-energy-2017/.

for the entire planet (The Guardian, 2016), and the Vietnamese national radio, Voice of Vietnam, has called for a halt to coal power plant construction (Voice of Vietnam, 2017).

According to the Air Quality Annual Status Report of the Green Innovation and Development Centre (GreenID), in the first six months of 2017 Hanoi suffered 139 days in which PM2.5 levels — the finest and most dangerous particulate matter — exceeded international standards (GreenID Vietnam, 2017). Over the course of a year, then, Hanoi's air quality is of an acceptable health standard for only two days a week. With an annual spend on public health costs due to air pollution of US$780 million (ibid.), Vietnam needs to create support for renewable energy development. It is essential for scientific research units to develop efficient, low-cost renewable energy equipment and to promote the renewable energy industry (Ngọc An, 2017).

According to research by Vu Quang Viet (2017), a former United Nations expert, the state should eliminate monopolies, privatize energy companies and raise electricity prices to international standards. Currently, electricity is so inexpensive that little attention is paid to clean energy competition. The average retail price of electricity in Vietnam in 2015 was 7.5 cents, compared to 8 cents in China and 13 cents in the USA. Raising electricity prices could also have an impact on polluting industries such as steel, as investors would no longer be able to take advantage of Vietnam's cheap electricity prices to run their factories. A study conducted by the World Bank found that 25 years of economic development in Vietnam had resulted in the exhaustion of natural resources, and in environmental pollution from industrial and urban waste water, leading to the risk of adverse health effects, especially for children and the elderly in large cities like Hanoi and Ho Chi Minh City (World Bank & Ministry of Planning and Investment of Vietnam, 2016).

5 Integration: Following the Path of the Four Asian Dragons

The lessons learned from Singapore and other Asian dragons South Korea, Taiwan, and Hong Kong have stimulated Vietnam and its international friends to look forward to a bright future as Vietnam continues its reforms (Dapice, Haughton, & Perkins, 1994). Vietnam has laid the foundations of an industrialization process in what might be called 'flying dragon orientation', but its potential has not yet been realized. Pham Do Chi and Tran Nam Binh (2002) note that the recommendations of many experts stress the importance of changing policy and improving the economic growth model to awaken the dormant dragon.

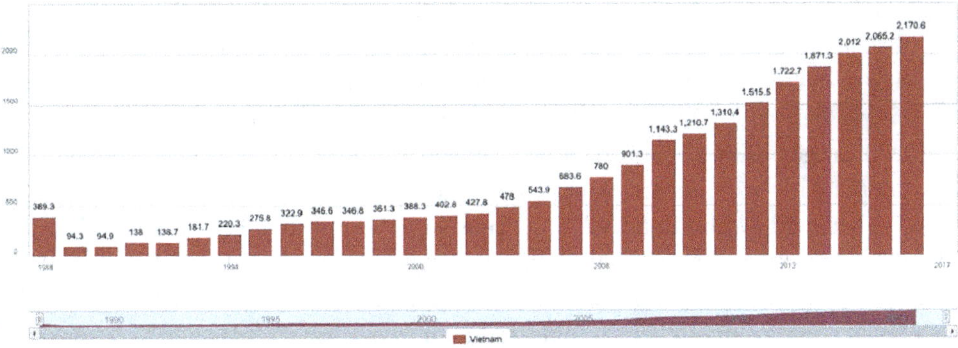

FIGURE 13.3 Vietnam's GDP per capita, 1988–2017
SOURCE: WORLD DEVELOPMENT INDICATORS: HTTP://DATABANK.WORLDBANK.ORG/DATA/REPORTS.ASPX?SOURCE=2&SERIES=NY.GDP.PCAP.CD&COUNTRY=VNM#

However, the industrialization process in Vietnam has squandered numerous resources and investments. Although the economy has steadily grown with an average annual GDP growth rate of more than 6% over 20 years, and although annual income per capita had increased to US$2,170.6 in 2016 (see Figure 13.3), the economy is held back by the government's ongoing support for state-owned enterprises (SOEs), which cause many problems. SOEs use substantial amounts of resources and capital, but their profit margins remain low. The failure of the shipbuilding group Vinashin — the result of rapid development, lack of financial transparency, and corruption due to the absence of state regulation — has led to a huge debt of VND86,031 billion (equivalent to US$4.2 billion) (Cooke et al., 2014). Such mismanagement has turned SOEs (commonly known as "steel fists") into a heavy burden of deficit and debt; the accumulated deficit of Vinaline, the Vietnamese shipping lines, is up to VND4,000 billion (Bạch Dương, 2016). Many have expressed concern for the future of Vietnam's economic development; the fear is that, if policy changes are not made, Vietnam will become a "little cat" rather than a "giant tiger" (Bằng, 2017).

In general, Vietnam has shown limited management capability and inadequate regulation over SOEs and corporations, although the number of SOEs with 100% state capital has decreased, from 12,000 in the 1990s to 5,655 in 2001 and 718 in October 2016. In the words of Pham Minh Dien: "In general, business efficiency and SOEs contribution are still low and inconsistent with the resources that government invested, not to mention the economic deficit and loss, corruption, wastes, and negative impacts" (quoted in Lưu Thủy, 2017).

Interest groups and crony relationships are now recognized as the major cause for deferring SOE reform, since the management of public property and

SOEs provides a great opportunity for profiteering (Quỳnh Nga, 2017a). This has led to a situation in which many individuals do not want change, resulting in delays to restructuring, especially of the management system. Another factor behind reform delays is the failure of local authorities to implement central government policies. Many industrial zones are built and operated inefficiently with an occupancy rate of under 30%, such as those in Sơn La, Bình Thuận, Ninh Thuận, Cà Mau, Hòa Bình, Thanh Hóa, Hà Tĩnh, and Cao Bằng provinces (Văn Bắc, 2015). In the early stages, in its eagerness to receive inflows of FDI, Vietnam was not concerned with such matters as the stability of the ecosystem. For example, industrial wastewater has had a disastrous environmental effect on Thi Vai River, with a major impact on the lives of locals (Hieu P. Nguyen & Huyen T. Pham, 2011). Plans for bauxite mining in Tay Nguyen gave rise to many controversial debates among scientists on the opportunity costs of economic development and environmental protection for future generations. Nonetheless, after a period of discussion, the government approved the plan for mineral resource exploitation in Tay Nguyen. The plan envisaged that companies in charge of the project would operate at a loss in the first four years; however, total losses have now reached VND3,690 billion, much higher than expected, while total revenue in 2017 from the mining companies was only about VND107 billion, VND330 billion lower than the previous analysis of economic viability had predicted (Thanh Hương, 2017).

One of many factors which has helped the Vietnamese economy to develop is the opening to foreign investment. Such investment is a good way for Vietnam to gain experience, transfer technology, seek markets outside of the country, and create jobs for labourers (Trần Đình Lâm, n.d.). Among the three engines of economic growth — foreign investment, state-owned enterprise, private enterprise — foreign investment is still the most efficient means. FDI accounts for about 50% of Vietnam's industrial production value, 20% of GDP and more than 70% of total export turnover. However, this advantage can be fully utilized only when domestic human resources, particularly at the top level of business, are sufficiently developed. Vietnam thus needs to adjust its policies to improve conditions for effective domestic competition, spurring economic growth in the long term. A World Bank report of September 2017, focusing on the experiences of domestic and foreign affairs, stresses the importance of the relationship between a dynamic FDI inflow and Vietnam's private enterprise in bringing together all the resources necessary for development (World Bank Group, 2017).

On the other hand, the inflow of foreign investment in the form of official development assistance (ODA) can also negatively influence Vietnam's economic growth if not properly regulated. On a small scale and with effective management, foreign aid has proved beneficial. However, as the amount of money increases, management system tend to fail to properly supervise spending, causing many losses and fostering opportunities for corruption. According to research by Burnside and Dollar (2016), foreign aid has positive impacts on developing countries that have good monetary policy and trade policy, but much less positive effects on economic growth in countries with inadequate policies. Indeed, data from the Ministry of Foreign Affairs of Denmark indicates that Vietnam received more than US$90 billion in aid from 1993 to 2014 (Lattanzio Advisory, 2017), but with very disappointing results. According to evaluations by the Ministry of Planning and Investment, there are at least five main reasons for this: (1) the perception of being 'subsidized' by government, especially in the case of ODA and government bonds; (2) the limited capabilities of contractors in areas that require foreign aid; (3) the wasting of investment capital, along with capital loss during operations; (4) tangled lines of responsibility between management and responsibility for public debt; and (5) lack of clear lines of responsibility for arranging loans and repaying debts as foreign aid is allocated and used (Ngọc Quang, 2017). Public debt levels reached 58.3% of GDP in 2015, as shown in Table 13.2, and were forecast to keep climbing. Table 13.2 also shows the quota of public debt, revenue, spending, and budget deficit for Vietnam in comparison to Indonesia in 2015.

Vietnam's public debt was estimated at US$95 billion in July 2017 and public debt per capita at around US$1,039 (The Economist, n.d.). Since Vietnam is now officially classified as a middle-income country, it will not be offered preferential credit and will have to look for credit in the open market. If Vietnam continues to promote growth through public investment, this 'fertile ground' for corruption could have devastating consequences for economic development. Kenichi Ohno (2009) suggests that, in order to escape the middle-income trap, Vietnam needs to select a team of technical professionals as well as

TABLE 13.2 Vietnam's 2015 government budget compared to Indonesia (percentages)

Fiscal policy	Vietnam	Indonesia
Public debt/GDP	58.3%	26.9%
Tax revenue/GDP	22.1%	13%
Expenditure/GDP	27%	15.6%
Budget deficit/GDP	−5.6%	−2.6%

SOURCE: ĐỨC MINH & HOÀNG LY (2017)

dedicated leaders who will cooperate with other countries around the world to figure out a sustainable industrial direction for economic development.

Economist Alex Tabarrok (2015), discussing the effects of geography on institutions, stresses that the powerful and rich do not want innovations for fear of losing position. By contrast, Vietnamese Prime Minister Nguyen Xuan Phuc has demonstrated a clear desire for innovation in his call for a second reform. He stresses that this reform must be based on the spirit of freedom, liberalization, innovation, and promotion of the role of enterprises in the private sector in every stage of the agricultural production value chain (Đức Tuân, 2017). As Vietnam joins the ranks of middle-income countries, its sustainable economic growth and development will require determination and enthusiasm for innovation across the whole government system. In order to achieve that goal, legal thresholds need to be consistently respected: no one is above the law. With lessons learned from Asian dragons like Singapore, South Korea, and Taiwan, Vietnam has many chances to carry out its second reform successfully. Furthermore, enforcing institutional regulations and fighting corruption will help to regain the trust of foreign investors and domestic enterprises.

The success of the Asian dragons stems from healthy competition between many well-educated individuals, protected by appropriate laws. Politicians work with talented technocrats for mutual development. The Vietnamese government is currently implementing a number of measures to improve the quality of the government system, but a competitive mechanism is needed to search for experienced advisors. These professionals are needed to formulate specific policies to help the government address deficiencies in SOEs, improve ODA management, and strengthen the relationship between FDI and domestic private enterprise, which will provide leverage for bigger changes and integration with other countries in the region. This is a precious lesson from Singapore's People's Action Party that Vietnam needs to study if it is to learn the key factors that led to the success of the sustainable industrialization process in that country.

6 Building Trust for Integration and Enrichment

Research by Zak and Knack (1998) suggests a strong, positive relationship between trust and economic growth. Vietnam's recent reform process resulted in some praiseworthy achievements in increasing the faith of the people in the government's determination to change. Nevertheless, Vietnam has to face the reality of corruption in its public administration and SOEs which reduces that trust in government. According to Pak Hung Mo (2001), the higher the level of corruption, the lower the rate of economic growth. This leads to political instability, which can further damage growth. The Vietnamese government is promoting the rule of law to eliminate corruption in an effort to increase the trust of the people.

There are many obstacles to industrialization, including corruption, a widening gap between rich and poor, and loss of public trust. Many large projects have suffered financial losses on a grand scale (VietnamNet, 2016). Interest groups have had an adverse impact on BOT[5] projects, all of which have appointed contractors and charge high fees, producing social tensions and injustices. Dr Nguyen Si Dung, former Vice-Chairman of the Office of the National Assembly, said that charging BOT fees, which is seen as theft, must be terminated (Mai Ha, 2017). Former Deputy Prime Minister Phan Van Khai also opposed BOT projects on old highways (VietnamNet, 2017b).

According to Daniel Treisman (2000), the best way to fight corruption is to protect the legal system; fighting corruption among government officials helps restore people's faith in the fair execution of the legal system. Another way to reduce corruption is through democracy. Acemoglu and Robinson (2012) argue that sustainable economic success can be achieved only with good, efficient and "inclusive" institutions. This issue is very pertinent to Vietnam, as a few examples from Vietnam's history illustrate. In 1977, Doan Xa commune in Northern Vietnam experienced the paradox that people were hungry while the fields were deserted because they did not support the cooperative policy of the state. The commune-level officials met, and decided to apply the "Khoán hộ" policy.[6] Productivity increased significantly because people actively produced and cultivated for themselves. Institutional changes made this community one of the wealthiest communes in the district (VietnamNet, 2017a).

The provinces of Southwest Vietnam changed the stagnant system of budget subsidies that had prohibited the circulation of goods under the state's motto: "Buy like a robber, sell like a giver", and "Agriculture must sacrifice for industrialization". Pricing does not, according to the market, act as a disincentive, leading to food shortages. Local officials decided to raise the purchase price based on the market price. The price increase stimulated production in

---

5 BOT (Build–Operate–Transfer) is a form of project financing, wherein a private entity receives a concession from the private or public sector to finance, design, construct, own, and operate a facility.

6 These Party members were criticized and were not given Communist Party cards for allegedly going against the Party Central Committee. This model was then replicated throughout the district and the whole city, instead of concentrating farming land on cooperatives.

these provinces, reaching a very high level of mobilization and even exceeding the standard quantity set by the Party Central Committee in some places.[7]

In September 1979, the Party Central Committee's 6th plenum discussed the removal of a number of old positions and institutions, criticized the weaknesses of the model, and decided to free up production, allowing an open mindset on economic matters. In 1986, an important milestone was reached with the Sixth National Party Congress, which officially changed the basic institution of the economy and recognized the commodity economy, the multi-sectoral economy. The biggest breakthrough was to invite overseas Vietnamese experts from Western countries to advise the government on the basic steps needed to develop the country under the market system (Đặng Phong, 2016b).

When institutions change as the social economy develops, lives improve and people become happier which, in turn, increases the value of trust. Vietnam has seen the importance of institutions. In this context, the 2011–2020 socio-economic development strategy states that Vietnam will continue to build institutions and enhance transparency and officials' responsibility to society.[8] The current social situation requires great determination in achieving institutional reform to build trust among the population. Fostering solidarity among the Party Central Committee, localities, and citizens is essential to build inclusive institutions which empower people to develop and monitor policies. The Ministry of Industry and Trade also needs to abolish many provisions that hinder production, and to allow enterprises to be more active and reduce corruption (Bạch Dương, 2017).

7 Continuing to Reform: towards the SDGs

Vietnam's economic reforms have resulted in many successes. After 30 years of reform, the average income has increased 3.5 times. International trade is

---

7 Officials raised the purchase price from 52 dong / kg under the direction of the state to 1,500 dong / kg and 2,000 dong / kg. An Giang province received 100,000 tons from the Party Central Committee, and mobilized 162,000 tons. At the same time, localities acquired and sold enough food to feed 3 million people in Ho Chi Minh City, so that they did not have to eat coix seed instead of rice.

8 This includes reforming administrative procedures in the process of institution building; summarizing practical experience and enhancing dialogue between the state, enterprises and people; expanding democracy and promoting the role of independent organizations and consultants; independently building institutional and national standards on administrative procedures; drastically reducing current procedures; publicizing standards and administrative regulations so that people can oversee their implementation; enhancing transparency and responsibility in public activities.

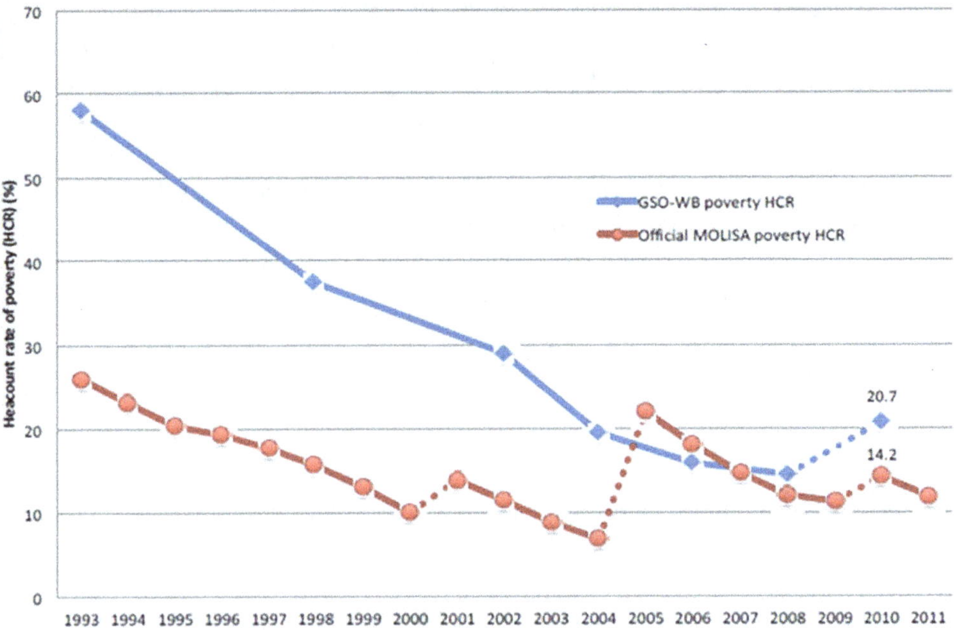

FIGURE 13.4 Progress at reducing poverty using GSO-WB and MOLISA monitoring systems
SOURCE: WORLD BANK (2012)

flourishing due to investments from over 100 countries with a total value up to US$257 billion (as of December 2017). Developmental achievements are shared across society. The official poverty rate has fallen significantly, from 60% in 1993 to 14.2% in 2010, taking the poverty standard in rural areas as US$1.03 per capita per day and in the cities as US$1.34 per capita per day. Using the World Bank's poverty standard of US$2.24 per capita per day, the poverty rate is at 20.7% (see Figure 13.4).[9] According to data from the General Statistical Office, income per capita in rural areas was only 36.9% of that in urban areas in 1995, but 52.8% in 2014 (World Bank & Ministry of Planning and Investment of Vietnam, 2016, p. 220). As of 2017, income in Vietnam has increased and Vietnam has been classified as a middle-income country.

9 In Vietnam, there are two main approaches to measuring poverty. The first is used by the Ministry of Labour, Invalids, and Social Affairs (MOLISA) to generate a classification used for determining those who are eligible for the national anti-poverty programme, as well as monitoring poverty over the short term. A separate approach has been used by the General Statistics Office and the World Bank (GSO-WB), principally to examine poverty changes over the long run. On the latter, see: http://blogs.worldbank.org/eastasiapacific/how-we-measure-poverty-vietnam.

However, Vietnam still faces many obstacles on its way towards sustainable development such as low productivity, low levels of centralization of agricultural land, and a large wealth gap. The minority ethnic groups face a widening gap between themselves and the majority groups; they account for 15% of the total population but make up half of the poor. Inequality still exists, especially with regard to access to health care and education. Economic growth without poverty reduction is not sustainable in terms of economy and society. Thus, policies targeting disadvantaged groups in order to reduce inequality will be a crucial factor for sustainable development.

8 Conclusion

Thanks to some major institutional changes, Vietnam has shifted from a planned economy to a successful market economy with continuous economic growth and is now classified as a middle-income nation. From being a war-torn country, it has moved towards a deeper dialogue and cooperative path with other Southeast Asian countries. However, evidence from other countries in the region shows that economic development is positive and sustainable only in combination with numerous other factors including social and environmental factors. Vietnam needs to pay attention to domestic resources, and create conditions for private enterprise to learn about management methods based on international standards regarding sustainable development models. SOEs should undergo an equitization process to avoid losing state assets, and private companies should be promoted within the national economy.

If it is to successfully implement a strategy of industrialization and development, based on the model of the Sustainable Development Goals, there are a number of things that Vietnam needs to focus on, including: improving resource efficiency; adapting to climate change; protecting ecosystems; developing clean energy policies; improving transparency; fighting corruption; closely managing public investment and foreign aid, while avoiding the potentially negative impact of interest groups; ensuring an effective competitive working environment; building a social welfare system in order to support ethnic minorities; and reducing income and wealth gaps and other forms of inequality.

In order to move towards sustainable economic development and improve the quality of its growth, Vietnam must develop renewable energy sources, reduce environmental pollution, create new consensus to build public trust, and continue to promote institutional reforms to meet people's expectations and needs. Looking outwards, it also needs to join hands with other countries to build a peaceful environment, respect the laws of other nations, unite with others to give voice to international issues, and work together for shared sustainable development and prosperity.

Acknowledgement

I would like to thank Prof. Scott Frey, Prof. Michael Lavin, and Prof. Ronald Holzhacker for helpful comments on drafts of this chapter.

References

Acemoglu, D., & Robinson, J. (2012). *Why nations fail. The origins of power, prosperity and poverty*. London: Profile Books.

Bạch Dương (2016, November 1). Lỗ lũy kế của Vinalines lên hơn 4,000 tỷ đồng. *Vneconomy*. Retrieved from http://vneconomy.vn/doanh-nhan/lo-luy-ke-cua-vina lines-len-hon-4000-ty-dong-20161101030834420.htm.

Bạch Dương (2017, September 22). Bộ công thương và quyết định chưa từng có trong lịch sử. *Báo mới*. Retrieved from https://baomoi.com/bo-cong-thuong-va-quyet-dinh-chua-tung-co-trong-lich-su/c/23346208.epi?utm_source=dapp&utm_campaign=share.

Bằng, L. (2017, June 27). Việt Nam mong trở thành 'hổ lớn' hay chỉ là 'mèo nhỏ'. *Vietnamnet*. Retrieved from http://vietnamnet.vn/vn/kinh-doanh/dau-tu/viet-nam-mong-tro-thanh-ho-lon-hay-chi-la-meo-nho-380565.html.

Bộ Công Thương (2017). Retrieved from http://vbpl.vn/bocongthuong/Pages/vbpq-van-ban-goc.aspx?ItemID=121589.

Burnside, C., & Dollar, D. (2016). Aid, policies, and growth. Policy research working paper 1777. Washington, DC: World Bank. Retrieved from https://papers.ssrn.com/sol3/papers.cfm?abstract_id=569252.

Cơ sở dữ liệu quốc gia về văn bản pháp luật (2001). Government's Prime Minister decision on approval of "Educational development strategy 2001–2010". Retrieved from http://vbpl.vn/bogiaoducdaotao/Pages/vbpq-print.aspx?ItemID=22027.

Công báo (1998). Luật cho phép tư bản nước ngoài kinh doanh ở Việt Nam.

Cooke F. M., Rosazman Hussin, Shi Xueqin, Fei Huang, Khoo Ying Hooi, & Tran Dinh Lam (2014). The state-owned enterprises reform in Vietnam. *Suvannabhumi*, 6(1), p. 56.

Đặng Phong (2016a). Phá rào cản trong đêm trước Đổi Mới. *Tri Thức*.

Đặng Phong (2016b). Tư duy kinh tế Việt Nam 1975–1989. *Trí Thức*.

Dapice, D. O., Haughton, J. H., & Perkins, D. H. (1994). In search of the dragon's tail: Economic reform in Vietnam. Cambridge, MA: Harvard Institute for International Development.

Đức Minh & Hoàng Ly (2017, September 15). TS Vũ Minh Khương: Thúc đẩy tăng trưởng bằng tăng cung tiền tệ, giảm lãi suất giống như thúc người áp huyết cao ăn nhiều thịt bò. *Cafebiz*. Retrieved from http://cafebiz.vn/ts-vu-minh-khuong-day-tang-truong-bang-tang-cung-tien-giam-lai-suat-giong-nhu-thuc-nguoi-ap-huyet-cao-an-nhieu-thit-bo-20170915112051964.chn.

Đức Tuân (2017, April 9). Thủ tướng yêu cầu phải đổi mới tư duy một lần nữa. *Báo chính phủ*. Retrieved from http://baochinhphu.vn/Tin-noi-bat/Thu-tuong-yeu-cau-phai-doi-moi-tu-duy-mot-lan-nua/302656.vgp.

Eckardt, S, Demombynes, G. & Chandrasekharan Behr, D. (2016). Vietnam systematic country diagnostic. Washington, DC: World Bank.

The Economist. The Global Debt Clock. Retrieved from https://www.economist.com/content/global_debt_clock.

Feyrer, J. (2011). Trade and income. Exploiting time series in geography. NBER working paper 14910. Cambridge, MA: National Bureau of Economic Research.

GreenID Vietnam (2017, September 21). Hà Nội trung bình mỗi tuần có ít hơn 2 ngày chất lượng không khí ở mức chấp nhận được. Retrieved from http://greenidvietnam.org.vn/notices/(tcbc)-ha-noi-/trung-binh-moi-tuan-co-it-hon-2-ngay-chat-luong-khong-khi-o-muc-chap-nhan-duoc.html.

The Guardian (2016, May 5). Plans for coal-fired power in Asia are 'disaster for planet' warns World Bank. https://www.theguardian.com/environment/2016/may/05/climate-change-coal-power-asia-world-bank-disaster.

Hieu P. Nguyen & Huyen T. Pham (2011). The dark side of development in Vietnam. Lessons from the killing of the Thi Vai River. *Journal of Macromarketing*, 32(1), 74–86. http://journals.sagepub.com/doi/abs/10.1177/0276146711423666.

Hương Quỳnh (2018, April 9). Thủ tướng: Trước khi gieo hạt, phải tính bán cho ai. *Vietnamnet*. Retrieved from http://vietnamnet.vn/vn/thoi-su/chinh-tri/thu-tuong-nguyen-xuan-phuc-truoc-khi-gieo-hat-phai-tinh-ban-cho-ai-440611.html.

Huỳnh Hải (2017, April 27). Bình Thuận khuyến cáo không dùng nước ngầm gần nhiệt điện Vĩnh Tân. *Newszing*. Retrieved from https://news.zing.vn/binh-thuan-khuyen-cao-khong-dung-nuoc-ngam-gan-nhiet-dien-vinh-tan-post741648.html.

Lattanzio Advisory (2017). Vietnam–Denmark: Transformation of a partnership. Synthesis report. Copenhagen: Ministry of Foreign Affairs of Denmark. Retrieved from http://www.netpublikationer.dk/um/vietnam_denmark_synthesis_report_jan2017/html/helepubl.html#kap05.

Ling Tong (2017, April 30). Vietnam fish death casts suspicion on Formosa steel plant. *Thediplomat*. Retrieved from http://thediplomat.com/2016/04/vietnam-fish-deaths-cast-suspicion-on-formosa-steel-plant/.

Lưu Thủy (2017, May 31). Doanh nghiệp nhà nước: Thua lỗ, thất thoát, tham nhũng. *VTC*, para.7. Retrieved from https://vtc.vn/doanh-nghiep-da-nop-thue-nuoi-ngan-sach-thi-chinh-phu-phai-chi-tieu-cho-hop-li-d326626.html.

Mai Ha (2017, September 8). TS Nguyễn Sĩ Dũng: Chấm dứt ngay thu phí BOT kiểu trấn lột. *Thanh niên*. Retrieved from https://thanhnien.vn/thoi-su/ts-nguyen-si-dung-cham-dut-ngay-thu-phi-bot-kieu-tran-lot-873590.html.

Ngọc An (2017, July 31). Còn điện than, còn ô nhiễm. *Tuổi trẻ*. Retrieved from https://tuoitre.vn/con-dien-than-con-o-nhiem-1361152.htm.

Ngọc Quang (2017, May 25). Nợ công, nợ chính phủ, nợ quốc gia đang ở mức nào. *Giáo dục*: para. 13. Retrieved from http://giaoduc.net.vn/Kinh-te/No-cong-no-Chinh-phu-no-quoc-gia-dang-o-muc-nao-post176886.gd.

Nguyễn Khắc Nhân (2017, August 21). Việt Nam có thể đạt 100% năng lượng tái tạo vào năm 2050. *Mekongcuulongblogpost*. Retrieved from http://mekong-/cuulong.blogspot.com/2017/08/viet-nam-co-at-100-nang-luong-tai-tao.html.

Ohno, K. (2009). Avoiding the middle income trap: Renovating industrial policy formulation in Vietnam. *ASEAN Economic Bulletin*, 26(1), 25–43.

Pak Hung Mo (2001). Corruption and economic growth. *Journal of Comparative Economics*, 29(1), 66–79.

Pham Do Chi & Tran Nam Binh (2002). *The Vietnamese economy: Awakening the dormant dragon*. London and New York: Routledge.

Phạm Thị Quý (2015, May 31). Từ khoán hộ đến đổi mới trong nông nghiệp. *Vanhoanghean*. Retrieved from http://www.vanhoanghean.com.vn/van-hoa-va-doi-song27/ong-kinh-van-hoa35/tu-khoan-ho-den-doi-moi-trong-nong-nghiep.

Pham Van Ninh (2003). Biển Đông 2: Khí tượng thủy văn động lực biển. *ĐHQG*.

Quỳnh Nga (2017a, June 29). Cổ phần hóa chậm do bị trì hoãn lợi ích cá nhân. *Tiền phong*. Retrieved from https://www.tienphong.vn/kinh-te/co-phan-hoa-cham-do-bi-tri-hoan-loi-ich-ca-nhan-1162509.tpo.

Quỳnh Nga (2017b, September 26). Phát triển năng lượng tái tạo: Rào cản từ cơ chế. *Baocongthuong*. Retrieved from http://baocongthuong.com.vn/phat-trien-nang-luong-tai-tao-rao-can-tu-co-che.html.

Tabarrok, A. (2015). The effect of geography on institutions. Development economics course material, MR University. Retrieved from https://www.mruniversity.com/courses/development-economics/effect-geography-institutions.

Thanh Hương (2017, August 21). Số thu từ bauxite gây thất vọng. *Báo đầu tư*. Retrieved from http://baodautu.vn/so-thu-tu-bauxit-gay-that-vong-d68392.html.

Thanh Phan, Tran Dinh Lam, & Van Mon (2011). The economic, cultural, and social life of Bahnar people sustainable development. Ho Chi Minh City: Vietnam National University Ho Chi Minh City Press.

Thu vien quoc gia Vietnam (2012). Quá trình Đảng lãnh đạo thực hiện cải tạo xã hội chủ nghĩa đối với công thương nghiệp tư bản tư doanh ở thành phố Hồ Chí Minh (1975–1986). Retrieved from http://luanan.nlv.gov.vn/luanan?a=d&d=TTcFlGyLKQmq2012.1.16#.

Trần Đình Lâm (n.d.) Foreign direct investment in Vietnam. Retrieved from https://www.bot.or.th/Thai/MonetaryPolicy/NorthEastern/DocLib_seminar56/Tran%20Dinh%20Lam--Foreign%20Direct%20Investment%20in%20Vietnam.pdf.

Treisman, D. (2000). The causes of corruption: A cross-national study. *Journal of Public Economics*, 73(3), 399–457.

Trung Chanh (2017, August 24). Chuyên gia: xây 14 nhà máy nhiệt điện ở DBSCL là quá nhiều. *The Saigon Times*. Retrieved from http://www.thesaigontimes.vn/163998/Chuyen-gia-xay-14-nha-may-nhiet-dien-o-DBSCL-la-qua-nhieu.html.

Tuổi trẻ (2016, August 24). Không đánh đổi kinh tế lấy môi trường gây ảnh hưởng cuộc sống. Retrieved from http://tuoitre.vn/khong-danh-doi-kinh-te-lay-moi-truong-gay-anh-huong-cuoc-song-/1160059.htm.

Tuổi trẻ (2017, July 31). Khiếu nại trong lĩnh vực đất đai chiếm hơn 95%. Retrieved from https://tuoitre.vn/khieu-nai-trong-linh-vuc-dat-dai-chiem-hon-95-1361299.htm.

Văn Bắc (2015, November 3). Kiên quyết xử lý các khu công nghiệp hoạt động kém hiệu quả. *Báo hải quan*. Retrieved from http://www.baohaiquan.vn/Pages/Kien-quyet-xu-ly-cac-khu-cong-nghiep-hoat-dong-kem-hieu-qua.aspx.

Vietnam Management Consultant (2017). Potential of Vietnam wind power. Retrieved from http://www.antconsult.vn/news/potential-of-vietnam-wind-power.html.

VietnamNet (2016, December 23). 12 dự án ngàn tỷ thua lỗ: 'ông lớn' dầu khí, hóa chất vô địch. Retrieved from http://vietnamnet.vn/vn/kinh-doanh/dau-tu/12-du-an-ngan-ty-thua-lo-ong-lon-dau-khi-/hoa-chat-vo-dich-347764.html.

VietnamNet (2017a, September 13). Chuyện 'khoán chui' của ông 'Chủ nhiệm khoai lang' ở 'đêm trước đổi mới'. Retrieved from http://vietnamnet.vn/vn/kinh-doanh/doanh-nhan/chuyen-khoan-chui-cua-/ong-chu-nhiem-khoai-lang-o-dem-truoc-doi-moi-398320.html.

VietnamNet (2017b, December 8). Triết lý về BOT của Nguyên thủ tướng Phan Văn Khải 22 năm về trước. Retrieved from http://vietnamnet.vn/vn/kinh-doanh/dau-tu/triet-ly-ve-bot-cua-nguyen-thu-tuong-phan-van-khai-22-nam-truoc-415566.html.

Voice of Vietnam (2017, September 8). Người dân Hà Nội đang phải hít thở không khí bẩn đến cỡ nào. Retrieved from http://vov.vn/tin-24h/nguoi-dan-ha-noi-dang-phai-hit-tho-khong-khi-ban-den-co-nao-/668331.vov.

Vu Quang Viet (2017, July 21). Không nên duy trì giá điện rẻ. *The Saigon Times*. Retrieved from http://www.thesaigontimes.vn/162679/Khong-nen-duy-tri-gia-dien-re.html.

Vu Thanh Tu Anh & Dam Quang Minh (2006, April 5). Năng lượng gió của Việt Nam: Tiềm năng và triển vọng. Fullbright. Retrieved from http://www.fetp.edu.vn/vn/bao-cao-chinh-sach/nghien-cuu-chinh-/sach/nang-luong-gio-cua-viet-nam-tiem-nang-va-trien-vong/.

Wikipedia (n.d.). Thuyền nhân Việt Nam. Retrieved from https://vi.wikipedia.org/wiki/Thuy%E1%BB%81n_nh%C3%A2n_Vi%E1%BB%87t_Nam.

World Bank (2010). Wind energy resource atlas of Southeast Asia. Retrieved from http://documents.worldbank.org/curated/en/252541468770659342/Wind-energy-resource-atlas-of-Southeast-Asia.

World Bank (2012). Well begun, not yet done. Vietnam's remarkable progress on poverty reduction and the emerging challenges. 2012 Vietnam poverty assessment. Hanoi: World Bank. Retrieved from http://documents.worldbank.org/curated/en/563561468329654096/2012-Vietnam-poverty-assessment-well-begun-not-yet-done-Vietnams-remarkable-progress-on-poverty-reduction-and-the-emerging-challenges.

World Bank Group (2016). Hệ thống đăng ký hộ khẩu ở Việt Nam. Hanoi: World Bank.

World Bank Group (2017). Việt Nam: Tăng Cường Năng Lực Cạnh Tranh Và Liên Kết Của Doanh Nghiệp Vừa Và Nhỏ. Hanoi: World Bank.

World Bank, & Ministry of Planning and Investment of Vietnam (2016). Vietnam 2035: Toward prosperity, creativity, equity, and democracy. Washington, DC: World Bank. Retrieved from https://openknowledge.worldbank.org/handle/10986/23724.

Zak, P. J., & Knack, S. (1998). Trust and growth. *The Economic Journal*, 111(470), 295–321.

CHAPTER 14

Legal Protection of Construction Workers and the Right to Safe Working Conditions: Lessons from Cambodia

Kimsan Soy

1 Contextual Background

Increased globalization and norm diffusion since the late 1980s mean that businesses have a growing responsibility for the safeguarding of human rights (Mutua and Howse, 2001). However, the lack of concerted efforts to control the surge of privatization, deregulation, and liberalization of trade reflects the inability or unwillingness of states to establish a responsive system of governance to effectively tackle the increasing impacts of business activities on human rights (ibid.). The primary objective of economic development should be society-wide improvements, creating equal opportunities and empowering all people, so that they can become involved in the process and benefit from it (Dipholo, Tshishonga, & Gumede, 2014). The success of economic development must not be defined simply by economic indicators such as Gross Domestic Product, but should be seen as a holistic process that incorporates "economic, social, political, cultural and environmental needs of people to promote improvement in the quality of life for all" (ibid., p. 2926).

The impact of globalization on human rights has also been felt in Cambodia. In that context, 1993 can be seen as a significant milestone, with the adoption of a new constitution aimed at promoting liberal democracy and economic development, in an attempt to put the country's violent past behind it. Economic and political development are closely linked: real economic growth is dependent on political stability, and high levels of political violence and human rights violations acted as a brake on Cambodia's economic growth at first. It is only from 2000 onwards that Cambodian politics has been relatively stable (Un, 2011). Since then, the country has maintained a steady growth, averaging 7.7% annually (ADB, 2017), with the garment, tourism, agriculture, and construction sectors the main drivers of the economy. Nevertheless, Cambodia remains one of the poorest countries in Asia, which raises the question: does the country's rapid growth actually serve the needs and interests of all people, especially those marginalized workers who participate in and contribute to

 | DOI:10.1163/9789004391949_015

the development process? The adverse effects of business activities on human rights have become a reality and have grown ever more acute as economic growth continues.

Since the early 2000s, the construction sector has been one of the main pillars of the Cambodian economy; investment in the sector has escalated sharply, with a 267% increase to about $1.65 billion in the first quarter of 2016, compared with $448 million in 2015 (Phnom Penh Post, 2016a). In 2016, there were 473 construction projects affecting 3.7 million square metres of land (ibid.), requiring a labour force of between 175,000 and 200,000 construction workers (Cambodia Daily, 2014). However, construction is not only one of the most important sectors for contributing to the economy of Cambodia; it is also widely considered to be one of the most dangerous, highly prone to work-related injuries and fatalities. As part of the Decent Work country programme for Cambodia, the International Labour Organization (ILO) estimated that in 2009, about 1,500 workers died as a result of accidents at the workplace, mostly on construction sites. Poor working conditions and the lack of safety equipment and safety standards on construction sites are generally the main cause of incidents, which include falling from buildings, building collapses, electric shocks, and so on. In the words of a construction worker quoted in the Cambodia Daily (2014): "I need a harness belt, boots, and helmet to protect my life". The risky nature of the work is arguably one of the main reasons why there has been a shortage of construction workers, in spite of their average earnings — between $6 and $7.5 per day — being higher than other industries, especially the garment sector (Khmer Times, 2015). According to Sovandeth, President of Cambodia Building Worker Confederation, "if there are suitable benefits and a safe job, there will be no shortage [of labour]" (Phnom Penh Post, 2012).

The government has recognized the appalling safety record of the construction industry and pledged to expedite the drafting of building codes, but it has not yet adopted a proper legal framework or monitoring mechanisms related to building and construction (Phnom Penh Post, 2016b). The government also acknowledges that even though building codes could help to protect workers' safety, the implementation of any new laws will not be a quick process (ibid.).

Corruption in the sector, particularly in obtaining construction permits, and the limited sanctions which construction companies face for regulation violations, are further areas of concern and factors contributing to the infringement of construction workers' rights. Most workers are paid by the day without an employment contract, a result of subcontracting labour for short periods, which makes regulating the sector extremely difficult (Cambodia Daily, 2014). "They stay at one site for a short time and then move to another", said a spokesman for the Ministry of Labour and Vocational Training (quoted in Khmer Times, 2015). Moreover, weak monitoring also results in accidents at construction sites often being kept hidden by construction contractors (ibid.).

This chapter examines the right of construction workers to safe working conditions under Cambodia's legal and policy framework as well as under relevant international instruments. It traces progress and challenges in the protection of these rights by looking at the effectiveness of monitoring and enforcement mechanisms, particularly within the ILO and Sustainable Development Goals (SDGs) frameworks. The roles of relevant stakeholders, particularly the government, non-governmental organizations (NGOs), and civil society organizations (CSOs) are examined. These actors are seen as key for responding to the immediate and urgent need for systematic, comprehensive, and coordinated actions to deal with violations of this important right, as well as the root causes of such violations.

2 Linking the Sustainable Development Goals, Human Rights, and ILO Conventions

This chapter focuses on the detailed legal and regulatory framework of Cambodia with regard to worker safety, especially in the construction industry. It demonstrates that the effective implementation of the SDGs by countries will often depend on strengthening longer-term processes which have legally defined requirements for stakeholders and existing state enforcement institutions. For example, while SDG 8 calls for "decent work for all", the implementation and realization of this broad goal is embedded in an ongoing process of creating and adapting Cambodian law to integrate it with existing human rights treaties and ILO conventions.

It is important to underscore the links between these various international agreements. For example, SDG 8 has a specific target (Target 8.8) which expresses the global goal towards worker safety in terms of the need to "Protect labour rights and *promote safe and secure working environments for all workers*, including migrant workers, in particular women migrants, and those in precarious employment" (UN, 2015, emphasis added). This SDG target is inherently linked to existing human rights and ILO treaties and agreements and existing processes of monitoring and action (see DIHR, 2018). It can be linked at the broadest level to the Universal Declaration of Human Rights (UDHR), which states: "Everyone has the right to work, to free choice of employment, to just and favourable conditions of work and to protection against unemployment" (UN, 1948, Section 23.1). In addition, the International Covenant on Economic,

Social and Cultural Rights (ICESCR) refers to the creation of "safe and healthy working conditions" (UN, 1967, Section 7.b).

The ILO's Labour Inspection Convention of 1947, No. 81, contains the most specific commitments to worker safety and implementation of appropriate standards by states. It reads: "This ILO priority governance Convention requires ratifying states to maintain a system of labour inspection in industrial workplaces, to secure the enforcement of legal provisions relating to conditions of work and the protection of workers" (ILO, 1947). The more recent (2006) ILO Promotional Framework for Occupational Safety and Health Convention, No. 187, states that "This ILO Convention requires ratifying states to promote continuous improvement of occupational safety and health to prevent occupational injuries, diseases and deaths, by the development of a national policy, national system and national programme" (ILO, 2006). In addition, human rights treaties provide for special protection with respect to woman and worker safety: the Convention on the Elimination of All Forms of Discrimination Against Women (CEDAW) provides for "The right to protection of health and to safety in working conditions, including the safeguarding of the function of reproduction" (UN, 1979, Section 11.1.f). Thus, while SDG 8 contains a commitment to the global goal of "decent work for all" by 2030, its realization and implementation are part of a decades-long process developing more precise legal and institutional instruments.

3 Decent Work: Overview of the International Legal Framework

Guaranteeing the safety of construction workers is a complex undertaking. There are many factors contributing to the health and safety of workers on construction sites, including the intrinsically dangerous nature of construction work, workers' behaviour and perceptions about safety, working procedures, workload, mental stress, worksite conditions, and safety management (Idrees, Hafeez, & Kim, 2017). For nearly a century, there have been global efforts, mostly led by the United Nations through its various specialized agencies, particularly the ILO, to develop and establish international legal instruments and policies in order to ensure decent work and safety for all workers.

In Cambodia, laws enacted at the national level apply to the whole Cambodian territory. Cambodia is a party to eight out of nine core human rights treaties, and to the eight fundamental ILO conventions, as well as the Employment Policy Convention, No. 122 (ILO, 1964). As a general rule, state parties to international treaties have an obligation not only to implement the treaties, but also to submit reports on implementation to the relevant international committees. Cambodia upholds the principles of direct application of the human rights treaties it has ratified.

3.1 *Decent Work and the ICESCR*

Article 7 of the ICESCR reads:

> The State Parties to the present Covenant recognize the right of everyone to the enjoyment of just and favourable conditions of work which ensure, in particular:
> (a) Remuneration which provides all workers, as a minimum, with:
> i. Fair wages and equal remuneration for work of equal value without distinction of any kind, in particular women being guaranteed conditions of work not inferior to those enjoyed by men, with equal pay for equal work;
> ii. A decent living for themselves and their families in accordance with the provisions of the present Convention;
> (b) Safe and healthy working conditions;
> (c) Equal opportunity for everyone to be promoted in his employment to an appropriate higher level, subject to no considerations other than those of seniority and competence;
> (d) Rest, leisure and reasonable limitation of working hours and periodic holidays with pay, as well as remuneration for public holidays. (UN, 1967)

Within the context of Article 7, this section focuses on remuneration, and safe and healthy working conditions, which are essential for construction workers. In principle, however, human rights are interdependent and interrelated; thus the realization of Article 7, and in particular Article 7(b) depends on the implementation of other provisions of the convention. The Committee on Economic, Social, and Cultural Rights (CESCR) states that the right to work does not mean only the availability and accessibility of employment, but also the acceptability and quality of the employment, which includes safe and healthy working conditions and environment (CESCR, 2006, para 12).

3.1.1 Remuneration

The ICESCR requires that all workers receive "fair wages and equal remuneration" for work of equal value without discrimination. Fair wages go beyond minimum wage and include supplements to wages where workers are engaged in "precarious contracts" such as construction work (CESCR, 2016, para 10).

Equal value can only be determined through "ongoing objective evaluation" of factors such as "skills, responsibilities and efforts required by workers, as well as working conditions" within one company or across companies, rather than trying to compare whether types of work are exactly the same (ibid., paras 12–13, 18). Remuneration also includes other allowances such as contributions to "health insurance, housing and food allowances, and on-site affordable childcare facilities" (ibid., paras 7, 10). The remuneration must be sufficient for workers to afford social security, healthcare and an adequate standard of living (ibid., para 18), all of which have an impact on the occupational health and safety of workers.

3.1.2 Safe and Healthy Working Conditions

The CESCR notes the importance of preventative measures to avoid occupational accidents and disease (ibid., para 25). Rest, leisure, limitations on hours of work, and paid holidays help reduce "work-related stress, accidents and disease", and must be ensured (ibid., para 34). The right to health is closely linked to safe and healthy working conditions (CESCR, 2000, para 11). Article 12.1 of the ICESCR "recognizes the right of everyone to the enjoyment of the highest attainable standard of physical and mental health" (UN, 1967). The convention requires companies to ensure good environmental and industrial hygiene at work sites, by adopting measures to prevent occupational accidents and diseases, minimize contact with harmful chemicals and substances, provide clean water and basic sanitation to workers, and reduce negative impacts on the environment (CESCR, 2000, para 15). The CESCR also takes the view that to prevent occupational disease, training is required on "behavior-related health concerns" and that "urgent medical care" should be available at work sites to deal with accidents (ibid., para 16). Construction workers can access available, acceptable, and quality healthcare outside the construction sites, only when they have economic resources to afford such healthcare (ibid., para 12). Workers must also be able to avail themselves of social security, especially when they are injured while working, regardless of the length of their employment (CESCR, 2008, para 17).

3.2 *Decent Work and the ILO*

The CESCR encourages states parties to incorporate the ILO conventions into their domestic laws in order to strengthen the "effectiveness of measures taken to guarantee the right to work" (CESCR, 2006, para 49). The purpose of this section is to suggest how international labour standards might positively impact the working conditions of construction workers. The ILO's fundamental conventions and other construction worker-related recommendations will be the primary guiding materials.

The ILO (2018b) lists some of the benefits of national compliance with international labour standards; these benefits include the creation of paths to decent work, improving country-specific economic performance, enhancing the global economy, and poverty reduction. The implementation of international labour standards can be especially useful to countries that lack certain labour laws of their own. Cambodia does not currently have sufficient laws protecting the rights of construction workers: it is therefore possible that the application of international labour standards could help to protect potentially vulnerable workers.

That said, the use of international labour standards has limitations. It can be difficult to force governments to uphold international standards — even those they have ratified. While international standards can help to shape national laws, they can sometimes act more as guidelines for social policy. However, there are countries in which international labour standards apply to cases on which national law is silent (ILO, 2018a).

It is also important to keep in mind that international labour standards will not apply exactly as written to every country. Individual nations have unique legal frameworks, which means that the application of international standards may require certain aspects to be amended. Likewise, international standards may need to be altered to better adhere to the diverse needs of a country's citizenry. While it is possible that international labour standards could positively impact the lives of Cambodian construction workers, caution is needed when suggesting that any government adopt standardized recommendations. Cambodia has a unique political climate and Cambodian construction workers also have diverse needs. It would be unrealistic to assume that every international standard related to construction work that has been applied successfully elsewhere could feasibly be implemented in Cambodia. The ILO itself recognizes that international labour standards must be treated flexibly when they are translated into national laws and practices (ILO, 2018a).

3.2.1 ILO and the Decent Work Agenda

Decent work has been defined as productive work for women and men in conditions of freedom, equity, security, and human dignity (ILO, 2008). During the UN General Assembly in September 2015, decent work was adopted as an element of the 2030 Agenda for Sustainable Development. The Decent Work Agenda consists of four pillars: employment creation, social protection, rights at work, and social dialogue. The ILO has published several resources related to decent work, including some that are country- and industry-specific.

The *Toolkit for mainstreaming employment and decent work* (ILO, 2008) provides a conceptual approach to the application of decent work, and gives a detailed overview of the four pillars, under the headings of (1) employment creation and enterprise development; (2) social protection; (3) standards and rights at work; and (4) governance and social dialogue. The Toolkit also includes a self-assessment checklist and guidelines for self-assessment.

One of the region- and industry-specific resources for the Asia-Pacific is the ILO's Regional model competency standards: Construction (ILO, 2015). These standards include skills and knowledge related to construction throughout the Association of Southeast Asian Nations (ASEAN) member states. Emphasis is placed on labour mobility and the need for skills recognition, as well as training systems and the need for improvement in the construction industry. These standards were developed to help identify the competencies needed in construction workplaces to enable the development of training resources. The standards can also be used as reference material for processes such as recruitment and the development of job descriptions. It is recognized that individuals will have varying needs according to personal and cultural circumstances, and so the standards must be flexible (ibid.). The standards include both generic competencies and vocational and technical competencies. While the standards define a general framework for the construction sector, the ILO does not consider them to be all-encompassing (ibid.).

The ILO has also published Cambodia-specific documents related to decent work, including the decent work country profile for Cambodia (ILO, 2012) and the Cambodia decent work programme (ILO, 2016). Cambodia's decent work programme focuses on three priorities: (1) improving industrial relations and rights at work; (2) promoting an enabling environment for decent employment growth and sustainable enterprises, with a particular focus on young people; and (3) improving and expanding social protection and occupational safety and health (ibid.). Although this programme was designed to focus primarily on decent work for garment workers, all of these goals are directly relevant to the Cambodian construction industry, and may help to promote the rights of construction workers.

MacNaughton and Frey (2011) examine the ways in which the ILO Conventions and Decent Work Agenda can work in tandem. Their article explores the potential of using a holistic human rights approach to achieve decent work for all. This approach reframes decent work as a set of specific legal obligations under both international and human rights laws in an effort to move beyond political committees. Three principles from the human rights framework embraced by the UDHR and the ICESCR are key: universality, interdependence, and the equality of all human rights. Applying these three principles to work rights would arguably serve to integrate the ILO's Conventions and Decent Work pillars into a holistic framework (MacNaughton & Frey, 2011). Such a framework could potentially be of use to countries such as Cambodia.

3.2.2 Safety and Health in Construction: A Code of Practice

In *Safety and health in construction* (ILO, 1992), the ILO provides recommendations that can be used by both the private and public sectors to help ensure safety and health in construction work. It is recognized that local circumstances and technical possibilities will determine how far it is practical to follow these guidelines. The ILO offers guide on both general and specific aspects of a variety of construction activities. It also provides "a free, comprehensive, international, and digital training package in occupational safety and health for the construction industry" entitled ILO Construction and OS&H.[1]

4 Cambodia's Legal and Policy Framework on Safe Working Conditions Related to Construction

It is important to state from the outset that the Constitution of Cambodia makes no specific provision for decent work for all; rather, the intention to promote decent work must be inferred from or read into the content of other provisions in the Constitution. A starting point is Article 31 of the Constitution, which can be construed to incorporate or recognize decent work.

4.1 *1997 Labour Law*

In principle, labour rights of Cambodian workers are protected under the Labour Law of the Kingdom of Cambodia adopted in 1997.[2] Specifically, this law sets the minimum standards for occupational safety and health (OSH) of workers that are required of private entities (Chapter VIII). For instance, Article 229 imposes obligations on private entities to maintain standards of hygiene and sanitation at the workplace while Article 230 focuses on appropriate measures to prevent accident risks such as falling, moving heavy objects, handling dangerous apparatus, and fire prevention. The Labour Law also demands that employers provide workers with primary health services (Articles 239–45). For enterprises employing at least 50 workers, a permanent infirmary

1 See https://www.ilo.org/wcmsp5/groups/public/---ed_dialogue/---sector/documents/instructionalmaterial/wcms_161768.pdf.

2 Labour Law of the Kingdom of Cambodia, 1997. Adopted on 10 January 1997 by the National Assembly during its 7th session of its first legislature and promulgated on 13 March 1997.

equipped with adequate materials, bandages, and medicines, and capable of providing emergency care in the event of accidents or illness at the workplace, is compulsory. The cost of providing such basic healthcare should be borne by the employer.

Although the Labour Law does not precisely specify legal consequences or liability incurred by employers for failing to adhere to health and safety standards, it does state that employers are liable for all work-related accidents regardless of the personal status of workers (Article 249). According to Article 248, a work-related accident is defined as one that "happens to a worker during the working hours, whether or not the worker was at fault …, inflicted on the body of the worker … in whatever capacity or whatever place". Every manager of an enterprise has a legal obligation to take appropriate measures to prevent occupational accidents (Article 250). The employer may be exempt from paying compensation in the event that the victim deliberately causes the accident, and compensation may be reduced "if it is proved that the accident was the result of an inexcusable mistake of the victim" (Articles 253–255). One can thus conclude that the Labour Law generally guarantees OSH of workers, and compensation in case of infraction.

However, it can be argued that the Law only provides protection to workers in the formal sector, particularly those in employer–employee relationships. This argument is based on Article 1, which broadly states the purpose and scope of the Law as follows: "this law governs relations between employers and workers resulting from employment contracts". The term "worker" is defined as "every person of any sex and nationality, who has signed an employment contract in return for remuneration, under the direction and management of another person, whether that person is a natural person or legal entity, public or private" (Article 3).

In practice, there are many informal economy workers in Cambodia, as employers resort to this form of labour relation in order to bypass costly bureaucracy and adherence to labour standards. Construction workers are reported as informal economy workers because they usually do not have employment contracts. They work and get paid by the day. For this reason, they are considered as being particularly vulnerable to workplace accidents and poor standards of OSH.

4.2 *Regulations Concerning Construction and Safe Workplaces*

The 1997 Labour Law entrusts the Ministry of Labour and Vocational Training (MoLVT) with the power to issue special regulations or sub-decrees (Prakas) relating to certain professions or certain types of work (Article 231). Although the MoLVT spearheads the campaign to regulate and govern OSH, other ministries are also authorized and obliged to issue special regulations to ensure OSH within their mandate and specialty. For instance, in relation to OSH on construction sites, the Ministry of Land Management, Urbanization and Construction issued two key regulations: Anukret No. 86 dated 19 December 1997 concerning construction permits; and Prakas No. 32 dated 20 March 2001 concerning the administration of construction sites, issued with the intention of maintaining public order and security and safety on construction sites.

In addition, the Ministry of Economy and Finance issued a joint Prakas with MoLVT — Prakas No. 659 on monetary fines for violations of the labour law, which is also applicable to workers in the construction sector. The annex to this joint Prakas consists of a list of 105 labour offences which are subject to fines, including absence of employment contract, exploitation by labour contractor, mass layoff without complying with legal procedures, failure to guarantee hygiene and safety of workers at the workplace, non-response to work-related accidents and compensation, restrictions on joining union, and lack of cooperation with labour and medical inspectors.

To be eligible to receive a construction permit, these regulations require construction companies to ensure public safety and well-being, including controlling environmental impact, on construction sites. During the construction period, inspectors from the municipal and provincial administration visit the sites: if the inspectors observe that the construction company is failing to meet its obligations to ensure public safety and well-being, the construction permit may be suspended or withdrawn completely (Anukret No. 86). However, there seems to be no regulation that holds the construction company accountable for failing to protect worker safety and well-being. Although there are some existing mechanisms regulating OSH standards in the construction sector, they are vague and confusing in nature. As a consequence, it is difficult to translate these ambiguous provisions from words into deeds. Clearly, a new law is needed which deals specifically with OSH of workers in the construction sector, and which sets out a range of specific conditions on construction companies which apply for a construction permit, such as providing safety gear, lifting equipment or proper safety training to workers. Such a new law is also needed to consolidate all relevant existing regulations and to standardize them into one comprehensive construction law. The government has acknowledged this shortcoming and has been working on a draft law on construction since 2014.

The enforcement of such a law, once enacted, will be another big challenge in terms of resources and time for implementation. First, the costs implicit in implementing the new law could become so high that contractors and developers are discouraged from complying, and instead continue to adopt unsafe practices. Second, the government must prepare a significant number of trained personnel or experts in a number of fields who can provide the necessary training about OSH standards not only to construction workers themselves but more importantly to employers or developers, to make sure that their practices comply with the law.

4.3 *Occupational Health Standard in the Construction Sector*

The MoLVT has the main responsibility for ensuring OSH standards in the workplace, and three departments are mandated with OSH roles under the MoLVT: Department of Occupational Health and Safety, Department of Labour Inspection, and Department of Social Security which has now become the National Social Security Fund. Between 2001 and 2005, it is reported that the three departments jointly conducted 48 investigations, including some involving construction sites, and found serious OSH concerns such as insufficient ventilation and hot dust in the workplace, injury by falling objects, and falling from buildings (MoLVT, 2011). Current data seem to be unavailable.

In cooperation with the ILO, the MoLVT also developed and established the first OSH Master Plan 2009–2013, which was officially launched on 28 April 2009. The Master Plan (MoLVT, 2009) aims to promote OSH standards by identifying six priority action points:

1. Strengthen national OSH systems
2. Improve safety and health inspection and compliance with the Labour Law
3. Promote OSH activities by employers' and workers' organizations
4. Implement special programmes for hazardous occupations
5. Extend OSH protection to small enterprises, and rural and informal economy workplaces
6. Promote collaborative actions in the fields of hazardous child labour and HIV/AIDS projects to achieve stronger compliance.

As a result of the Master Plan, a number of ministerial regulations relating to OSH have been issued under MoLVT, four of which specifically focus on OSH standards on construction sites:

- Prakas No. 077 dated 30 March 2011 concerning information at the construction site
- Prakas No. 075 dated 30 March 2011 concerning sanitation at the construction site
- Prakas No. 076 dated 30 March 2011 concerning the prevention of risks associated with changing weather conditions at the construction site
- Prakas No. 078 dated 30 March 2011 concerning storage, waste management and cleanliness at the construction site.

In addition, a Department of Labour Inspection was created under the 1997 Labour Law (Chapter VIII, Section 2), with a key role in ensuring good working conditions and providing training concerning health and safety of workers. There are 91 OSH inspectors (19 based in the ministry and 72 based in municipalities and provinces) who undertake inspection visits (MoLVT, 2011). Whereas in the past, regular inspection visits were made only in Phnom Penh City, since 2010 OSH inspections have been extended nationwide. During these visits, the role of the OSH inspector is to:

- conduct technical inspections on industrial hygiene (level of lighting, noise, dust, vibration and other harmful physical hazards)
- build an OSH network in the company
- promote medical check-ups for workers
- provide OSH protection to workers and prevent workplace accidents and occupational diseases
- advise employers and workers on OSH-related regulations and ministerial orders (ibid.).

If labour inspectors identify potential health and safety concerns and standards which do not meet the requirements of the Labour Law, they can issue a prior notice as a warning to managers of the company in question to address the problems (1997 Labour Law, Article 233). The notice should contain details of the specific violations of health and safety standards and demand that the employers address such infractions within a given deadline. The OSH inspector will write up an official report to the MoLVT for further action against the employers if they fail to respond by the deadline (Article 236). If the concern is particularly serious or requires immediate action, the inspector can issue an official report in writing directly without serving prior notice (Article 234). It is the mandate of the National Social Security Fund to hold employers accountable for occupational accidents and injuries. To regulate this matter, the MoLVT issued Prakas No. 243 dated 10 September 2002 concerning the notice for work-related accidents, formula for compensation and degree of disability.

4.4 *Cambodia's Obligations under the SDGs to Guarantee Decent Work for Construction Workers*

The term 'decent work' was coined by the ILO in 1999 as a strategic approach to fight poverty by improving working conditions and providing more social protections for all workers including those in the informal economy (ILO, 1999). Simply put, decent work is indispensable to a quality life with human dignity, which allows workers to meet basic needs for food, education, healthcare, housing, and clothing (Frey and MacNaughton, 2016). This newly emerging norm failed to make its way into the eight Millennium Development Goals (MDGs) established in 2000, according to which leaders from 189 countries pledged their collective commitment to protect human rights, promote global peace, and halve poverty globally by 2015. Despite significant success in reducing poverty (Sachs, 2012), it was strongly felt that the MDG framework missed a key piece of poverty's jigsaw puzzle, namely full employment and decent work. The ILO led a campaign to include decent work in its fight against poverty by establishing a monitoring and reporting system within the MDGs which adopted indicators for decent work. As a consequence, the decent work norm was eventually adopted within the SDGs — the successor of the MDGs — as Goal 8 on decent work and economic growth (Frey and MacNaughton, 2016).

Cambodia committed itself to the MDGs and realized four of the eight targets in child care, communicable diseases, maternal health, and partnership, which was considered a remarkable achievement given its struggle to transition from decades of civil wars. Although Cambodia's poverty goal of 19.5% was not met, there was a steady improvement and poverty was reduced from 47.8% in 2007 to 19.8% in 2011 (RGC, 2013). However, it is not possible to see how this poverty reduction benefited vulnerable groups such as construction workers, since there were no specific indicators to show this.

Given its success with the MDGs, Cambodia fully committed to the SDGs. As with other countries, the transition from the Cambodian Millennium Development Goals to the Cambodian Sustainable Development Goals (CSDGs) is an ongoing and challenging process. The main challenge is the localization process, determining how each SDG is integrated into national and local policies, plans, and strategies, which requires significant resources, and complex coordination between line ministries and local authorities. Poch Sovanndy, Deputy Director General of Planning at the Cambodian Ministry of Planning, acknowledged the complexities of the task in 2017: "It has been a long and challenging process but we are making good progress. We had a lot of coordination with line ministries to review the goals and targets for their respective sectors but the main difficulty is the availability of data to establish the baselines and targets" (quoted in Khmer Times, 2017).

In 2015, the Cooperation Committee for Cambodia (CCC), an umbrella NGO bringing together many international and local civil society organizations, prepared a report on Cambodia's preparations for implementing the SDGs, noting that the challenge for the country is to adopt SDGs that "are feasible for the local context and that do not pose significant barriers to growth" (CCC, 2015, p. 3). NGOs have organized numerous events around implementation of the Cambodian SDGs; the Asia–Europe Foundation, for example, organized a workshop in Phnom Penh in 2015 on the SDGs and national implementation challenges. Tin Ponlok, Secretary General of the National Council for Sustainable Development in the Cambodian Ministry of Environment, spoke about the importance of sustainable development in Cambodia's future national planning.

Various ministries took part in the event, including the Ministry of Planning and the MoLVT (ASEF, 2015).

However, the localization process of CSDGs may be delayed or complicated further by other factors such as changing political and socio-economic priorities. The realization of the CSDGs requires considerable improvements in human rights, good governance, rule of law, corruption, transparency, and accountability — a common problem in many transitional countries such as Cambodia. Having escaped from protracted civil wars and mass killings, maintaining peace and security and achieving economic growth have been the overriding priorities of the Cambodian government (CCC, 2015).

With regard to the prosperity-related SDG targets, including Goal 8 on decent work and inclusive growth, approximately 70% of the targets have been addressed in the National Strategic Development Plan (NSDP). However, with regard to Goal 8, the priorities in the NSDP focus mainly on Target 8.5 to ensure gender equality between women and men and equal treatment between them in the workplace, and Target 8.7 to eradicate forced labour, modern slavery, child labour, and human trafficking. Improving working conditions and labour protection for workers in the informal economy sector, such as construction, do not feature in the NSDP or in other national policies. Given that the deadline for achieving the SDGs is 2030, it could be argued that it is too early to make an assessment of the Cambodian government's performance concerning Goal 8 as it might relate to construction workers; it is possible that this sector may be seen as more relevant and become a government focus at some future stage. That said, Cambodia faces many practical challenges if it is to realize all the SDG targets, including issues around the rule of law, governance, weak capacity of institutions and human resources, poor cooperation and strained relationships among development partners and stakeholders and, most importantly, the lack of localization of the SDGs (CCC, 2015).

5 Challenges Facing Cambodia in Guaranteeing Decent Work for Construction Workers

The continuing uncertainty about the ability of the labour law, regulations, and policies to protect labour rights of construction workers, described in the above analysis, has been exacerbated by a serious lack of concerted effort and attention from major actors — the government, NGOs, civil society, and donors. A retrospective view of growth in Cambodia since the early 2000s shows that the construction sector has contributed substantially to the country's economic transformation. Yet it is equally clear that regulating the sector to protect construction workers has failed to keep pace with the growth. It is no exaggeration to say that construction workers receive virtually no attention from the key actors to address their poor working conditions.

5.1 *Power Relations and the Lack of Political Interest in Improving Decent Work in the Construction Industry*

In these decades of steady economic development, there is a risk that technical solutions to fix loopholes in laws, regulations, and practices to protect workers from getting harmed are thwarted by the political reality in Cambodia. Economic growth often breeds a human rights 'governance gap', especially in a developing economy with institutionalized corruption like Cambodia, where the government shows no real interest in imposing more regulations and obligations on businesses to guarantee decent work (CCHR, 2010). Furthermore, in their pursuit of national economic development, policy makers in any developing economy tend to adopt laws and regulations that attract investment and employment, while they also try to ensure that legal frameworks protect workers' rights, directly or indirectly, to decent work and a safe working environment (ibid.). In Cambodia, the government seems to lack the capacity and the will to develop and apply such legal obligations for fear of losing investment; businesses are able to take advantage of this fear and exercise influence over the government and its policies (ibid.). This section therefore argues that the government's economic approach and its relations with Cambodian business are the major bottlenecks in developing rigorous standards and laws that could protect construction workers from dangerous working conditions.

By 2012, four-fifths of the construction companies in Cambodia were owned by Cambodian investors while the remainder belonged mostly to investors from ASEAN and China, with 5.5 per cent and 2.7 per cent from the Republic of South Korea and European countries, respectively (Bruni, Luch, & Kuoch, 2013, p. 24). This raises serious questions as to the connections between the Cambodian construction industry and political power.

Despite the outward adoption of liberal democracy since 1993, in practice Cambodia is generally recognized as an authoritarian regime which maintains control through a powerful patronage system. According to John D. Ciorciari, in its ideal form, a patron–client relationship confers mutual benefit on both parties — a patron of higher authority and status whose influence is used to give protection and benefits to the client, in exchange for various forms of support and assistance (Ciorciari, 2013). In assessing the overall quality of democracy in Cambodia, Un Kheang asserts that democratization has been derailed and "has evolved from unstructured competitive authoritarianism toward an authoritarianism characterized by the presence of a stable hegemonic party system", following a coup d'état in 1997 and subsequent wins for the party in the 1998 general election (Un, 2011, p. 546). Kelsall and Heng (2014) argue that the ruling Cambodia People's Party (CPP) has conducted a campaign for power consolidation, achieving almost absolute control over the country by establishing a political settlement between the government and business (see also Laws & Leftwich, 2014). This politico-economic strategy has enabled poverty reduction by providing mass employment, thereby gaining regime legitimacy through economic growth and political stability, but at the expense of human rights.

The CPP has built a patronage system involving close ties with an inner circle of local business associates, military generals, top politicians, and relatives and family members of the Prime Minister Hun Sen, and an outer circle of development partners and foreign investors (Kelsall & Heng, 2014). In some situations, it is difficult to distinguish between government and business figures, not only because of their close financial relationships (CCHR, 2010), but because the two roles are often played by the same individuals. Un Kheang (2015) contends that this patronage regime leads to corruption, as business tycoons (as clients) are expected to contribute financially to their patrons, usually top officials in the CPP, or to support government development projects in various forms such as donations for humanitarian causes, while in fact they are courting influence, favour, and protection for their businesses. The same applies to government officials at all levels of the hierarchical power structure, as they contribute part of their income — normally earned through corrupt practices — to the party, while trying to keep as much as possible for themselves. In this way, they become loyal protectors of the party: the CPP receives their political support, and they are rewarded in return. The CPP can thus maintain the status quo by using elements of the socio-political patronage system. In response to the trenchant criticism of enduring widespread human rights violations in the country, the CPP asserts that it has brought peace, stability, and rapid economic development. Un Kheang describes the political economy of the CPP as a balancing act performed by the party in order to just keep the democratic institutions functioning, while the party retains its grip on power (ibid.).

This view is further reinforced in a recent controversial report of the NGO Global Witness on the wealth of Cambodia's ruling family (Global Witness, 2016). The report reveals that the ruling family has steadily accumulated its wealth based on "a huge network of secret deal-making, corruption and cronyism" (ibid., p. 3) in a wide range of business sectors such as trading, finance, tourism, energy, mining, media, transport, construction, and real estate, which helps to strengthen its political stronghold and hegemony. The ruling family has interests in seven major construction companies, which are often involved in government development projects (ibid., p. 10). High-ranking government officials and ruling party law makers are also closely involved with big business, including the construction industry (ibid., p. 16). There are currently about 700 business tycoons in the country, compared to just 20 a decade ago. A business tycoon ("oknha") is an honorary title given to business people who make donations of at least $100,000 to the government. Ideally, this donation should be used to contribute to the country's development and acts as a display of commitment from the business tycoon to run ethical and responsible businesses in a way that will help develop the country and its people. However, the role of business tycoons is more aspirational than actual; the majority of them are calculating and opportunistic business people who are seeking to secure government contracts or protection for the operation of their businesses (Global Witness, 2016).

In light of the above observations, the sluggish progress in developing rigorous regulations in the construction sector can be understood by a closer examination of the power relationship between the ruling party and the industry. It is axiomatic that strict construction laws will not be in the interests of the businesses and those in authority.

5.2 *Lack of Involvement from NGOs and Weak Civil Society*

Despite their widespread occurrence, violations of labour rights in the Cambodian construction sector remain largely undocumented and receive virtually no attention from local advocacy NGOs and CSOs. This raises a number of questions. What drives advocacy NGOs and CSOs to prioritize one particular human rights issue over another? Why do labour rights of construction workers (or the lack thereof) seem to be widely ignored? Do advocacy NGOs decide independently on the issues that should be addressed in the context of the priorities and demands of local people? This section argues that the selection of particular human rights issues by NGOs and CSOs is very much influenced and steered by the preference of donor countries or international NGOs (INGOs), which results in a widespread neglect of, or lack of interest in, issues perceived as less important — including the labour rights of construction workers.

NGOs play a crucial role in upholding key international human rights treaties through local empowerment, advocacy, and monitoring of implementation. Since the 1960s, the United Nations — through the Economic and Social Council — has officially granted consultative status to NGOs due to their increasingly significant role in providing information concerning human rights violations across the world (Stamatopoulou, 1998). NGOs can also mobilize social movements against governments and for democracy and human rights.

Franz (1987) draws the distinction between governments reaching society at the macro level, and NGOs reaching it at the micro level. In other words, governments prepare national policies and measures for society as a whole, whereas NGOs represent the needs and interests of specific groups of civil society. Governments are highly susceptible to ineffectiveness, inefficiency, and subjectivity in addressing the issues of certain social groups (ibid.), whereas NGOs are in a better position to closely interact with local communities, especially in remote areas, and to understand the heart of their problems. As a consequence, NGOs are able to access more objective information at local level by using participatory methods, enabling them to carry out projects more effectively and efficiently compared to government agencies (ibid.).

Given the limited capacity of the government, due to its poor human resources, systems, and institutions, the reconstruction and rehabilitation process undertaken by Cambodia since 1993 has been donor driven in nearly all aspects, involving international development cooperation and assistance (Curtis, 1998). There are a number of studies that explain power relations between donors and NGOs and how they affect the credibility and accountability of NGOs' work, which could shed some light on the lack of attention to labour rights of construction workers in Cambodia. In general, advocacy NGOs and CSOs in Cambodia are highly donor driven, which undermines their autonomy in setting an agenda that responds to particular human rights violations. As a consequence, they tend to be more assimilated into the demands of donors rather than the actual needs of the local population (Chambers and Pettit, 2004). In assessing the accountability of local NGOs, Johnson and Prakash (2007) argue that some NGOs and CSOs conceive projects which follow the donors' agenda and resources, as these are strategically vital to the continued existence of the organizations and their activities. This attachment to donors reinforces the government's negative perception of the role and work of NGOs, as representing the donors' agenda (Marcussen, 1996), imposing foreign views, and overstating human rights issues in order to secure financial support from donors (Un, 2006). In turn, this affects the realization of human rights that depend heavily on a good relationship between NGOs and the government.

The vast majority of construction businesses in the country (80%) are owned by Cambodians, and the rest are owned by foreigners who are mostly Chinese and Korean. There is very little involvement in the construction sector from donor countries, making it very different to the garment sector in which many big brands from donor countries are involved, bringing extraterritorial obligations from their home countries to conduct socially responsible business. Data from the Ministry of Land Management, Urban Planning and Construction suggest that the construction industry employed approximately 260,000 workers a day in 2017, an increase of about 30% compared to the year before (Phnom Penh Post, 2018b). Despite its importance as the second-largest driver of the country's economy, its high employment rate, and the dangerous nature of the work, the construction sector has failed to attract global attention for the improvement of working conditions as compared to the garment sector, which currently employs some 800,000 workers (Phnom Penh Post, 2018a).

The lack of general interest in improving working conditions on construction sites was highlighted by a protest of about 50 construction workers who demanded that the government and general public "pay more attention" to their "terrible conditions". They also called on the government to ensure regular safety inspections at construction sites and "to treat all workers fairly", referring to the different levels of attention paid by the government, NGOs, and donors towards other sectors, particularly the garment sector (Khmer Times, 2015). This protest yielded no substantial changes in the attitude of the government or other stakeholders, perhaps due to the small number of protesters taking part. Their concerns seemed peripheral to workers in other sectors, particularly the garment sector that could easily mobilize hundreds of workers, thereby attracting more attention. Finally, weak and disorganized labour unions in the construction sector have also contributed to the lack of effective representation of construction workers' rights and interests. This problem was extensively discussed in the report of a fieldwork assessment on the leading federation of construction unions, the BWTUC (Building and Woodworkers Trade Union Federation of Cambodia) which saw a fall in membership from 23 affiliated unions in 2010 to just 12 by 2011 (LO-Norway, 2012).

6 Conclusion

This chapter began with a brief overview of the existing legal protection related to occupational health and safety of construction workers, and institutional mechanisms for labour inspection and monitoring in Cambodia. These laws and institutions are critically important for the achievement of SDG 8, decent work for all. One clear lesson from the above analysis is the need to fully recognize construction workers as part of the formal economy with strong employment relations. NGOs and civil society remain weak in this area, with almost no groups advocating rights for construction workers or giving voice to their concerns.

The chapter has examined international labour and human rights standards in order to highlight some ways in which the health and safety of Cambodian construction workers may be more adequately protected. There is a range of options that the government could take, from regulating foreign businesses and other employers to helping promote employee unionization. However, it is not only the Cambodian government that needs to take action. Workers, unions, governments, and businesses all need to come together to help implement health and safety standards within the Cambodian construction sector.

Finally, while it is possible that international labour and human rights standards could positively impact the lives of Cambodian construction workers, it is important to be cautious when suggesting that any government adopt standardized recommendations. Cambodia has a unique political climate and Cambodian construction workers also have a variety of diverse needs. It would be unrealistic to assume that every international standard related to construction work could feasibly be implemented in Cambodia. The goal of this chapter is to help promote the rights of Cambodian construction workers, not to suggest the implementation of standards that are inappropriate or that could end up hurting rather than protecting construction workers.

Acknowledgements

This study was conducted as part of the 2nd Regional Human Rights Research Initiative programme, organized by the Raoul Wallenberg Institute of Human Rights and Humanitarian Law. The research was also partially supported by Pro Bono Students Canada (PBSC) law students from the University of Victoria, namely Madeline J. Adam and Suzy Flader and Ratana Ly, led by Victor V. Ramraj.

References

ADB (2017). Member fact sheet: Cambodia. Asian Development Bank. Retrieved from https://www.adb.org/sites/default/files/publication/27757/cam-2017.pdf.

ASEF (2015). Cambodia on implementation of Sustainable Development Goals. Asia–Europe Foundation. Retrieved from http://www.asef.org/projects/themes/sustainable-development/3676-workshop-on-sustainable-development-goals-national-implementation-challenges/3714-cambodia-on-implementation-of-sustainable-development-goals.

Bruni, M., Luch, L., & Kuoch, S. (2013). Skills shortages and skills gaps in the Cambodian labour market: Evidence from employer skills needs survey. ILO Asia-Pacific Working Paper Series. Geneva: International Labour Organization.

Cambodia Daily (2014, November 20). Construction workers' lives hang in the balance. Retrieved from https://www.cambodiadaily.com/news/construction-workers-lives-hang-in-the-balance-72679/.

CCC (2015). Analysis of Cambodia's preparedness for the implementation of the Sustainable Development Goals: Challenges, opportunities and financing. Cooperation Committee for Cambodia. Retrieved from http://www.ccc-cambodia.org/en/download?file_id=807&action=view&view_file_id=91325d9cb47415f97c31bac2f6af3954b87834be.

CCHR (2010). Business and human rights in Cambodia: Constructing the three pillars. Report Summary. Phnom Penh: Cambodian Center for Human Rights.

CESCR (2000, August 11). General comment no. 14: The right to the highest attainable standard of health (Art. 12 of the Covenant). E/C.12/2000/4. Committee on Economic, Social, and Cultural Rights.

CESCR (2006, February 16). General comment No. 18: The right to work (Art. 6 of the Covenant). E/C.12/GC/18. Committee on Economic, Social, and Cultural Rights.

CESCR (2008, February 4). General comment no. 19: The right to social security (Art. 9 of the Covenant). E/C.12/GC/19. Committee on Economic, Social, and Cultural Rights.

CESCR (2016, April 27). General comment No. 23: The right to just and favorable conditions of work (Art. 7 of the Covenant). E/C.12/GC/23. Committee on Economic, Social, and Cultural Rights.

Chambers, R., & Pettit, J. (2004). Shifting power to make a difference. In L. Groves & R. Hinton (Eds), *Inclusive aid: Changing power and relationships in international development* (pp. 137–162). London and Sterling, VA: Earthscan.

Ciorciari, J. (2013). China and Cambodia: Patron and client?. International Policy Center Working Paper Series No. 121. Ann Arbor, MI: Gerald R. Ford School of Public Policy, University of Michigan.

Curtis, G. (1998). Cambodia reborn: The transition to democracy and peace. Geneva: United Nations Research Institute for Social Development.

DIHR (2018). Human rights and the SDGs. Copenhagen: Danish Institute for Human Rights. Retrieved from https://www.humanrights.dk/our-work/sustainable-development/human-rights-sdgs.

Dipholo, K. B., Tshishonga, N., & Gumede, N. (2014). A double-edged sword? Reflection on the development of the San community of Botswana. *Mediterranean Journal of Social Sciences*, 5(20), 2926–2933.

Frantz, T. R. (1987). The role of NGOs in the strengthening of civil society. *World Development*, 15(S1), 121–127.

Frey, D. F., & MacNaughton, G. (2016) A human rights lens on full employment and decent work in the 2030 Sustainable Development Agenda. *Journal of Workplace Rights* (online). https://doi.org/10.1177/2158244016649580.

Global Witness (2016). Hostile takeover: The corporate empire of Cambodia's ruling family. London: Global Witness. Retrieved from https://www.globalwitness.org/en/reports/hostile-takeover/.

Idrees, M.D., Hafeez, M., & Kim, J.Y. (2017). Workers' age and the impact of psychological factors on the perception of safety at construction sites. *Sustainability*, 9(5), 745.

ILO (1947). ILO convention no. 81. Labour inspection convention. Geneva: International Labour Organization. Retrieved from http://blue.lim.ilo.org/cariblex/pdfs/ILO_Convention_81.pdf.

ILO (1964). ILO convention no. 122. Employment policy. Geneva: International Labour Organization. Retrieved from https://www.un.org/ruleoflaw/files/Employment%20Policy%20Convention,%201964%20(No.%20122).pdf.

ILO (1992). *Safety and health in construction*. Geneva: International Labour Organization.

ILO (1999). Report of the Director General: Decent work, ILC 87. Geneva: International Labour Organization.

ILO (2006). ILO convention no. 187. Promotional framework for occupational safety and health convention. Geneva: International Labour Organization. Retrieved from http://blue.lim.ilo.org/cariblex/conventions_10.shtml.

ILO (2008). *Toolkit for mainstreaming employment and decent work*. Geneva: International Labour Organization.

ILO (2012). *Decent work country profile: Cambodia*. Geneva: International Labour Organization. Retrieved from https://www.ilo.org/wcmsp5/groups/public/---dgreports/---integration/documents/publication/wcms_185267.pdf.

ILO (2015). Regional model competency standards: Construction. Geneva: International Labour Organization.

ILO (2016). Kingdom of Cambodia decent work country programme 2016–2018. Geneva: International Labour Organization.

ILO (2018a). How international labour standards are used: Models and targets for labour law. Retrieved from http://www.ilo.org/global/standards/introduction-to-international-labour-standards/international-labour-standards-use/lang--en/index.htm.

ILO (2018b). The benefits of international labour standards. Retrieved from http://www.ilo.org/global/standards/introduction-to-international-labour-standards/the-benefits-of-international-labour-standards/lang--en/index.htm.

Johnson, E., & Prakash, A. (2007). NGO research program: A collective action perspective. *Policy Science*, 40(3), 221–240.

Kelsall, T., & Heng, S. (2014). The political settlement and economic growth in Cambodia. ESID Working Paper No 37. Manchester: University of Manchester.

Khmer Times (2015, December 30). Construction workers take to the streets. Retrieved from https://www.khmertimeskh.com/news/19306/construction-workers-take-to-the-streets/.

Khmer Times (2017, July 17). SDGs localisation in Cambodia. Retrieved from https://www.khmertimeskh.com/5075180/sdgs-localisation-cambodia/.

Laws, E., & Leftwich, A. (2014). Political settlements. DLP Concept Brief 01. Birmingham: Developmental Leadership Programme, University of Birmingham. Retrieved from http://publications.dlprog.org/PolSet.pdf.

LO-Norway (2012). Evaluation report of Building and Wood Workers International BWI/BWTUC in Cambodia. Oslo: Norad. Retrieved from https://www.norad.no/en/toolspublications/publications/ngo-evaluations/2013/evaluation-report-of-building-and-wood-workers-international-bwibwtuc-cambodia/.

MacNaughton, G., & Frey, D. F. (2011). Decent work for all: A holistic human rights approach. *American University International Law Review*, 26(2), 441–483.

Marcussen, H. S. (1996). NGOs, the state and civil society. *Review of African Political Economy*, 23(69), 405–423.

MoVLT (2009). The first occupational safety and health master plan 2009–2013. Ministry of Labour and Vocational Training. Retrieved from http://ilo.org/wcmsp5/groups/public/---asia/---ro-bangkok/---sro-bangkok/documents/policy/wcms_112931.pdf.

MoVLT (2011). Overview of occupational safety and health in Cambodia. Ministry of Labour and Vocational Training. Retrieved from https://www.ilo.org/wcmsp5/groups/public/---ed_protect/---protrav/---safework/documents/policy/wcms_187746.pdf.

Mutua, M. W., & Howse, R. L. (2001) Protecting human rights in a global economy: Challenges for the World Trade Organization. In H. Stokke & A. Tostensen (Eds), *Human rights in development. Yearbook 1999/2000: The millennium edition* (pp. 51–82). The Hague: Kluwer Law International/Nordic Human Rights Publications.

Phnom Penh Post (2012, September 13). Cambodia sees shortage of builders. Retrieved from https://www.phnompenhpost.com/real-estate/cambodia-sees-shortage-builders.

Phnom Penh Post (2016a, May 5). Construction investment skyrockets. Retrieved from https://www.phnompenhpost.com/post-property/construction-investment-skyrockets.

Phnom Penh Post (2016b, May 12). Construction safety training centre floated for Phnom Penh. Retrieved from https://www.phnompenhpost.com/post-property/construction-safety-training-centre-floated-phnom-penh.

Phnom Penh Post (2018a, April 5). A sector too big to fail?. Retrieved from https://www.phnompenhpost.com/business/sector-too-big-fail.

Phnom Penh Post (2018b, May 10). Construction booms but salaries fall behind. Retrieved from https://www.phnompenhpost.com/post-property/construction-booms-salaries-fall-behind.

RGC (2013). Development cooperation & partnership strategy 2014–2018. Royal Government of Cambodia. Phnom Penh: Cambodian Rehabilitation and Development Board and Council for the Development of Cambodia.

Sachs, J. D. (2012). From Millennium Development Goals to Sustainable Development Goals. *Lancet*, 379, 2206–2211.

Stamatopoulou, E. (1998). The development of the United Nations mechanisms for the protection and promotion of human rights. *Washington and Lee Law Review*, 55(3), 687.

Un, K. (2005). Patronage politics and hybrid democracy: Political change in Cambodia, 1993–2003. *Asian Perspective*, 29(2), 203–230.

Un, K. (2006). State, society, and democratic consolidation: The case of Cambodia. *Pacific Affairs*, 79(2), 225–245.

Un, K. (2011). Cambodia: Moving away from democracy?. *International Political Science Review*, 32(5), 546–562.

UN (1948). The Universal Declaration of Human Rights. New York: United Nations.

UN (1967). International Covenant on Economic, Social and Cultural Rights. New York: United Nations. Retrieved from https://treaties.un.org/doc/treaties/1976/01/19760103%2009-57%20pm/ch_iv_03.pdf.

UN (1979). Convention on the Elimination of All Forms of Discrimination against Women (CEDAW). New York: United Nations. Retrieved from http://www.un.org/womenwatch/daw/cedaw/text/econvention.htm.

UN (2015). Transforming our world: The 2030 Agenda for Sustainable Development. New York: United Nations. Retrieved from http://www.un.org/ga/search/view_doc.asp?symbol=A/RES/70/1&Lang=E.

CHAPTER 15

The Bumblebee Doctrine: Small and Medium-Sized Enterprises as a Force for Achieving the Sustainable Development Goals in Myanmar

Thuta Aung

1 Introduction

As a Burmese person educated in the West, I have experienced both the accumulated wisdom of prestigious business schools and the challenges of applying that wisdom to a developing economy. When Myanmar opened up to democracy in 2010, I decided to return to the country of my birth, to make a contribution to modernizing the economy in a way that respects its unique culture and ecology. As an advisor to government and aid agencies and the proprietor of my own consulting firm, I have had the privilege of testing out numerous developmental and economic theories in the context of Myanmar. I believe strongly that, as Myanmar aims to meet the United Nations' Sustainable Development Goals (SDGs), the country's small and medium-sized enterprises (SMEs) represent a blueprint that will allow it both to meet the SDGs and to prosper well into the 21st century.

This chapter primarily addresses two of the SDGs: Goal 8 (Decent Work and Economic Growth) and Goal 9 (Industry, Innovation, and Infrastructure). Fulfilling, well-paid work is one of the fundamental building blocks of a thriving commercial sector. In an expanding economy like that of Myanmar, it is not enough to ask simply: is there work? A more pertinent question is: is there *decent* work? Are workers paid fairly and treated with respect? Do firms operate an inclusive recruitment policy? This is the heart of Goal 8.

Other essential aspects of a developing economy are hard work that leads to success, creativity that leads to innovation, and investment that enables infrastructure growth. For many people, the word 'innovation' conjures an image of expensive consultants who bring complex ideas and concepts from far-flung places. When they consider infrastructure, they assume that the costs will be high and the technology complex. However, this need not be the case.

In this chapter, I intend to show that the best results often emerge from small-scale, appropriate technologies, many of which can be installed at little or no cost. I shall argue that, in many sectors in Myanmar, we should not be

 | DOI:10.1163/9789004391949_016

seeking to emulate the Western development model wholesale. Myanmar is a unique country with the potential to thrive on its own terms. It would be a mistake to think of Myanmar as a second Vietnam or China. Indeed, I believe that the country has the potential to reach the SDGs in a unique manner, meeting the needs of stakeholders without copying from another developing country.

I call this approach the Bumblebee Doctrine, based on the legend that the bumblebee's wings are too small to raise its plump body into the air. Yet, seemingly in defiance of physics, the bumblebee flies. Something similar can be said of Myanmar. Looked at from the outside, the country appears poor and troubled. It seems impossible to imagine that it could become a leader in sustainable development. I contend, however, that it is this very aspect of Myanmar that gives it so much potential. Unfettered by ingrained dependence on aid, unattached to a Western development model, Myanmar has the freedom to plot an inspiring new flight path. The country's SME owners are the key to attaining this unlikely flight. They are the bumblebees whose determination and resilience allow them to access the opportunities scattered around Myanmar's developing economy and bring those opportunities to fruition. The Bumblebee Doctrine evolved from my book, *The Bumblebee Takes Flight*, a critical yet hopeful look at Myanmar's evolution towards democracy (Thuta Aung, 2017). The doctrine itself is a response to the question of how Myanmar can innovate to meet the SDGs.

It is important to note that, while this chapter will concentrate on Goal 8 and Goal 9, many of the SDGs are interconnected. Goal 1, for example, is the elimination of poverty. A country with a growing economy and plentiful decent work is a country with low levels of poverty. Goal 10 is the reduction of inequality, while Goal 11 is the establishment of sustainable cities and communities. Again, decent work implies reduced inequality, because it entails fair pay and comfortable living conditions. SMEs which utilize innovative employment and procurement practices can play a direct role in achieving this aim. Similarly, cities built on the growth created by SMEs and infrastructure tailored to meet their needs will undoubtedly be moving in the direction of sustainability. Goal 8 and Goal 9 should be seen as pathways towards additional goals.

Within the scope of this chapter, however, it would be impossible to address all 17 goals. It will therefore focus on Goal 8 and Goal 9, with the understanding that their effects can ripple out to other sectors. I should also note that there may be other routes towards meeting the SDGs — even Goal 8 and Goal 9. My expertise lies in the field of international development and government liaison, so that will be the focus of this chapter's attention. This should not be seen as an exhaustive survey of every possible means by which SMEs can meet the SDGs, but as a snapshot of what can be achieved through the judicious engagement of SMEs and their owners.

2 The Current Situation in Myanmar

It is undeniable that — as of 2018 — Myanmar remains a poor country, at least in terms of per capital income. This does not mean, however, that the country is a blank canvas. Taking infrastructure as an example: although much of it requires an upgrade, there is more infrastructure in place than might first appear. There are, for instance, many airports and a vast rail network. The problem is that the infrastructure is relatively old, meaning that rail journeys take a long time in comparison with similar journeys in neighbouring countries. Some parts of the country, especially the south, are not connected to the national grid. This means that people must rely on diesel generators to produce electricity, an expensive and environmentally damaging process. Nonetheless, major projects involving sustainable energy technologies, such as solar energy, are underway. Myanmar's infrastructure may be limited, but it is improving.

There is a risk that development professionals observing Myanmar may conclude that they have a golden opportunity to intervene, bringing their pet projects or exercising their own agendas. I would argue that it is essential — both for the country's prosperity and for the attainment of the SDGs — that Myanmar proceeds along a path of its own making. The challenges it faces are borne of inefficiency and low levels of technological development, not a total lack of infrastructure. Furthermore, Myanmar is a place of opportunities, many of which are overlooked both by SME owners and political figures. Government efforts to promote the country, for example, usually focus on exports. This is an important *facet* of growth, but it tends to be overemphasized. Similarly, SME owners are often distracted by the promise of export markets, asking themselves how to serve clients in Australia, Singapore, or Thailand. They forget about the growing markets in Moulmein and Shan State.

Despite its weaknesses, Myanmar is one of the most democratic countries in the ASEAN. That it is often criticized for not moving faster towards democracy is perhaps understandable. When State Counsellor Aung San Suu Kyi took office in 2015, she was seen as a symbol of hope. When her party, the National League for Democracy (NLD) won the 2015 elections, some people dreamed of a perfect democracy. The State Counsellor was perceived as a figurehead of truth and justice, almost saintly. Some people are disappointed to discover that she is also a politician. Since her election, Myanmar has taken steps towards democracy, but has not undergone the wholesale transformation some onlookers expected. Aung San Suu Kyi is working very hard, but she faces many obstacles, both internally and externally. The dream of democracy is coming up against the hard realities involved in getting the job done. In addition, the recent crisis in western Rakhine State has significantly lowered the State Counsellor's stock in the eyes of the Western world.

Looked at from a different perspective, however, Myanmar's progress has been remarkable. Just 10 years ago, in 2008, the country was wracked by Cyclone Nargis, the damage from which was so severe that it prompted the military dictatorship to open up the country and allow international relief teams to deliver aid. This opening was itself a first step towards democracy. A year earlier, in 2007, Myanmar had been the scene of the Saffron Revolution, in which Buddhist monks and clergy were subject to violence on the streets of Yangon.

While Myanmar is still far from perfect, the change in a decade has been phenomenal. Citizens of the country are more optimistic: business owners, students, factory workers, waiters — people from all walks of life — almost all feel that life is getting better. People have more disposable income than they did a decade ago. They feel that the horizons of their lives have broadened, and they believe in the possibility of future positive change. If it were possible to bring a Burmese person from 2008 to 2018, they would be stunned at the extent of the country's evolution.

As household incomes increase, the consumer market in Myanmar is growing at a tremendous pace. People want to consume more, and they have the means to do so. There is great potential for SMEs willing to take on the challenge of serving local markets more efficiently. Nonetheless, SMEs that focus on the local market should not assume that they have a captive audience: Neighbouring ASEAN countries are bringing businesses into Myanmar, many of which compete with local businesses. Myanmar is a rice-producing country, but as people become richer they may be willing to spend increasing amounts of money on imported rice varieties from India or Pakistan.

In addition to economic challenges, Myanmar faces undoubted political challenges, including an escalating conflict in the west of the country. This presents both internal and external difficulties, as SME owners wrestle both with the question of how to address the problem themselves and how to communicate with international partners. At present, engaging commercially with Myanmar is often seen as a reputational risk. The troubles in Rakhine State may be interpreted as an accurate benchmark of the country's attitude to all refugees, and therefore sufficient reason to boycott the country.

There are many aspects to the problem in Rakhine State, some of which are neglected by Western media. While this is not the place for an extensive discussion, it is important to provide a brief summary. The Western narrative paints the conflict as a direct battle between the Buddhist majority and the Muslim minority. In reality, it is more complicated, involving Buddhists, Muslims, and Hindus. Some of the Muslims — especially the refugees — refer to themselves as Rohingya, while others — primarily ethnic Burmese — identify as Kaman. While it is true that Buddhists are attacking Muslims, it is also true that Muslims are attacking other Muslims and Hindus. Of all the groups, the Hindus probably suffer the most. Clearly, this is a humanitarian disaster, with many people dying needlessly and painfully. It is also true that the Burmese government has responded to the crisis in a way that has exacerbated the situation. It is important that we recognize this as a problem, rather than trying to brush it under the carpet and pretend that no issue exists. For international partners, there is no dodging the issue, and it represents a major ethical concern when deciding whether to do business with SMEs in Myanmar. Later in this chapter, I will discuss in more detail ways to improve communication with international partners and, ultimately, how to build a Myanmar in which religious discrimination is no longer a concern.

Despite its difficulties, Myanmar remains a country with enormous potential. The typical international development landscape in Southeast Asia is somewhat depressing: aid agencies and non-governmental organizations (NGOs) work to promote a Westernized model of development with limited engagement from host populations. Many aid workers seem more concerned with progressing their own careers than understanding the unique challenges of the countries in which they work. Myanmar can break this cycle, becoming a model of development in which local businesspeople learn how to help themselves, working towards independence from foreign aid. It can become a leading example of genuine development, incorporating practical, hands-on solutions. This may be bad news for consultants and aid agencies, but it will be good news for entrepreneurs and the people of Myanmar. SDG Goal 8 and Goal 9 are the perfect mechanisms to achieve this.

3 Achieving the SDGs: Myanmar's Unique Opportunity

Since Myanmar has opened up only recently, it is of particular interest to researchers. Given its relatively low level of development, Myanmar offers a place where fresh ideas can be tested, as well as an opportunity to embed lessons learned from other countries. Unfortunately, this interest does not always translate into positive outcomes. Many foreign development professionals content themselves with conducting multiple interviews that rarely seem to lead anywhere useful. As a consequence, Burmese entrepreneurs may suffer from 'researcher fatigue', while the researcher misses the opportunity to undertake productive research with local SMEs — a combination of practical work and reflection. With leadership from national entities such as government agencies, Myanmar could be a test case for innovation and appropriate technology. Let aid agencies apply a potential solution, assess its effectiveness, and iterate rapidly to ascertain what works and what doesn't. Naturally, mistakes will be made, but they will be working towards positive goals. The focus of the research will be the benefit it can provide to Myanmar, with everyone involved in the process having a stake in the success of the country.

Additionally, Myanmar benefits from an enviable geographical position. It is the only country to share a land border with both India and China, the two most populous countries in the world. This gives Myanmar a land border with over 2 billion people, more than a quarter of the global population. India is keen to become a key player in the ASEAN, so Myanmar's position represents an opportunity to exercise influence. Some initiatives by Indian NGOs and business associations include working with Myanmar SMEs as a pilot programme for reducing the gap between the more developed ASEAN countries and the CLM nations.[1]

While Singapore is much admired as a model of development, it is a small island with a relatively limited population. The whole of Singapore is only about one third the size of Yangon, the commercial capital of Myanmar. Managing a diverse, resource-rich country is a very different task from managing a city-state. As Myanmar pursues economic development, it must remain mindful of the impact on the environment and on society as a whole. However, there is also a role for pragmatism: if the government refuses all potential development, the country will remain extremely poor, and Burmese people will be forced to cross international borders as migrant workers, most likely facing poorly paid jobs and inadequate living conditions.

The world is evolving: populations are growing, energy needs are increasing, and the demand for food will continue to grow. To meet these challenges, Myanmar needs to develop intelligently, converting more of the countryside into arable land and perhaps employing hydropower or other sustainable energy generation techniques. For many countries, being resource-rich has proven to be a curse. It has led them on a path towards inequality and exploitation, harming both the environment and workers. Myanmar has the opportunity to be a resource-rich country that sidesteps the resource curse and models a healthy way to develop; the SDGs can play a key role in helping to plot this

1 The CLM countries are Cambodia, Lao PDR and Myanmar.

path. Put into action on a large scale, this could have powerful knock-on effects, bringing electricity to clinics, schools, and universities. Some pilot projects have already taken place, with potential for both local and foreign SMEs to play a role in scaling up these operations.

Those who do not know Myanmar well may imagine that the country is full of problems. There are certainly challenges, but now is a great time to make a difference. Both local and foreign SME owners have a chance not only to increase their own wealth, but to help lift people out of poverty, providing valuable services that will set Myanmar on a path towards achieving the SDGs.

4 Resilience in the Face of Shocks

As suggested above, Myanmar is a place of both challenge and opportunity. SMEs wishing to operate there will need certain characteristics if they are to prosper. In the country's evolution, some SMEs will surely perish, wiped out by foreign competition. One of the greatest skills SME owners can develop in this environment is *resilience*, the capacity to absorb and respond to unexpected shocks. This is not simply a matter of preparing for a particular challenge, but of preparing for any one of a number of unspecified challenges. They must be mindful of myriad risks and multiple scenarios in order to create contingency plans for these risks. If they are suddenly without buyers, what will they do? If the supply chain is interrupted, what will they do?

Burmese SME owners need to strike a delicate balance. They must be willing to seek out existing but overlooked opportunities in the local market, while also understanding that the tastes of that market are evolving. They must engage with foreign markets where appropriate, while understanding that the greatest opportunities may lie closer to home. They must be aware of the risks to their businesses, while putting in place plans to mitigate those risks and to grow.

The question, then, is how the owners of SMEs should develop the resilience that they will surely need. The first step is to improve the way they communicate, especially with international partners concerned by the ethics of doing business with Myanmar. Traditionally, Myanmar has been a place where minimal communication was seen as a positive trait. Burmese people working in close proximity to each other develop shared values and a shared understanding of their work, with little need for communication. Juniors are expected to understand aspects of their roles without being told, an attitude they subsequently adopt towards *their* juniors as they grow in seniority. Most Burmese business owners are strong, decent people, but they tend to under-communicate. When working with international partners, they need to buck this trend, making a point of communicating their values explicitly.

As well as effective communication, business owners must remain calm and exercise the wisdom of the stoic philosophers, who took a practical approach and believed in virtue as the only good. It is tempting for business owners to see the negative consequences of the crisis in Rakhine State and become frustrated. If they ask instead how they can exercise a positive influence, within both the local and the international spheres, they will be equipped to succeed. The key is to remain calm and emotionally equable. SME owners cannot bring an immediate halt to the violence in the west of Myanmar, but they can operate a policy of inclusiveness towards people of all ethnic backgrounds. Equally, farmers cannot guarantee that their crops will not be affected by a cyclone, but they can take out an insurance policy to protect against lost income.

Despite the best efforts of SME owners, there may be times when key buyers or partners are resistant to the idea of working in Myanmar. If this occurs, and there is no way to salvage the relationship, it is important to take a stoical approach, continuing to present a positive face of the company and the local area. In practice, this probably means parting as amicably as possible, with the intention of reconnecting when the situation allows. SME owners must respect the constraints their partners are working under, yet balance that with the intention to be helpful and engaging. If, for example, the partner is facing pressure from media outlets in their home country, it might be possible to set up a media engagement activity to ease the pressure.

A resilient businessperson should always be asking how they can build relationships with partners. If, ultimately, their efforts come to naught, they must be willing to adapt and move forwards. To share one illustration of this: in the late 1990s, a friend of mine ran a thriving media company with major global brands as partners. When Myanmar became subject to sanctions from the American government, her international partners had no choice but to walk away. This seemed like a massive blow to her company, but it became a huge opportunity. Just as she was losing key partnerships, many talented Burmese — who were working for Western companies — became suddenly unemployed. My friend took the opportunity to employ some of the brightest people in Myanmar. Before long, she had rebuilt her company as the number one national media business in Yangon.

5 Inclusivity: a Powerful Tool for Peace and Economic Development

The centre of Yangon is a remarkable place. Within close proximity you can find a Buddhist pagoda, a Christian church, a Muslim mosque, a Hindu temple, and a Jewish synagogue. Taken together, these places of worship form a powerful symbol of religious tolerance that businesspeople can use as a guiding light. There may be discrimination in the west of the country, but SME owners have a responsibility to demonstrate that they do not discriminate. They also have a responsibility to match their words with actions. Some may argue that talk is cheap. In an environment of confusion and concern, however, clear communication is an essential first step towards rebuilding trust. For that trust to take root, it is vital that good communication is backed up.

Myanmar consists of fourteen states. The troubles in Rakhine State, disturbing though they are, affect only one part of one of those states. Rather than focusing on the negative and applying it to the entire country, SME owners have an opportunity to focus on the positive and use it to promote specific regions. Shan State in the east, and Kayin State and Mon State in the southeast, all produce unique goods. None has any issues with violent discrimination. Some of Myanmar's towns and cities, notably Mandalay and Yangon, carry an air of romance that has existed since colonial times. Businesspeople can share the values and the products of their own regions, showcasing the best of Myanmar to the world rather than attempting to fight the Western narrative of affairs in Rakhine State.

This is more than merely symbolism. Establishing a tolerant, inclusive atmosphere in other parts of the country can send a message of tolerance to the more troubled areas. Myanmar can choose to be a country of diversity and, in so doing, may be able to shift perceptions and encourage acceptance in the most troubled areas. The conflict may initially appear to be purely religious, but there are also economic elements. In poor areas, the fight for resources is constant. Many believe that they do not have enough, and this leads to resentment of other groups. Rakhine State is the second poorest in the country, and many of those involved in violence live on the poverty line.

If we can address the economic elements of the conflict, it may be possible to improve the situation considerably; SMEs can play a significant role in doing this. For example, plans are afoot to build an industrial zone in Rakhine State, close to the site of the conflict. Refugees could take roles in factories, giving them an economic footing in the country and easing the tension that comes with hunger. It may also be possible for them to work in agriculture, enabling them to produce their own food and reduce their dependence on external aid. If Rakhine State could become prosperous, I believe the problems would largely disappear. People with a stake in their communities are motivated to work and improve their lives, not to fight.

It is essential that the international community does not abandon Myanmar, and instead chooses to engage more mindfully. For reform and development to take place in Rakhine State, and for the Bumblebee Doctrine to take root, the government must feel secure. Does this mean it should be immune from criticism? No. Does it mean that other countries should not seek to engage in constructive dialogue? Absolutely not. What it does mean is that the international community should resist the temptation to call for boycotts or sanctions against Myanmar. These will only increase insecurity, creating a siege mentality and making it harder for the country to remain open and connected.

From the perspective of investors, it is easy to believe that the whole of Myanmar is subject to conflict. It is therefore worth repeating: Myanmar is a large country, and only a portion of one state is embroiled in violence. Investors may be concerned that their stock prices will be negatively impacted if they are associated with Myanmar. Buyers, too, may feel pressure to divest from the country. Their good intentions, however, may have the opposite effect to the one they intend. SME owners may lose business, impacting their livelihoods and their ability to serve their communities.

While it is completely understandable that people reading about the crisis in western Rakhine State want to do something to help, isolating Myanmar risks making the situation worse. Sadly, there is no easy solution to the problems in Rakhine State. If international partners choose to boycott Myanmar as a whole, they forgo the opportunity to play a part in bringing the country into the global value chain and to exercise their leverage to demand better treatment of people and the environment. Some sectors of the Myanmar economy have already achieved that goal, while others are moving in that direction. If they are cut off from buyers, suppliers, and sources of benefit, the chances are high that they will turn to unscrupulous partners from other countries. While Myanmar remains part of the global value chain, people at every level of the economy can be incentivized to pursue employment policies with equality at their heart.

In the late 1990s, a factory in Yangon manufactured women's undergarments for a major Western brand. The workers received decent wages, the factory was air-conditioned, and the work hours were reasonable. When sanctions came into force, the factory lost its Western buyer. Another buyer, from a country with less regulation, took its place. The new buyer's first move was to shut off the air-conditioning, closely followed by increasing work hours without improving pay. The intentions behind sanctions were positive, but the people who were most affected were ordinary citizens, whose lives were made harder.

The original buyer was constrained by sanctions, and had no choice but to pull out of the country. Nowadays, there are no official sanctions against Myanmar, only the risk of reputational damage. What if, instead of operating a punitive boycott, foreign partners said that they would purchase more goods from companies that could prove a certain percentage of their board members were female, or that could demonstrate a commitment to hiring ethnic minorities? What if they did the same for companies that paid a living wage? This is a powerful opportunity to bring more people into decent work, and it will be squandered if the international community opts merely to boycott Myanmar. This principle applies equally to large supermarkets in Yangon. By offering to reward companies for good practices, instead of punishing them for the country's existing deficiencies, change becomes far more achievable.

If Myanmar's SMEs are to reach their full potential, it is essential to confront the conflict in western Rakhine State directly. Some people try to minimize the issue and label it an immigration problem, which is not helpful and presents a negative face of Myanmar to the world. Much better to be honest and acknowledge that it is a problem for all Burmese people to address. Indeed, it will likely be a generational issue that takes decades to truly resolve. We may wish to distance ourselves from the problem, but for anyone living in Myanmar, that is not possible. As a comparison, consider South Africa. Individual white people in the country may claim that they do not discriminate against black people, and indeed they may not. Nonetheless, the country clearly has problems with racial inequality, even decades after the fall of the apartheid regime. At the same time, it is important to recognize that not every Burmese person is directly responsible for the troubles, and to acknowledge and applaud the efforts being made by young people from the area to alleviate the suffering. Many are volunteering to deliver humanitarian aid to refugees of all ethnicities. They want to be part of the solution, not part of the problem.

Myanmar is a very large country and each state has its own unique identity. SME owners can help to highlight this by describing the heritage of their own state and region. Those who are based in Mandalay can talk about Mandalay. Those in Shan State can discuss Shan State. In this way, they offer international partners a chance to understand the diversity of Myanmar, just as they already understand the distinct regions of other countries. Manchester, for example, is quite different from London. It has its own cultural identity. Indonesia consists of thousands of islands, each with its own heritage. Let the businesspeople of Myanmar show the world what each of its regions can offer.

In summary, Myanmar faces a grave reputational problem, driven by the violence in the western part of Rakhine State. This is a significant issue and must not be underplayed by political and business leaders. Nonetheless, the way to achieve the SDGs is not to abandon Myanmar but to work in partnership with responsible representatives of the country, who wish to see a diverse, inclusive, prosperous nation. Turning away from Myanmar will only create more poverty, making it harder for SME owners to generate decent work or engage in innovation. Partnership built on a foundation of inclusivity can lift the overall wealth of Myanmar, develop the social fabric, and — in time — allow people from poverty-stricken segments of Burmese society to share in those benefits.

6 The Bumblebee Doctrine in Action

This section will concentrate on practical examples of what can be achieved in Myanmar with the skilful application of intelligence and resources. In many cases, far from relying on hefty aid budgets, the potential improvements require minimal investment or even no investment at all. E.F. Schumacher, whose 1973 book *Small Is Beautiful* is an iconic economics text, made his career working in Burma — as it was then known — in the 1950s and 1960s. He was a big advocate of appropriate or intermediate technology, and also coined the term 'Buddhist economy' to describe an economy that created prosperity while remaining in harmony with others and with the environment. Since Schumacher's ideas have had a considerable influence on the Bumblebee Doctrine, it is useful to catalogue some of the ways in which they can lead to palpable improvements in the lives of Burmese SME owners. These are relatively straightforward ways of boosting productivity, increasing the value of products, or otherwise taking steps forward. The focus here is on appropriate technology, not on high-tech solutions.

It is a cliché that economic development is best served by aiming to pluck the 'low-hanging fruits'. Our first example *literally* involves low-hanging fruits. Myanmar's climate is ideal for growing many different crops, among them mangos and avocados. Unfortunately, there is often a lot of wastage. When crops are bounteous, many succulent fruits never make it to market, instead rotting on the trees or in storage facilities. Yet there are simple and straightforward solutions to these problems, which have the potential to make a huge difference to the incomes of farmers. Mangos, for example, can be turned into mango puree. Very little investment or complex machinery is required to make puree, yet it can fetch good prices.

Going further, a little more investment could allow farmers to install cold storage units or freeze dryers at some locations. Freeze drying fruit can preserve freshness and allow mangos that would otherwise be wasted to become assets. This may be too expensive for individuals, but farmers who are willing to join forces with their neighbours may find that purchasing cold storage units and freeze dryers is within their purview. Alternatively, investors could reap a solid return by providing this equipment, while simultaneously helping farmers to boost incomes.

Another simple way to make mango farming more lucrative is diversifying which varieties are grown in a particular locality. Myanmar's climate allows numerous types of mango to be grown, each of which produces fruit at a slightly different time of year. In most districts, a single variety is preferred, leading to a bumper crop followed by a period of scarcity. This results in fluctuating income for farmers and a paucity of employment for workers. Pilot schemes are currently underway in which three different species of mango, each with a slightly different life cycle, are grown in close proximity to one another. This allows farmers to stagger their harvests, providing more consistent income for them and work for their employees.

Our second example comes from another field — no pun intended. Many Burmese villagers manufacture potato chips. On the open market, they lose out to competition from Thailand. Although in terms of taste and nutrition they are easily the equals of their Thai counterparts, they lose out aesthetically to the Thai chips, which are lighter in colour and look more appetizing. The Thai chips contain chemicals that give them their pale colour. One might imagine that the only way for Burmese chips to compete on the open market would be to adopt the same methods as the Thai manufacturers, but there is another, much simpler solution. If the chips were removed from oil thirty seconds sooner, they would still taste good, but they would be lighter and appear more appetizing, making them more appealing to consumers.

Farmers — for example those who harvest potatoes — may rely on monocropping. That is, they plant the same crops repeatedly, impoverishing the soil and ultimately leaving them dependent on chemical fertilizers. This problem could be easily resolved by intercropping, for example by alternating a crop of potatoes with a crop of onions, and subsequently leaving the land fallow for some time to recover. Potatoes and onions require different nutrients and, crucially, leave different residues in the soil. This makes them complementary, keeping the soil fertile and reducing farmers' dependence on chemical fertilizers.

A similar principle operates in fisheries. Small particles of uneaten fish food sink to the bottom of lakes, decompose, and produce nitrogen, altering the composition of the water. Fish excrement also contains nitrogen and, when fish die, their decomposition adds to the nitrogen content of the water. When nitrogen levels in a given pond exceed 1 part per million (ppm), the water is unable to sustain aquatic life, causing mass die-off. The solution, however, is straightforward. Nitrogen-rich water can be used to irrigate neighbouring rice fields. Rice thrives on the nitrogen in the water and recycling the water saves resources for poor farmers. Meanwhile, refreshing the water regularly keeps fish ponds thriving.

Fish ponds represent another environmental conservation opportunity, too. Hyacinth is a fast-growing plant that thrives in dirty water and readily absorbs toxins. Water sanitation is expensive, but planting hyacinth can ease the burden, cleansing water expelled from factories. In addition, hyacinth stalks have numerous uses, including the potential to be woven into furniture (Su Phyo Win, 2018).

Although Myanmar is one of the world's largest bamboo producers, much of the bamboo used in the country is imported from Thailand. If it were grown locally, it would be eagerly received by the existing Burmese market. In general, the government focuses primarily on import substitution and export promotion, in an attempt to bring Myanmar further into the global economy. This attitude, however, is somewhat outdated. In today's interconnected world, it is impossible to stop imports from other countries. On the other hand, the climate in Myanmar makes it easy to grow some of the most delicious fruit on the planet. Similarly, the export market is challenging, requiring farmers to achieve consistency of supply and quality and meet numerous regulations. In many cases, it would make more sense for them to focus on local markets.

The examples continue. Myanmar is an exporter of avocados. The fruit is highly prized and fetches comparatively good prices, but it is also fragile and easily spoiled. Many otherwise usable avocados go to waste, when they could be making their way onto the dinner tables of consumers. If they could be turned into oil, their freshness would be preserved. Avocado oil is a premium product, not to mention a nutritional powerhouse. Delivering avocado oil to local or export markets could be a lucrative and satisfying move for farmers. Again, the equipment required to press avocados into oil is relatively simple: it can fit into a 40 ft^2 shed. A group of local farmers, working together, could combine their resources and put together a local, small-scale operation pressing and supplying avocado oil.

This is a model that has worked successfully in other countries. New Zealand, for example, has a number of Maori businesses, which are partly owned by the community. Instead of one individual entrepreneur reaping the benefits of their labours, everyone in the community receives a share. It should be recognized, however, that this is a sensitive issue in Myanmar. Prior to the advent of socialism in the country, there was a thriving cooperative movement. It was destroyed so completely that the concept of cooperatives has become a taboo. The concept may be reborn, but it will need to be rebranded to gain widespread acceptance in Myanmar. If this can be done, there is potential for a mutual or co-ownership movement to cross ethnic boundaries. In some areas, shared ownership of a business — or an asset — has the potential to bring together different sectors of the community, forging a sense of belonging and fostering a desire to contribute towards shared goals.

If Myanmar is to meet the SDGs, one area in which it must make progress is electricity generation. There is also a major commercial opportunity here: investors in sustainable energy production can be assured that they will always have paying customers to serve. Myanmar has the third-largest bamboo reserves in the world. Sustainably-managed bamboo plantations can supply bamboo chips, which can be used as biomass for electricity production. This is a perfect example of a simple innovation that could bring prosperity to investors, while serving local people and supporting a sustainable environment.

Another practical way to boost both the productivity and inclusivity of Burmese SMEs is matching skilled workers with available jobs. The typical Burmese business may overlook non-Burmese speakers. This is exacerbated by the fact that people from some areas of the country, such as Dawei and Myeik, in the south of Myanmar, have very strong regional accents, making it hard for more orthodox speakers to understand them. Many SMEs in Yangon complain that skilled workers are scarce. Yet when I visit universities and colleges in Dawei and Myeik, the young people there possess valuable skills. Their home cities do not have a lot of industry, so they move to Yangon or Mandalay. There are far more jobs in those cities, but workers from Dawei and Myeik may find themselves subject to prejudice, unable to participate in the labour market, and liable to become the butt of jokes. This is a lose-lose situation. They lose the opportunity to do productive work and the companies which might otherwise employ them lose the opportunity to benefit from their skills. Both the supply and demand are present, but there is a resistance on the part of SMEs to employing people from Dawei and Myeik. With sufficient entrepreneurial will, it would be a simple matter to match the workers with appropriate roles.

Another challenge comes in the form of religion. Some SMEs may wish to avoid employing people of certain religions. If they can be encouraged to move towards diversity, their companies can be strengthened, the social ties between people of different religions can be improved, and skills can be more effectively shared.

In summary, the SMEs of Myanmar are sitting on many opportunities that could move the country in a positive direction and assist in the task of meeting the SDGs. The country is highly fertile, with numerous crops that could command a high market value with minimal processing. There are millions of people acquiring skills that could be put to good use if SME owners could be encouraged to see beyond regional and religious prejudices. The best news is that, if these hurdles can be surmounted, SME owners will benefit not only themselves, but also their communities and their country.

7 Conclusion

The key recommendation of this chapter is that policy makers support SME owners in the activities that will build prosperity, industry, and innovation in Myanmar, generating decent work and sustainable infrastructure. The bumblebees — SME owners — have the capacity to make a significant contribution both to the wealth of the country and its social fabric. Occasionally they may require active encouragement, but often all they really need is the freedom to make a difference.

While Myanmar has become far more open in recent years, it is still subject to relatively high levels of bureaucracy. In Denmark and New Zealand, for example — two countries which make it easy for people to start small businesses — a lot of bureaucracy can be done online. In Myanmar, prospective business owners must fill out several sets of paper forms before they can open a business, a procedure that is cumbersome and time-consuming. If aid workers and development economists wish to boost the productivity and prosperity of Myanmar, they could make a genuine contribution by supporting the installation of more efficient systems for founding and taxing small businesses. They could use their experience in developed market economies to teach Burmese government officials how to streamline the system in Myanmar. Unfortunately, many seem uncomfortable with this kind of contribution. While they profess a belief in market economies, their work in Myanmar reveals that many of them are closet planned economists. Instead of lobbying to lower the barriers for SME owners, they generally campaign on fringe issues such as making the tourism sector more inclusive or bringing more women into business.

This is a sensitive issue and it is important to be totally clear. Inclusivity is positive, and it is vital that the Burmese economy develops along inclusive lines. When NGOs push for inclusivity *as a good in itself*, however, they ignore the realities of business for the average Burmese. Is the business profitable? Is it expanding? Earlier in this chapter, we discussed the potential for partners to link inclusivity with additional purchases. By doing this, they can create an association between inclusivity and successful business development, so that it becomes a by-product of effective business management. Many foreign aid workers demand inclusivity but ignore business fundamentals — an approach akin to icing a cake that has not yet been baked. To truly serve the economy of Myanmar, they might push for improved tourist infrastructure in areas of the country that could welcome visitors. By encouraging the government to make it easier to start businesses, they could stimulate the building of guest houses and the foundation of tour operators. In time, these businesses might become stable enough to welcome moves towards inclusivity. While they continue to push their own agenda, however, they miss the chance to make a positive difference and may actually have a negative impact. How? Every time an overstretched government department signs up to an NGO programme, vital resources are diverted away from the key activities that can bring prosperity to Myanmar.

SME owners do not require protectionism. They only require a level playing field. Some incentives to start businesses may be appropriate, but there should be no question of rescuing businesses that do not establish themselves successfully. If they fail, they fail, and their owners will hopefully learn from those failures and achieve success the next time they start an enterprise. On the other side of the coin, however, SME owners should not be punished for their success. Large businesses often lose the tax advantages that SMEs enjoy, stunting the growth of organizations or incentivizing managers to engage in creative accounting — perhaps by starting an additional separate business, as opposed to growing their existing concerns. A government that rewards smallness will discourage the growth of businesses.

The aim of this chapter was to show that it is entirely possible to achieve the UN's SDGs in Myanmar, especially Goal 8 and Goal 9. The Bumblebee Doctrine tells us that, even when it seems impossible, business growth is achievable. The key is to focus on low-cost or no-cost innovation, promoting projects that bring maximum benefit for minimum outlay. To support the bumblebees — the SME owners — we must resist the urge to overcomplicate matters. Instead, government representatives should look to simplify processes and lower the barriers to wealth creation. For the international aid community, the challenge is to see clearly what Myanmar's SME owners truly need, instead of getting lost in a world of frameworks and reports. If we can coordinate all these activities, we can create a healthy environment in which bumblebees can flourish, turning the pollen of opportunities into the honey of thriving, sustainable businesses. When this is achieved, we will all share in the sweetness.

References

Schumacher, E. F. (1973). *Small is beautiful: A study of economics as if people mattered.* New York: Harper & Row.

Su Phyo Win (2018, March 26). Forestry industry hopes hyacinth can save trees, create jobs. *Myanmar Times.* https://www.mmtimes.com/business/24200-forestry-industry-hopes-hyacinth-can-save-trees-create-jobs.html.

Thuta Aung (2017). *The bumblebee takes flight: A Burmese insider's guide to the rapidly developing business landscape of Myanmar.* Yangon: HamsaHub Consulting Pte Ltd.

PART 7

SDGs, Agriculture, and Community Development through Partnerships

CHAPTER 16

Bluewashing, Green Coffee, and the Sustainable Development Agenda in Southeast Asia

Amador IV Peleo and Titus C. Chen

1 Introduction

This chapter examines the presumed incompatibility between the state-led project to attain the Sustainable Development Goals (SDGs) of the UN's 2030 Agenda for Sustainable Development (UN, 2015), on the one hand, and the working conditions in Southeast Asian agricultural communities, on the other. Both the SDGs and the notion of universal human rights were introduced to Southeast Asian societies by non-Southeast Asians whose interests in the region did not necessarily match the material culture and normative beliefs in the region. The same is true of the Southeast Asian coffee culture. The 'third wave' trend of coffee consumption that has emerged in the region coincides with an increasing awareness among Southeast Asians that "specialty and good quality" coffees with attributable origins are aspirational products (Project Café 2016 South East Asia, 2015). The prevailing global perception is that Southeast Asia is capable of providing only anonymous 'commodity' coffee but also that coffee farming may yet be disassociated from the exploitative conditions of plantation economies.

In contrast, the engagement of the Association of Southeast Asian Nations (ASEAN) member governments with the SDGs appears to reflect a status quo mindset in the governments and societies in the region. That is, for the ASEAN constituent governments, the concept of 'sustainability' extends to the defence of governing regimes against criticism that inequality, poverty, and criminality are structural rather than symptomatic problems. The ASEAN political security agenda is juxtaposed with the "Peace, Justice and Strong Institutions" goal of Agenda 2030 (SDG 16). The ASEAN organization and its constituent governments have the prerogative to determine which of the SDGs may be analogues of the outputs that extant programmes have already produced. This chapter argues that the harmonization of ASEAN with international norms such as the SDGs has not occurred directly. This is because the 'ASEAN Way' of non-confrontational inter-governmental interaction enables corporate enterprises, especially those not headquartered in Southeast Asia, to maintain

 | DOI:10.1163/9789004391949_017

cross-border operations across the region and thereby collectively serve as a proxy regionalizing mechanism. Through their ability to engage all Southeast Asians, regardless of nationality and socio-economic strata, such corporations are able to embed slogans ostensibly promoting global values in the language associated with the production, distribution, marketing, and consumption of products and services. But if the ASEAN member governments were to effectively 'outsource' the attainment of the SDGs to the private sector, this would not only reify the notion of profit motive as a determinant of effective global and national governance: it would also allow corporations to promote their corporate social responsibility (CSR) projects, which constitute only a small proportion of their operations, as significant contributions to the well-being of the communities in which they operate.

2 Problematization of Bluewashing

The UN itself has described the SDGs as a means to "bring together world leaders, the business community, civil society and citizens around shared challenges to make change happen … for communities and businesses to thrive everywhere, now and in the future" (UN Global Compact, 2017e, p. 6). It explicitly associates the SDGs with the existing UN Global Compact — a voluntary initiative for businesses and NGOs that hope to gain promotional benefits from the disclosure of their activities in a UN-approved format. Not only does the UN present the 17 SDGs as interests that businesses may include in their promotional campaigns (UN Global Compact, 2017f), but it also presents the 10 principles of the Global Compact as compatible with the monitoring and evaluation of all businesses, regardless of the industry (UN Global Compact, 2011, pp. 12, 14). The SDGs are thus presented as "part of the sound values and principles that the UN Global Compact is built upon" (Karbassi, 2018). More directly, "the 2030 Agenda for Sustainable Development, including the Sustainable Development Goals (SDGs), presents an historic opportunity for the UN Global Compact and the corporate sustainability movement writ large", as the UN Global Compact is "the main interface between the United Nations and business" (UN Sustainable Development Knowledge Platform, 2018, p. 1).

Periodically, businesses are de-listed from the Global Compact's partnership programme either for failing to communicate progress on the SDG commitments, or at the request of the businesses themselves; this gives the appearance that the UN Global Compact remains a set of standards to which businesses conform. Nonetheless, CSR-sensitive businessmen still expect to thrive in a network of similarly inclined businesses, regardless of their listing with the Global Compact (Larsen, 2015, p. 31). Small companies with limited resources to meet the UN Global Compact standards, and companies that already operate according to strict government standards, are not likely to benefit from alignment with the programme (Knudsen, 2011). Of particular note for this chapter, all of the companies which deal mainly with coffee and remain listed with the UN Global Compact were already extensively networked with the relevant international supply chains prior to their application, such as Starbucks Coffee (UN Global Compact, 2017d) and Sithar Coffee (UN Global Compact, 2017c). They are also likely to have access to international sources of finance and other industrial inputs; this applies, for example, to Farm Mountain Coffee which, although headquartered in Denmark, has its operations in Uganda (UN Global Compact, 2017a) and to Java Mountain Coffee (UN Global Compact, 2017b).

'Bluewashing', or the alignment of corporate interests with the development goals of the United Nations (Bruno & Karliner, 2000), is not regarded as entirely objectionable by the UN, notwithstanding the criticism of anti-globalist activists that the failure of bluewashed corporations to meet UN standards in their operations may damage the credibility of the UN as an institution (The Economist, 2004). Because the promotion of the SDGs benefits from the resources, expertise, and communications networks of businesses with international value chains, concerns about bluewashing are always assessed relative to the ease of diffusing SDG-related information through corporate channels. "Partnering with SMEs (small- and medium-sized enterprises) usually results in higher transaction costs due to the lack of partnership expertise, the need to translate UN agendas into local contexts and higher costs of due diligence processes" (UN Global Compact, 2013, p. 14). However, the case of UN–corporate cooperation highlights a dilemma, in which the attainment of a development agenda reinforces the structural inequality that necessitated the creation of development agendas in the first place. Bluewashing conflates the desirability of non-profit institutions, which effectively implement social welfare programmes, with the social acceptability of corporations that retain the profit motive as a core value. The issue is not easily explained away as a matter of the 'haves' helping the 'have-nots'. It is not hard to imagine that even minor contributions from corporations headquartered in the industrially developed countries will significantly improve the well-being of persons in the developing world (Galatsidas, 2015). In practice, however, corporate managers who seek to ensure the viability of the profit motive invariably locate the contribution of their corporations to 'development' at the end of what has been termed an 'immiseration process'. That is, the process of "social development is a matter of tidying up after the market ... a polarising mode of economic growth [is] followed up by social impact studies to assess its pauperising impact" (Nederveen Pieterse, 1997, p. 144).

Even assuming that, for instance, certain UN-partnered corporations are able to attain SDG targets to reduce inequalities[1] in some communities through the incorporation of goods and services from impoverished individuals into their value chains, this benefit is likely to be incidental and therefore not reliable in the long term. The alignment of "a company's social and environmental activities with its business purpose and values" (Rangan, Chase, & Kahel, 2015) may indicate that businessmen are aware of their operational environment. But unless their organizations are social enterprises, such an alignment may be a concession to particular non-corporate interests. Any shift in these interests may yet allow corporations to re-assert that the values of profitability and wealth generation are at least as important to the society-at-large as the values attached to environmental sustainability and societal well-being. Bluewashing may even be regarded as a sub-function of CSR operations in general, and may likewise be categorized as "a firm-centric activity, and as such, a narrow, voluntary, and discretionary activity incapable of delivering sustained comprehensive and equitable development" (Sagebien & Whellams, 2010, p. 506). Against the claim that "big and small" can "co-exist perfectly well" (Bhagwati, 2007, p. 223), a counter-claim may be made that larger, capital-rich corporations will inevitably have the advantage.

By implication, a bluewashed corporation may claim that activities contribute to the development of the environment in which it operates, beyond its tax and other payments. However, the presence of inequality and underdevelopment already constitutes part of the market opportunities from which the corporation is intended to benefit. Although a bluewashed company may be easier to promote, the absence of the bluewash is not automatically disastrous for a company that has already stabilized its operations and value chain.

3 Operationalization of Bluewashing in ASEAN

The point above is not intended to delegitimize either the need of businessmen to remain operable in their political geography or the effectiveness of CSR as a means to popularize the operations and outputs of corporations. Rather, the critical focus is on what amounts to an 'outsourcing' of the SDGs to the

1 SDG Target 10.1: "By 2030 progressively achieve and sustain income growth of the bottom 40% of the population at a rate higher than the national average" (UN, 2015).

private sector by governments and by an intergovernmental organization. In this case, three issues may be associated with the institutionalization of corporate values across international borders. First, because the operationalization of the abstract SDGs ultimately depends on bureaucratic harmonization between states and corporations, it is likely that the SDGs will be particularized as justifications of extant practices of states and corporations. The coffee industry in Southeast Asia demonstrates that corporate values already have a normative effect on Southeast Asian societies regardless of the ability or willingness of Southeast Asian coffee farmers and consumers. Cases such as this, where conditions on the ground have a normative influence parallel to that of states and international institutions, indicate that the final form of global values may actually, if not intentionally, be determined by globally capable non-state organizations.

Second, the reconfiguration of the SDGs as part of the CSR of corporations has implications for the 'false consensus effect' associated with CSR-related messages. Individuals falsely believe that the operations of companies are always in accordance with the values that the companies present to the general public (Golob, Lah, & Jancic, 2008, p. 86). Even with a system for vetting companies that bluewash with false promises, governments and international organizations that have endorsed the CSR of certain companies will nonetheless be subject to criticism if the CSR of these companies is proven misleading. The SDGs have been openly criticized not only for "offering nothing new" to debunk the notion of a state–corporate collusion but also as a programme that passes off the "business as usual" behaviour that is incapable of meeting the SDGs in the short-term as nonetheless capable of attaining them in the long term. "[The SDGs] lock in the global development agenda for the next 15 years around a failing economic model that requires urgent and deep structural changes, and they kick the hard challenge of real transformation down the road for the next generation to deal with — by which time it may be too late" (Hickel, 2015).

In particular, the issue of equitable benefits from natural resources is clearly linked to the willingness of powerful governments to be perceived as working for the benefit of corporations. At the request of the government of the United States, SDG targets 2.5 and 15.6[2] were altered from their original form

2 SDG Target 2.5: "By 2020, maintain the genetic diversity of seeds, cultivated plants and farmed and domesticated animals and their related wild species, including through soundly managed and diversified seed and plant banks at the national, regional and international levels, and promote access to and fair and equitable sharing of benefits arising from the utilization of genetic resources and associated traditional knowledge, as internationally agreed"; SDG Target 15.6: "Promote fair and equitable sharing of the benefits arising from the utilization

which would have "ensured" rather than merely "promoted" environmentally responsible behaviour amongst "rich nations — whose corporations and research institutions extract the vast majority of world's natural biodiversity" (Muchhala, 2015). By implication, even if a consensus cannot be formed on whether a company is broadly socially or environmentally beneficent, a consensus can nonetheless be formed on its currently profitable condition as well as on its prospects for strategic expansion. The condition of the global coffee industry reflects this consensus problem. Although the worldwide demand for Rainforest Alliance 'certified-as-sustainable' coffee has grown by 45% from 2011, only about 4.5% of coffee consumed worldwide is sustainably produced (Rainforest Alliance, 2013). Moreover, statements from Nestlé and Jacobs Douwe Egberts, which together account for 39% of the global coffee market, confirm that coffees produced at plantations where human rights abuses are known to take place are likely to have been included in their supply chains (Hodal, 2016). Notwithstanding claims from both companies regarding their sustainable sourcing goals, only about 3.3% of the coffees purchased by both companies qualify as environmentally sustainable (Craves, 2015).

Third, the failure of governments to meet the SDGs in even the medium term indicates that the public has in practice accepted the economic and socio-political power imbalances created by national governments to benefit large multinational corporations. Although corporate operations are capable of creating similar operational conditions in different geographic contexts, the general public will nonetheless perceive these effects as permitted by governments through their sovereign rights. In the ASEAN environment, the coordinated exercise of sovereign prerogative has taken the form of a "norm brewery": that is, even non-Southeast Asian states and institutions who engage with ASEAN member states are able to "influence participants' interests and policies" and to thereby leverage "incentives and pressure in an ideational or moral sense" so long as all participants defer to the goal of facilitating high-level international cooperation (Katsumata, 2006, p. 195).

The region-specific issue of whether ASEAN 'integration' can be meaningful beyond increasing intergovernmental coordination co-occurs with a broader global issue of how to manage the "quiet international importance" (Stratfor, 2014) of Southeast Asia. Southeast Asian societies are located in an area where great powers have sought military and economic access as well as political and cultural influence but have not consolidated total hegemonic control. The

of genetic resources and promote appropriate access to such resources, as internationally agreed" (UN, 2015).

confident rhetoric associated with the 2015 ASEAN Community programme, Vision 2025 (ASEAN Secretariat, 2015a, p. 13) seems to imply that Southeast Asians are not, or are no longer, ambivalent about the importance of their governments, economies, and cultures to their region and the world. However, the existential condition in Southeast Asian societies does not appear to be underpinned by an unambiguous notion of Southeast Asian-ness. Several counterclaims have been made that regional incoherence remains a strong tendency in Southeast Asia (Dent, 2010) and that, paradoxically, non-normativity is the only normative feature of Southeast Asian regionalism (ADB, 2015).

The features traditionally ascribed to the developing societies of Southeast Asia — comparatively low wages and a high growth rate for large populations — now appear to be conjoined to comparatively high GDP growth rate and low production costs. This combination may be regarded as an advantage for international manufacturing and trade, and has been the basis for the claim that the "workshop of the world" term may eventually refer to Southeast Asia as well as to China (Ho, 2015). Whether the "ASEAN Way" of "patient consensus building" (Severino, 2001, p. 2) between state officials (Ewing-Chow & Tan, 2013, pp. 4–5) may be a form of soft power remains disputable. ASEAN Way values such as informality, non-confrontational behaviour, and non-interference are regarded by Southeast Asian officials as crucial to the attainment of "the end goal of ASEAN, [which] is globalisation, not just regional integration" (ACCA, 2016). Nonetheless, the accommodating behaviour of officials from Southeast Asian states in conflict resolution or other diplomacy-related environments may still be interpreted as the result of "structural" limits of societies that are prone to political, economic, and social turbulence (Masilamani & Peterson, 2014) rather than as indicative of a viable template for regionalism for the Asia-Pacific or elsewhere (Langhammer, 2016).

Conversely, the non-confrontational ASEAN approach may be necessary when dealing with the 'development gap' amongst the ASEAN members. Claims that favourably portray the rapid growth rate of the economically disadvantaged ASEAN members, for instance, are helpful to moderate discontent within ASEAN based on other inequalities such as unemployment, GDP per capita, and trade (ASEAN Secretariat, 2015c, pp. 32, 45, 63). 'Servicification', which broadly refers to the increasing importance of the services sector to national economies, has been identified as an indicator that restrictions to the flow of goods and services across the borders of ASEAN member states may be easing (ASEAN Secretariat, 2015b, pp. xix, 108). This is because the services sector throughout Southeast Asia has generally followed what has recently been observed as a global trend, "that many services are embodied in the production of goods that are then exported, and hence the services content of goods trade is much higher when accounting for all the value-added originating in the services sector" (OECD, 2014).

If true, then this is likely to be an operationalization of the 'developmental non-confrontationalism' in ASEAN. ASEAN-style non-confrontational community building implies promoting the sort of outlook that links worker satisfaction, labour and resource mobility, company profitability, and increasing regional market demand throughout the region. By implication, the goals of ASEAN regional integration and sustainability are not very different from those of extant multinationals already operating in the region, even assuming the validity of the claim that, within ASEAN, "economic growth is not an end in itself, but rather a tool for achieving development goals" (Wu, 2013). Consequently, the entire ASEAN Economic Community 2025 Consolidated Strategic Action Plan appears premised on the idea that the attainment of development goals on the one hand, and of a region-wide increase in entrepreneurial behaviour on the other, are practically identical (ASEAN Secretariat, 2017).

4 Development and Coffee in Southeast Asia

The case of coffee production illustrates how existential commonalities have emerged in various communities throughout Southeast Asia without any direct involvement of either ASEAN-level officials or the members of the communities themselves in production and distribution systems other than their own. Coffee is excluded from the computation of food production indices of coffee-growing countries because, although edible, coffee is defined by the UN's Food and Agriculture Organization (FAO) as having "practically no nutritive value" (FAO, 2015, p. 224). However, coffee, like food staples such as soybeans, wheat, corn (maize), and rice is traded in international commodity markets. Unlike food staple commodities, coffee is currently priced at a significantly higher level to reflect a high level of demand. As of September 2017, soybeans were priced at US$0.342/kg, wheat at US$0.155/kg, corn (maize) at US$0.124/kg, rice at US$0.268/kg, and coffee at US$2.836/kg.

Globally, coffee is produced and consumed at levels much lower than food staples. In 2016, just over 9 million metric tonnes (mmt) of coffee were produced, compared to about 313 mmt of soybeans, 754 mmt of wheat, 1,067 mmt of corn (maize), and 483 mmt of rice. In the same year, global per capita consumption of coffee ranged from 1.2 kg to 9.6 kg, compared to wheat at about 67 kg, corn at 27 kg, and rice at 55 kg. However, the correlation of the trading price of coffee with the production and consumption volumes of coffee shows that populations with even low to moderate levels of coffee consumption

(Bernard, 2017) could be expected to pay as much for coffee annually as for the staples needed for survival and nutrition. In 2016, the per capita payment per kg for coffee ranged from US$3.4 to US$27.2; wheat was US$10.36, corn was US$3.41, and rice was US$14.61. This regard for coffee as a 'virtual staple' implies that, in principle, the development of this agricultural industry could have a direct impact on coffee farming communities and thereby contribute to the promotion of sustainable agriculture and of the SDGs in general. However, the image of coffee as a means to promote equitable and sustainable development is not mainly promoted by the farming communities themselves or by the governments that tax agricultural revenue. Rather, the promotion of this image remains within the prerogative of non-state organizations that, whilst committing to make coffee "the world's first 100 percent sustainable agricultural product" (Wenick, 2016), nonetheless depend on value chains that are neither entirely transparent nor structurally determined to eventually becoming sustainable (Conservation International, 2017).

Furthermore, differentiation of the coffee varietals further complicates the issue of coffee sustainability. Although most of the coffee produced in the world is of the Arabica variety, most of the production in Asia is of the Robusta variety (Silva, 2016). In 2016, Robusta accounted for 87.6% of the production in Indonesia, 100% in Malaysia, 68.73% in the Philippines, 100% in Thailand, and 96.2% in Vietnam. However, estimates of Robusta production, and of coffee production in general, differ widely not only across national sources,[3] but also across international monitors, particularly the International Coffee Organization (2017), the UN Comtrade database (2017), and the US Department of Agriculture (USDA Foreign Agricultural Service, 2017). There is also some disagreement about the commercial singularity of the Robusta species. The USDA describes Malaysian and Thai production as comprised entirely of Robusta beans. However, 90% of Malaysian production is comprised of Liberica (*Coffea liberica*) (Nor & Abdul Wahab, 2016) that, although taxonomically distinct from Robusta (*Coffea canephora*), is regarded, like Robusta, as commercially less desirable than Arabica (*Coffea Arabica*) (Davis, Govaerts, Bridson, & Stoffelen, 2006). In Northern Thailand, Chiang Mai, Chiang Rai, Mae Hong Son, and Phetchabun are areas where the Catimor cultivar has been introduced since 1972 by the UN Crop Replacement and Community Development Project and the Royal Project Foundation of Thailand as a means to deter farmers from growing opium (Angkasith, 2002, pp. 3–4). However, Catimor is a

3 For examples of data on Southeast Asian production, refer to the Ministry of Agriculture Indonesia (2014), the Philippine Statistics Authority (2016), the Assumption University of Thailand (Angkasith, 2002), and the East Asia Forum (Pham & Tien, 2014).

Robusta and Arabica hybrid, the popularity of which was determined more by its resistance to disease (Specialty Coffee Association of America, 2017) rather than by favourable 'flavour profile commercial beverage characteristics (Coffee Research, 2017).

There is a significant price difference between Robusta (about US$1.8 per kg) and Arabica (nearly US$3 per kg), which is usually brewed in specialty cafes (Verdin, 2016). Currently, this difference is attributable to the "rough and bitter taste" of the "beast" that is Robusta, which is "not very appealing for coffee connoisseurs and not as appreciated by the market as its 'gentler brother' that is the Arabica 'beauty'" (Penarredonda, 2017). The chief consideration for the widespread cultivation of Robusta has been disease and pest resistance as well as a shorter time to reach peak productivity (two years) compared to Arabica (five years). For Southeast Asia in particular, the epidemic of coffee leaf rust (*Hemileia vastatrix*) throughout the region between 1870 and 1900 ruined Arabica production and continued to have residual effects throughout the 20th century (Mishra & Slater, 2012). The disease practically destroyed the coffee industry in the Philippines and contributed to the drastic decline of coffee production in the established plantations in present-day Malaysia and East Timor (Bulbeck, Reid, Tan, & Wu, 1998, pp. 156–158). However, the Robusta plants introduced in Indonesia from what is now the Democratic Republic of Congo towards the end of the epidemic thrived, to the extent that coffee production by 1940 more than doubled compared to pre-epidemic levels (McStocker, 1987, p. 42). This was the combined output of the Robusta-rehabilitated plantations in Java and Sumatra as well the numerous Arabica-planted smallholdings in the east of the archipelago (Bulbeck et al., 1998, p. 155). After the rehabilitation of the Indonesia plantations, the price of coffee in the Netherlands, and in Western Europe by implication, was determined by the Robusta crop (ibid., p. 175) and the comparatively high price reflected scarcity of supply rather than flavour distinctions.

However, co-optation of the Southeast Asian coffee industry into large-scale Robusta production occurred during the Cold War, because of American government interests in assuring the supply of cheap coffee beans for industrial processors and in preventing coffee farmers from shifting to other crops in the face of an oversupplied international coffee market. Prior to World War II, consumers in the US were accustomed to drinking cheap coffee: a cup brewed at home cost about US$0.17, whilst a coffee in a restaurant cost about US$1.73. Prices through WWII remained at comparable levels because of government rationing (US$0.13 at home/US$2.01 at restaurant) (Elliot, 2017). The price stabilization was also made possible by the Inter-American Coffee Agreement between the US and 14 Latin American countries, whereby a quota system divided the US market among the Latin American treaty members and effectively prevented the US from importing coffee from non-member countries. The agreement led to a rise in coffee prices (Bilder, 1963b, p. 330), but also caused a dependency effect amongst Latin American farmers. The Agreement lapsed at the end of WWII. Despite the stabilization of consumption patterns in the US after the war, the cost of home-brewed coffee continued to rise (US$0.27 in 1955) whilst restaurant prices actually decreased (US$0.90 in 1955). This condition co-occurred with the significant increase in the global supply of coffee caused from revitalization of production in Kenya, Uganda, Ivory Coast, and what is now the Democratic Republic of Congo.

By 1955, producers in Africa were capable of exporting around 6 million bags (about 360 million kg), 80% of which was Robusta. The combination of the large African crop and the revitalized production of Robusta in India and Southeast Asia (chiefly Indonesia and Vietnam) resulted in a cheap resource that nonetheless remained mostly inaccessible to consumers in temperate, non-coffee-growing Europe and the US. This favoured not only restaurants that offered coffee as a value-added service but also US manufacturers (General Foods, Standard Brands, Folger's, Hills Brothers, and A & P) that invested heavily in the production of instant coffee, which was marketed on convenience rather than flavour. By 1958, some instant coffees were made from 100% Robusta, whilst the predominance of the large coffee manufacturers not only caused a drastic decline of smaller independent producers but also likely caused the widespread acceptance of the Robusta in the form of both brewed and instant coffee (Pendergrast, 2010, pp. 178–179). This was a significant problem for Arabica coffee producers in Latin America who traditionally supplied the US market; with a shift of the market to Robusta, they risked impoverishment and — a key concern for the US government — political alignment with the communist Soviet Union.

The 'coffee-communism' threat to the US was highlighted by two highly publicized incidents. First, whilst on a goodwill tour of Latin American countries, Vice-President Richard Nixon was heckled with shouts of *Muera Nixon* (death to Nixon) and was assaulted on 7 May 1958 by what were presumed by US officials to be communist student activists in Lima, Peru, which was and remains a coffee-producing country (Ambrose, 1988, pp. 467–468). Secondly, Cuba, which in 1961 reached its peak production of about 60,000 tonnes (Castellanos, 2016), was the site on 16–28 October 1962 of the US–Soviet Union military confrontation that later became known as the Cuban Missile Crisis. The US State Department adhered to the belief that:

> large publically-held corporations such as General Foods [the largest North American coffee processor] were aware of the consequences for US security if the market should break and Latin America went Communist: ... [it] would not only have been ruinous for American diplomatic interests in the Western Hemisphere, but many in the industry thought that [it] might [cause] enormous difficulties in obtaining supplies.
>
> FISHER, 1972, p. 22

With 1939 as an index price year, the price of coffee in the largely unregulated post-war global market dropped in 1955 (−19.16%), 1958 (−12.78%) and 1959 (−19.07%) but recovered to −3.59% in 1962 (US Bureau of Labor Statistics, 2017). Nonetheless, the issues of "short-run market instability" and "long-term overproduction" rather than political alignment became the basis for the International Coffee Agreement drawn up by the United Nations Coffee Conference on November 1962 (Bilder, 1963b, p. 333). The agreement became the basis for the creation of the International Coffee Organization (ICO), headquartered in London, UK, and the creation of a quota system whereby signatory coffee-producing countries would only supply the world market with specified volumes of coffee at prices based on "estimated levels of world demand" (Bilder, 1963a, p. 890). However, it also reinforced a categorization of coffees into 'Brazils' (Arabicas produced in Brazil), 'milds' (Arabicas produced in other Latin American countries), and 'Robustas' (coffees produced in Africa and elsewhere) that favoured Latin American producers, promoted the evaluation of coffee desirability based on geographic origin, and defined Robustas as the least commercially desirable of the lot.

Moreover, each of the participants in the negotiations tended to congregate at one of seven blocs divided along differences of ethnicity, wealth, productive capacity, purchasing capability, and administrative acumen. The complexity of managing the "constantly shifting pattern of coalition, pressure, and counter-pressure" ended with the majority accepting a "Latin American" proposal to allocate quotas based solely on country attribution instead of an "African" proposal, which also considered "consumer preference" and thereby also the peculiarities of the coffee beans (Bilder, 1963b, pp. 331, 345–347, 357–358). These eventualities put Indonesia and Philippines, the two Southeast Asian signatories to the agreement, in a peculiar position. The recovery of their respective coffee industries was thenceforth to depend on the production for export of Robusta not only because of the circumstances associated with the coffee leaf rust epidemic of the late 19th and early 20th centuries, but also — and more importantly — because the treaty essentially implied that the limited production of less profitable Robusta was preferable to the violation of treaty rules, which would mean having no commercially viable coffee crop at all.

The quota system was withdrawn and replaced several times across the 1960s through the 1980s, and was suspended on 4 July 1989. Prior to its indefinite suspension, it was renegotiated in 1968, 1976 and 1983. "The agreement was plagued by many difficulties, including the inability to deal with the structural causes of oversupply, the failure to do little more than dampen the unpredictable swings of the coffee cycle, and the persistent conflict among the signatory nations over the quota system" (Fridell, 2012, p. 291). The poverty of the coffee farmers in South America and Africa is associated with the end of the higher purchase price regime that the quota system represented. However, a key feature of this issue had been the attempt of the world's largest coffee producers and exporters, particularly Colombia and Brazil, to prevent challenges to the status quo by lobbying the world's single largest market, the US, to pressure the ICO to re-instate the quotas (International Trade Center Secretariat, 2017).

Nonetheless, the conditions particularly of Robusta producers in Southeast Asia would have proven attractive for coffee processors, who intended to circumvent the politics of "high prices and limited supply" of the ICO. In the 1970s and 1980s, not only did productivity per hectare of farms in Indonesia and Philippines match or exceed that of Latin American and African farms, but farm-gate prices for coffee paid directly to farmers were also lower than those in Latin America and Africa (Pieterse & Silvis, 1988, pp. 9–10). Moreover, producers in Indonesia, Philippines, and the "rest-of-Asia" were projected to significantly benefit from the removal of the quota system, compared to the large producers in Latin America and Africa (Akiyama & Varangis, 1989, pp. 11, 49–50).

It was into this environment that the Nestlé food company introduced a new political economy of coffee. Nestlé supplied the soluble coffee and chocolate rations of US military personnel who were deployed throughout the region during and after WWII. The resulting popularity of these and other products with the local population, combined with an aggressive expansion strategy in the region, resulted in a network of 30 factories that spanned Indonesia, Malaysia, Philippines, Singapore, and Thailand by the end of the Cold War (Pritchard & Fagan, 1999, pp. 12–14). Until about the mid-1970s, Nestlé was known in Southeast Asia mainly through retail operations that sold Nestlé-branded products, which were manufactured elsewhere. In 1960, a partnership between Nestlé and San Miguel Corporation in the Philippines resulted in the formation of Nutritional Products, Inc. (Nutripro). In 1962, Nutripro, which is the present-day Nestlé Philippines, Inc., started to manufacture

Nestlé's Nescafé brand coffee at its first plant in Alabang, Muntinlupa City, Metro Manila (Nestlé Philippines, 2017). In 1974, a factory was constructed in Chachoengsao, near the coast of Central Thailand to produce Nescafé coffee. It is presently advertised as "one of the largest and most advanced instant soluble coffee plants in Nestlé's world" (Nestlé Thailand, 2017). In 1979, a factory was opened in Panjang, Lampung, Indonesia to produce Nescafé coffee. This was the second factory constructed in Indonesia after a milk processing plant was constructed in Waru, East Java in 1972 (Nestlé Indonesia, 2017).

A key feature of Nestlé's coffee procurement is a simplified and direct purchasing system. The direct buying policy allows farmers to bring their harvests to Nestlé buying stations where the green dried and unroasted Robusta coffee beans are openly sorted, graded, and priced. A farmer who consents to the sale receives the payment within eight "banking hours" (ABS-CBN News, 2013). This system, which was in place when the worldwide supply of coffee surged towards the end of the global quota, allowed Nestlé to benefit from new and existing coffee producers in Southeast Asia, particularly in Vietnam (International Coffee Organization, 2016; Nestlé, 2004, pp. 6–7). In 1995, Nestlé Vietnam Co. Ltd. was established; the construction of the Nescafé plant at Tri An in Dong Nai Province, Southern Vietnam was completed in 2013, and a coffee decaffeination plant worth US$80 million was subsequently inaugurated in 2015 (Nestlé Vietnam, 2017). Through the Vietnam leg of the company-wide 'Nescafé Plan' launched in 2011, Nescafé staff had, by 2012, trained over 19,600 farmers who produced more than 15,000 tonnes of coffee beans and who, by 2013, had received around 2 million coffee seedlings for planting (Nestlé Global, 2017).

The regionalizing effect that the coffee operations of Nestlé had on Southeast Asia was consolidated by the ASEAN Industrial Joint Ventures (AJIV) programme of 7 November 1983, which was superseded by the ASEAN Industrial Cooperation (AICO) programme of 27 April 1996. Both programmes were similarly "aimed at increasing industrial production through resource pooling and market sharing by ASEAN firms" (Bello & Bullard, 1997), but most of the companies that took advantage of the programmes were automobile and automobile parts industries (Musungu, Villanueva, & Blasetti, 2004). Thailand in particular benefited from the export of Japanese automobile parts provided through the Keidanren-Japanese Federation of Economic Organizations to assembly and dealership operations throughout Southeast Asia until that AICO-supported operation was ended in 2010 (Chia & Plummer, 2015, pp. 71–73; Inama & Sim, 2015, pp. 170–173). Nestlé was the only food company that applied to the AJIV and AICO programmes, which benefited its instant coffee operations in Lampung, Indonesia. Most of the AJIV and AICO projects of Nestlé were not specifically linked to the production of instant coffee. Regardless, the corporate governance–state governance link is clearly inferable regardless of interpretation: the ASEAN system was instrumental in increasing the regional reach of Nestlé, and the corporate expansion of Nestlé occurred on the basis of market demand to which the ASEAN bureaucracy was incidental.

The production of the commodity from which Nestlé's instant coffee is derived in Southeast Asia appears to have had a normative effect on Robusta-growing communities. Not only is there an appreciable difference between Robusta and Arabica growing areas in terms of socio-economic development, but also a difference in the levels of socio-economic development across different coffee-growing states. These discrepancies are reflected in the Human Development Index (HDI) scores of the different areas. There is a clear Arabica–Robusta divide in Indonesia: the average HDI score for Arabica-growing areas is about 0.6952 whilst Robusta-growing areas score 0.6766, with Lampung, the centre of Nescafé production, scoring the lowest in HDI (0.6695). The difference is more clearly seen in the Philippines: predominantly Arabica-growing Benguet had an HDI score of 0.718, whilst Robusta-growing communities averaged 0.6189. In other places, accounting for the divide requires more qualification. Vietnam has shown significant variance across the different coffee-growing areas, notwithstanding the predominance of Robusta in all areas. This may be explained by the results of the Doi Moi economic reforms of 1986. The government of Vietnam not only sought large-scale foreign, mostly European, investment into the Robusta coffee industry, but also decollectivized agricultural production and gave incentives for lowland migrants to participate in the coffee industry in the highlands, particularly around Dak Lak. Although most of the coffee is currently produced by smallholders, the Vietnamese government claims that these smallholders are "well resourced, with a high level of tenure security, good access to finance, and high levels of ownership of farm equipment", through state policies and purchasing agreements with the US, Germany, and other European states (Neilson et al., 2011, pp. 191–193).

The relation appears to be inverted in Thailand, where Robusta-growing southern areas (average score 0.6391) are more developed than the northern Arabica-growing areas (average score 0.5936). Notwithstanding the debate on whether the Catimor Arabica–Robusta hybrid grown in the north produces a crop that may be priced at Arabica levels, a key issue in this case is the persistent involvement in the narcotics trade of the Hmong and other minority peoples in Phetchabun and Mae Song Hon. A prevailing theory is that the introduction of coffee and other crops into Northern Thailand dissuaded these peoples from growing their traditional opium crop. However, the network of the narcotics trade that spans Northern Thailand, Burma, Laos, and Southern

China remained intact, and resulted in a rise of the heroin trade in the areas where opium was previously cultivated (Chen, 2015; Minorities at Risk, 2010).

With some qualification, the coffee industry in Southeast Asia seems to indicate that the creation of an 'underclass', through the production of the coffee beans necessary for the processes of transnational cooperation, coincides with the regionalizing goals of the governing 'upper class'. Moreover, if the core values of regionalization become operationalized as the creation of structures that normalize behaviour across previously diverse societies, then the 'unity in diversity' maxim that identifies ASEAN may come to stand for: unity of system, diversity of the marginalized.

5 Sustainable Development: Uninterrupted Productivity or Endless Assistance?

The Human Development Index and SDGs both represent an attempt at "delinking of economic growth and human development" in order to emphasize that "development is an issue for every society, and that it is as much about human rights, the environment, and relationships as it is about economics" (Slim, 1995, pp. 145, 146). There is nonetheless a link between the present affluence of a society and the ability of that society to meet the broadened range of development criteria (Halle, 2016, p. 8). Moreover, as the case of coffee in Southeast Asia shows, any "intervention in a systemic process that attempts to control the process by using mechanistic tools" (Du Plessis, 1999, p. 380) also creates a structure of relations that accrues benefits to the intervener that remain largely beyond the ability of the intervened to attain.

This points to a dilemma that is not readily appreciated on account of the ostensible benefits that small-scale coffee farmers obtain through government-approved corporate programmes such as Nestlé's Nescafé Plan (The Philippine Star, 2016). On the one hand, sustainable development can mean that farmers and other small-scale participants in a region-wide corporate system are to be assured their place in a value chain that intends to maintain uninterrupted productivity. On the other hand, the lowest-level participants in the value chain remain unable to reconcile their persistent status as objects of corporate policy and practices with the seemingly ubiquitous need for socially and environmentally responsible growth. By presuming that the well-being of suppliers and consumers are externalities of commercial processes, corporate operatives can continue to present their products and services as intrinsically valuable regardless of the effects these have on the condition of the broader environment. In this case, the issue is exemplified by the growing worldwide popularity of instant soluble coffee.

Robusta-producing countries in Southeast Asia are themselves significant consumers of coffee-derived products, as Indonesia, Malaysia, Philippines, Singapore, and Thailand are amongst the top 20 importers of extracts, essences, and concentrates of coffee, which are mostly made from Robusta. African societies do not appear to be significant consumers, even if African plantations also produce significant amounts of Robusta. The Philippines, which is the fifth largest importer, is significant not only because Filipinos consume more Robusta than their own farmers are able to produce but also because the high level of consumption indicates an alignment with the promotional activities of industrial coffee refiners that are headquartered in temperate, non-coffee-producing areas. As the consumption of coffee-derived products is led by industrialized states, it would be reasonable to expect that the upward trend of consumption is not likely to cause a re-evaluation of coffee industry status quo. Instant soluble coffee, which is made by processing cheaper Robusta beans through large-scale industrial systems, is forecast to have a compound annual growth rate (CAGR) of 6% until 2020 (Technavio, 2016a). In contrast, the global roast and ground coffee market, which is sustained by Arabica production, is estimated to have a CAGR of 5% in the same period (Technavio, 2016b). It may be to the interest of Southeast Asian coffee farmers for corporations to promote the consumption of more expensive and more widely-produced Arabica coffee.

In practice, the instant coffee industry could introduce structural reforms, such as the propagation of Arabica or hardy hybrids such as Catimor in the crop renewal or replacement cycles of existing farms, or it could alter costing levels to allow for higher purchase prices of smallholder-sourced coffee. The goal would be to link the profit sustainability of the instant coffee industry to the social sustainability of coffee farming. An assurance to farmers that their coffee crops are not only profitable but also socially desirable, on account of these being composed of beans with superior flavour, would broaden the appeal and appreciation of coffee farming amongst the broader public. This approval, in turn, would highlight the continued reconfiguration of coffee farming, and of the global agricultural industry, as crucial to the attainment of the sustainable development goals. Grading standards for coffee beans have been compiled by the ICO (International Coffee Association, 2017), and the Specialty Coffee Association of America (SCAA) has released a set of criteria by which an Arabica crop may be assessed as "specialty grade" (Specially Coffee Association of America, 2009). However, beyond these sorting protocols, a unified standard for 'good coffee' has not yet emerged. "The reality is that there is no global consensus among consumers, about what a good quality cup of coffee really is ..., different forms of coffee and methods of preparation are preferred in different parts around the world making the task of standardising almost impossible" (Feria-Morales, 2002, p. 363).

Certain industrial applications can quantify the otherwise subjectively identified characteristics of beverages, and may serve as reference points for customer preference.[4] Nonetheless, the circumstances surrounding the production and consumption of coffee beverages, as well as the promotion of coffee drinking and growing, likewise determine the accessibility and social desirability of coffee in general, and of Robusta-based instant soluble coffee in particular. If the current forms of Robusta production can be both bluewashed and greenwashed by corporations, then not even governments are likely to disrupt the systems whereby Robusta farmers, though gainfully employed, are structurally limited from producing a better Arabica crop. This form of sustainable underachievement is seen in the "coffee statecraft" of Vietnam where production of Robusta was promoted by the state as a key export commodity during the Doi Moi economic renewal of 1986 (Fridell, 2014, pp. 415–416). The Vietnamese government sought "to develop and expand the rules and regulations demanded by a capitalist logic (creating vulnerable, export-dependent smallholders on semi-privatised land) while at the same time seeking to fulfil its territorial logic through economic growth and social stability (finding viable livelihoods for new farmers and profitable avenues for the new capitalist class)" (ibid., p. 21) As uninterrupted productivity, sustainable development would likely imply the defence of status quo prerogatives as well as the continuation of structural inequalities.

Conversely, sustainable development can mean that corporations and governments will still focus on high-level "broader political transformations and the underlying strategies of exclusion" (Frangie, 2011, p. 1183) whilst claiming that the concessions given to low-level participants are costs that the governing and corporate systems can afford. This "willingness to pay" is generally termed the neoliberal economic approach to development assistance. It also, however, is essentially a form of damage-control governmentality: "sustainable" would become operationalized as measures to justify economic viability, which take the form of acceptable levels of "costs" to the environment, natural resources and society. Bluewashing in this case would be the management of a "complex network of relations among ecosystems and socioeconomic systems"

4 See, for example, the Taste Sensing System TS-5000Z of Intelligent Sensor Technology, Inc., Japan, http://www.insent.co.jp/en/products/TS5000Z_web_E_0512.pdf.

that need to be regulated and governed by corporations and governments in tandem (Gomiero, Pimentel, & Paoletti, 2011, pp. 9–13). It also implies that the SDGs are regarded as categories through which the entire humanity may be classified into a global North that sets the regimes for global consumption and production and a global South whose consumption and production patterns are ultimately determined by dispensations such as capacity-building initiatives and developmental assistance (Frangie, 2011, p. 1188). However, it also implies that corporations are ascribed a key role in "commodifying morality", because socially desirable conditions such as clean environments, fair wages, and democratic decision making can only be "produced" by large-scale organizations that are capable of aggregating and operationalizing social aspirations into "market demand". "Consumers are expected to 'change the world' through their buying habits, and producers are expected to join collectives that donate a portion of their income to infrastructure projects that in the past would have been paid for by states (e.g., roads, sewers, schools, parks, and hospitals), and taxes paid by the wealthy" (Robbins, 2013, pp. 258–259).

The net effect of this hegemonic–corporatist approach to development is that, in exchange for dispensations from the corporate elite, the low-level stakeholders are expected to quickly adapt to market conditions that are not of their making. Ironically, the increased demand for sustainable coffee, for instance, has compelled tropical farmers and primary processors to think of their product in terms of specialty categories in alignment with foreign, typically European and North American, markets. "Specialty coffees are not precisely defined but cover a wide range of somehow differentiated coffees, such as organic, fair trade, bird-friendly ... and single-origin coffee: ... because the term single-origin is not precisely defined, single-origin coffees can originate in one country, one region, or one estate or farm" (Teuber, 2010, p. 278). "Shade-grown coffee" has also become a preferred product, presumably because coffee consumers are led to believe that the living conditions of farmers whose coffees are grown under jungle canopies are better than those who grow coffees in un-shaded monoculture plantations (Toledo, 2012).

In the 1970s, the United States Agency for International Development (USAID) and other development agencies pressured the same farmers to grow more productive varieties of coffee without shade and with heavy chemical inputs. "Shade-grown certification was ... conceived to solve a problem primarily articulated by people in the United States, and its primary goal is conservation rather than livelihood stability" (Lyon, 2009, p. 227). The culmination of corporatist coffee in Southeast Asia appears to be the organization of the ASEAN Coffee Federation (ACF) that was established in 2010. None of its objectives explicitly refer to the protection of coffee farmer welfare despite its stated vision "to be the key industry representative of ASEAN's coffee industry". And although its membership is presumably open to all corporations and associations that deal with the Southeast Asian coffee industry at all levels, its leadership appears to be composed of individuals who represent organizations that seek to raise the value and profile of Southeast Asian coffees rather than the well-being of Southeast Asian coffee farmers (ASEAN Coffee Federation, 2017).

6 Conclusion

Bluewashing may yet prove to be a way around the threefold problem posed by the 17 SDGs and the corresponding 169 targets. Not only do (1) the SDGs condition the public to perceive the value of businesses in terms of charitable work rather than profit, but (2) SDG-compliant businesses are still a collective "pocket to be picked" (Frantz, 2017) by governments for tax. Moreover, (3) as the SDGs are "country owned" in terms of monitoring and review (UN Office of Internal Oversight Services, 2017), the official SDG documents do not provide practical guidance for businesses that may be able to outperform governments in terms of SDG attainment.

Notwithstanding the ability of corporate governance to serve the ends of public governance, the policy implications of the SDGs are numerous, and not altogether positive. What becomes of governance if businesses are more efficient, more profitable, more humanitarian, and more environmentally friendly than governments? Conversely, what becomes of the SDGs if businesses are not able, or willing, to do much better than governments that regard the SDGs as attractive, but not essential, outputs of governance? Should the global public regard the SDGs like coffee; that is, should status quo governance and unremarkable coffee be regarded as acceptable because, only occasionally, examples of good governance and ethically produced coffee may be found? If the SDGs were created to benefit the extremely impoverished, oppressed and vulnerable, then does the successful pursuit of the SDGs imply upsetting the rest of the global public that, despite not being composed mostly of exploiters, have lifestyles that tacitly affirm forms of environmental degradation and social marginalization? And is there not a contradiction in making the attainment of the SDGs dependent on the largesse of government leaders and other politically privileged persons, whose elite status is accepted by a majority of persons mainly because they are structurally unable to gain similar levels of income, education, health, and social desirability?

The following recommendations are intended to help governments and businesses to derive the most benefit from the SDGs campaign. First, SDG-related programmes should, as far as possible, specifically and publicly identify the beneficiaries of these programmes as well as the nature of the benefits brought by these programmes. Such a practice may seem trivial, considering that the documentation of many welfare and CSR projects is often released for public scrutiny. However, public endorsement by institutions helps democratize access to mass media promotion. This is important because governments and businesses that are capable of conducting SDG-related activities gain promotional advantage even before any improvement in the lives of the programme beneficiaries occurs. Not even SDGs no. 3 (Health and well-being), no. 10 (Reduce inequality between and among countries), or no. 17 (Partnerships for the goals) include measures to actively reduce the practices of stigmatization, ridicule, or social exclusion of persons who by themselves are unable to resolve their poverty or other socially undesirable circumstances. The SDGs are the result of what has been described as the largest consultation programme in the history of the UN (Ford, 2015). However, the broad issue of social prejudice, with the narrow exception of SDG no. 5 (Achieve gender equality and empower all women and girls), nonetheless persists as a background issue of 'developed' and 'underdeveloped' categories that may be resolved only incrementally and indirectly.

In the case of the coffee industry, not only should farmers be sustained as producers of a socially desirable product, but also agriculture-based industries and economies should be more widely acknowledged, and more generously paid, as guarantors for humanity's existential condition. More broadly, this will entail acknowledging the importance of "food sovereignty", which assigns food self-sufficiency as a criterion for political legitimacy, whilst accepting that agricultural production is globally assessed in competitive terms and purchasing power (Lamy, 2011) rather than in terms of sustenance (Institute for Agriculture and Trade Policy, 2003, p. 4), information, waste, safety, or diversity (World Food Summit, 2017, p. 7).

Second, SDG programmes should be formulated with clear reference to a particular developmental model. This means identifying not only the particular SDGs to be attained, but also the broader political agenda that the developmental assistance, which funds the projects, likely serves. Although the SDGs are largely a UN initiative, they are also in alignment with the operations of the World Bank (World Bank, 2017), the International Monetary Fund (IMF, 2017), the Organization for Economic Co-operation and Development through its Development Assistance Committee (OECD, 2017), and numerous agencies of developed countries that work with their counterparts in developing countries. The SDGs are therefore qualified as a North–South cooperation project, the format of which has been current internationally since the end of WWII.

"Donors' legitimate leadership (donorship) over recipient countries was the final condition for the consolidation of international development as a field of practices" (Esteves & Assunção, 2014, p. 1778). Alternatives to this hegemonic model of donor-driven international development have since been created and have come into prominence since the global financial crisis of 2008. An example of these is the 2011 conference in Busan, South Korea that aimed at "mapping, documenting, analysing and discussing evidence on the synergies between the principles of aid effectiveness and the practice of South–South Cooperation" (Abdenur & Marques Da Fonseca, 2013, p. 1480).

In the case of the coffee industry, sourcing the funds for SDG-related projects to benefit coffee farmers or communities in the global South does seem to imply that the largest markets for coffee in developed countries should be made to pay more for their consumption. But it does not exclude the possibility of increasing the demand for a higher-quality product in the countries of origin, particularly if this implies empowering local producers to add value to their own coffees. On the one hand, it remains to be seen if consumers of products that have been processed, marketed, and thereby popularized by firms headquartered in developed countries would be willing to alter their consumption patterns for the attainment of the SDGs. On the other hand, if the SDGs are to be asserted as metrics of global, rather than hegemonic, development, then these must be promoted in tandem with other "policies and practices [that] embody a reinvigoration of South–South cooperation" (Muhr, 2016, p. 632).

Third, any programme intended at attaining specific SDGs in specific environments needs to be justified relative to other co-occurring programmes that are not directly development-related. This account of the political economy and history of the coffee trade shows that corporate value-creation activities are either located in developmental gaps that governments are not singly able to fill, or are complements of state programmes that have not been able to resolve status quo inequalities. Despite their appearance as discrete developmental milestones, the SDGs are structurally linked as ideal features of state and corporate governance, and are therefore subject to actual contingencies of governance in which the well-being of the marginalized is incidental, such as inter-elite struggle for political control, financial volatility, 'race to the bottom' capitalism and cut-throat competition, wars, and other forms of politically motivated violence.

Although the SDGs are promoted with the tagline "leaving no one behind", the pursuit of other state and corporate goals is likely to cause quite a number to be "left behind". Ironically, the failure of governments to attain most of the SDGs would not constitute a failure of the UN system through which the goals are articulated, or of the corporate system of business that is capable of subsuming the development agenda within a greater framework of profit generation. Thus, the least that can be expected of governments is a detailed explanation to their constituents that both the pursuit and neglect of the SDGs come at costs that need to be borne by all, especially by corporate businesses. Whether or not bluewashing can remove the stubborn stains from corporatism on the coffee trade remains to be seen.

References

Abdenur, A. E., & Marques Da Fonseca, J. M. (2013). The north's growing role in south–south cooperation: Keeping the foothold. *Third World Quarterly*, 34(8), 1475–1491.

ABS-CBN News (2013, March 13). Nestle expands network of coffee bean buying stations in PH. ABS-CBN News. Retrieved from http://news.abs-cbn.com/business/03/13/13/nestle-expands-network-coffee-bean-buying-stations-ph (accessed 1 September 2016).

ACCA (2016, April 11). ASEAN Way of integration may lack bite, but it works. Association of Chartered Certified Accountants. Retrieved from http://www.accaglobal.com/lk/en/discover/news/2016/april/acca-president-debate.html.

ADB (2015, December 29). ASEAN Economic Community: 12 things to know. Asian Development Bank. Retrieved from http://www.adb.org/features/asean-economic-community-12-things-know.

Akiyama, T., & Varangis, P. N. (1989). Impact of the International Coffee Agreement's export quota system on the world market. World Bank. Retrieved from http://documents.worldbank.org/curated/en/721281468739224011/pdf/multi-page.pdf.

Ambrose, S. E. (1988). *Nixon, Vol. 1: The education of a politician, 1913–1962*. New York: Simon and Schuster Inc.

Angkasith, P. (2002). Coffee production status and potential of organic Arabica coffee in Thailand. Assumption University of Thailand Journal of Technology. Retrieved from http://www.journal.au.edu/au_techno/2002/jan2002/article3.pdf.

ASEAN Coffee Federation (2017). ACF vision, objectives and mission. ASEAN Coffee Federation. Retrieved from http://aseancoffee.org/who-we-are/.

ASEAN Secretariat (2015a). ASEAN Community Vision 2025. Association of Southeast Asian Nations. Retrieved from http://www.asean.org/storage/images/2015/November/aec-page/ASEAN-Community-Vision-2025.pdf.

ASEAN Secretariat (2015b). ASEAN integration report 2015. Association of Southeast Asian Nations. Retrieved from http://asean.org/storage/2015/12/ASEAN-Integration-Report-2015.pdf.

ASEAN Secretariat (2015c). ASEAN statistical yearbook 2015. Association of Southeast Asian Nations. Retrieved from http://asean.org/storage/2012/05/ASEAN-Statistic-Yearbook-2015_r.pdf.

ASEAN Secretariat (2017). ASEAN Economic Community 2025 consolidated strategic action plan. Association of Southeast Asian Nations. Retrieved from http://asean.org/storage/2012/05/Consolidated-Strategic-Action-Plan-endorsed-060217rev.pdf.

Bello, W., & Bullard, N. (1997). APEC and the environment: A report to the Rio+5 conference. Focus on the Global South. Retrieved from http://focusweb.org/publications/1997/APEC%20and%20the%20Environment.htm (1 September 2013).

Bernard, K. (2017, April 25). Top 10 coffee consuming nations. WorldAtlas. Retrieved from http://www.worldatlas.com/articles/top-10-coffee-consuming-nations.html.

Bhagwati, J. (2007). Why multinationals help reduce poverty. *The World Economy*, 30(2), 211–228.

Bilder, R. B. (1963a). The International Coffee Agreement, 1962. *The American Journal of International Law*, 57(4), 888–892.

Bilder, R. B. (1963b). The International Coffee Agreement: A case history in negotiation. *Law and Contemporary Problems*, 28(2), 328–391.

Bruno, K., & Karliner, J. (2000). Tangled up in blue: Corporate partnerships at the United Nations. CorpWatch. Retrieved from http://www.corpwatch.org/article.php?id=996#intro.

Bulbeck, D., Reid, A., Tan, L. C., & Wu, Y. (1998). *Southeast Asian exports since the 14th century: Cloves, pepper, coffee, and sugar*. Singapore: Institute of Southeast Asian Studies.

Castellanos, D. (2016, June 15). The decline of Cuban coffee. Diario De Cuba. Retrieved from http://www.diariodecuba.com/cuba/1465976977_23086.html.

Chen, S. A. (2015, January 6). Singaporean moved by the plight of the Hmong. Catalyst Asia. Retrieved from https://catalystasia.wordpress.com/2015/02/04/singaporean-moved-by-the-plight-of-the-hmong-2/.

Chia, S. Y., & Plummer, M. (2015). *ASEAN economic cooperation and integration: Progress, challenges and future directions*. Cambridge: Cambridge University Press.

Coffee Research (2017). Arabica coffee bean varietals. Coffee Research. Retrieved from http://www.coffeeresearch.org/agriculture/varietals.htm.

Conservation International (2017, September). Sustainable coffee challenge. Conservation International. Retrieved from http://www.conservation.org/stories/sustainable-coffee-challenge/pages/about-the-challenge.aspx.

Craves, J. (2015, February 19). More corporate coffee consolidation. Coffee and conservation. Retrieved from http://www.coffeehabitat.com/2015/02/corporate-consolidation/.

Davis, A. P., Govaerts, R., Bridson, D. M., & Stoffelen, P. (2006). An annotated taxonomic conspectus of the genus Coffea (Rubiaceae). *Botanical Journal of the Linnean*

Society. Retrieved from https://academic.oup.com/botlinnean/article-lookup/doi/10.1111/j.1095-8339.2006.00584.x#.

Dent, C. M. (2010, June 1). Regionalism and East Asia: Which way forward? *World Politics Review*. Retrieved from http://www.worldpoliticsreview.com/articles/5664/regionalismandeastasiawhichwaysforward.

Du Plessis, C. (1999). Sustainable development demands dialogue between developed and developing worlds. *Building Research & Information*, 27(6), 378–389.

The Economist (2004, June 17). Corporate social responsibility: Bluewashed and boilerplated. *The Economist*. Retrieved from http://www.economist.com/node/2771522.

Elliot, M. (2017). What a cup of coffee cost each president since 1920. Culture Cheat Sheet. Retrieved from https://www.cheatsheet.com/money-career/historical-look-at-the-cost-of-a-cup-of-coffee.html/?a=viewall.

Esteves, P., & Assunção, M. (2014). South–south cooperation and the international development battlefield: Between the OECD and the UN. *Third World Quarterly*, 35(10), 1775–1790.

Ewing-Chow, M., & Tan, H.-L. (2013). The role of the rule of law in ASEAN integration. Centre for International Law, National University of Singapore. Retrieved from http://cil.nus.edu.sg/wp/wp-content/uploads/2010/10/Ewing-Chow-TanHsien-Li-The-Role-of-the-Rule-of-Law-in-ASEAN-Integration.pdf.

FAO (2015). *FAO statistical pocketbook 2015: World food and agriculture*. Rome: Food and Agriculture Organisation of the United Nations. Retrieved from http://www.fao.org/3/a-i4691e.pdf.

Feria-Morales, A. M. (2002). Examining the case of green coffee to illustrate the limitations of grading systems/expert tasters in sensory evaluation for quality control. *Food Quality and Preference*, 13(6), 355–367.

Fisher, B. S. (1972). *The International Coffee Agreement: A study in coffee diplomacy*. New York: Praeger.

Ford, L. (2015, January 19). Sustainable development goals: All you need to know. *The Guardian*. Retrieved from https://www.theguardian.com/global-development/2015/jan/19/sustainable-development-goals-united-nations.

Frangie, S. (2011). Post-development, developmental state and genealogy: Condemned to develop?. *Third World Quarterly*, 32(7), 1183–1198.

Frantz, D. (2017, March 14). The business case for Sustainable Development Goals. The Huffington Post. Retrieved from https://www.huffingtonpost.com/entry/the-business-case-for-the-sustainable-development-goals_us_58c6ffa5e4b03400023f4a47.

Fridell, G. (2012). Coffee and commodity fetishism. In D. Brock, R. Raby, & M. P. Thomas, *Power and everyday practices* (pp. 277–298). Toronto: Nelson Education.

Fridell, G. (2014). Coffee statecraft: Rethinking the global coffee crisis, 1998–2002. *New Political Economy*, 19(3), 407–426.

Galatsidas, A. (2015, January 19). Sustainable development goals: Changing the world in 17 steps — interactive. *The Guardian*. Retrieved from https://www.theguardian.com/global-development/ng-interactive/2015/jan/19/sustainable-development-goals-changing-world-17-steps-interactive.

Golob, U., Lah, M., & Jancic, Z. (2008). Value orientations and consumer expectations of Corporate Social Responsibility. *Journal of Marketing Communications*, 14(2), 83–96.

Gomiero, T., Pimentel, D., & Paoletti, M. G. (2011). Is there a need for a more sustainable agriculture? *Critical Reviews in Plant Sciences*, 30(1–2), 6–23.

Halle, M. (2016). Part three: Means of implementation. IISD Perspectives on the Development Agenda for Sustainable Development. Retrieved from http://www.iisd.org/sites/default/files/publications/sustainable-development-goals-iisd-prespectives.pdf.

Hickel, J. (2015). Five reasons to think twice about the UN's Sustainable Development Goals. The London School of Economics and Political Science. Retrieved from http://blogs.lse.ac.uk/africaatlse/2015/09/23/five-reasons-to-think-twice-about-the-uns-sustainable-development-goals/.

Ho, D. (2015). Alternative production bases in Southeast Asia: An introduction. Hong Kong Trade Development Council. Retrieved from http://economists-pick-research.hktdc.com/business-news/article/Research-Articles/Alternative-Production-Bases-in-Southeast-Asia-an-Introduction/rp/en/1/1X000000/1X0A32YW.htm.

Hodal, K. (2016, March 2). Nestlé admits slave labour risk on Brazil coffee plantations. *The Guardian*. Retrieved from https://www.theguardian.com/global-development/2016/mar/02/nestle-admits-slave-labour-risk-on-brazil-coffee-plantations.

IMF (2017, October 11). The IMF and the Sustainable Development Goals. The International Monetary Fund. Retrieved from https://www.imf.org/en/About/Factsheets/Sheets/2016/08/01/16/46/Sustainable-Development-Goals.

Inama, S., & Sim, E. (2015). *The foundation of the ASEAN Economic Community: An institutional and legal profile*. Cambridge: Cambridge University Press.

Institute for Agriculture and Trade Policy (2003). Towards food sovereignty: Constructive alternatives to the World Trade Organization's agreement on agriculture. Institute for Agriculture and Trade Policy. Retrieved from https://www.iatp.org/files/Towards_Food_Sovereignty_Constructive_Alternat.pdf.

International Coffee Association (2017). Grading and classification of green coffee. Retrieved from http://www.ico.org/projects/Good-Hygiene-Practices/cnt/cnt_en/sec_3/docs_3.3/Grading%20&%20class.pdf.

International Coffee Organization (2016). History, International Coffee Organization. Retrieved from http://www.ico.org/icohistory_e.asp?section=About_Us (1 September 2016).

International Coffee Organization (2017). Infographics on the global coffee trade. Retrieved from http://www.ico.org/coffee-trade-statistics-infographics.asp?section=Statistics.

International Trade Center Secretariat (2017). World coffee trade: Key events in the history of the International Coffee Agreement. International Trade Center. Retrieved from http://www.intracen.org/coffee-guide/world-coffee-trade/key-events-in-the-history-of-the-International-Coffee-Agreement/.

Karbassi, L. (2018). Advancing sustainable development. UN Global Compact. Retrieved from https://www.unglobalcompact.org/what-is-gc/our-work/sustainable-development.

Katsumata, H. (2006). Establishment of the ASEAN Regional Forum: Constructing a 'talking shop' or a 'norm brewery'. *The Pacific Review*, 19(2), 181–198.

Knudsen, J. S. (2011, November 24). Which companies benefit most from UN Global Compact Membership? *The European Business Review*. Retrieved from http://www.europeanbusinessreview.com/which-companies-benefit-most-from-un-global-compact-membership/.

Lamy, P. (2011). Trade is vital for food security, Lamy tells agricultural economists. World Trade Organization. Retrieved from https://www.wto.org/english/news_e/sppl_e/sppl203_e.htm.

Langhammer, R. J. (2016). What lessons can ASEAN learn from the EU? University of Pennsylvania Wharton School. Retrieved from http://knowledge.wharton.upenn.edu/article/what-lessons-can-asean-learn-from-the-eu/.

Larsen, M. L. (2015). The delisting dilemma. UN Global Compact International Yearbook 2015. Retrieved from http://www.keepeek.com/Digital-Asset-Management/oecd/united-nations/the-united-nations-global-compact-international-yearbook-2015_8d539869-en#.WaJnOtERU2w#page1.

Lyon, S. (2009). 'What good will two more trees do?' The political economy of sustainable coffee certification, local livelihoods and identities. *Landscape Research*, 34(2), 223–240.

Masilamani, L., & Peterson, J. (2014, October 15). The 'ASEAN Way': The structural underpinnings of constructive engagement. *Foreign Policy Journal*. Retrieved from http://www.foreignpolicyjournal.com/2014/10/15/the-asean-way-the-structural-underpinnings-of-constructive-engagement/.

McStocker, R. (1987). The Indonesian coffee industry. *Bulletin of Indonesian Economic Studies*, 23(1), 40–69.

Ministry of Agriculture Indonesia (2014). Tree crop estate statistics of Indonesia: 2013–2015 coffee. Directorate General of Plantation, Ministry of Agriculture. Retrieved from http://ditjenbun.pertanian.go.id/tinymcpuk/gambar/file/statistik/2015/KOPI%202013%20-2015.pdf (1 September 2016).

Minorities at Risk (2010). Chronology for northern hill tribes in Thailand. Retrieved from http://www.mar.umd.edu/chronology.asp?groupId=80003.

Mishra, M. K., & Slater, A. (2012). Recent advances in the genetic transformation of coffee. National Center for Biotechnology Information, US National Library of Medicine. Retrieved from https://www.ncbi.nlm.nih.gov/pmc/articles/PMC3437269/.

Muchhala, B. (2015, August 13). Last-minute lack of transparency weakens sustainable development goals. *The Guardian*. Retrieved from https://www.theguardian.com/global-development-professionals-network/2015/aug/13/lack-of-transparency-sustainable-development-goals-negotiations-united-nations.

Muhr, T. (2016). Beyond 'BRICS': Ten theses on south–south cooperation in the twenty-first century. *Third World Quarterly*, 37(4), 630–648.

Musungu, S., Villanueva, S., & Blasetti, R. (2004). Utilizing TRIPS flexibilities for public health protection through south-south regional frameworks. Essential Medicines and Health Products Information Portal. Retrieved from http://apps.who.int/medicinedocs/en/d/Js4968e/8.2.2.html (1 September 2016).

Nederveen Pieterse, J. (1997). Equity and growth revisited: A supply-side approach to social development. In C. Kay (Ed.), *Globalisation, competitiveness and human security* (pp. 128–149). Oxford and New York: Frank Cass.

Neilson, J., Arifin, B., Gracy, C., Tran, N. K., Pritchard, B., & Soutar, L. (2011). Challenges of global environmental governance by non-state actors in the coffee industry: Insights from India, Indonesia and Vietnam. In S. Lockie, & D. Carpenter (Eds), *Agriculture, biodiversity and markets: Livelihoods and agroecology in comparative perspective* (pp. 175–200). London and New York: Earthscan.

Nestlé (2004). The Nestlé coffee report: Faces of coffee. Retrieved from http://www.nestle.com/asset-library/documents/reports/csv%20reports/agriculture%20and%20rural%20development/faces_coffee_english.pdf (1 September 2016).

Nestlé Global (2017). Facts about Nestlé in Vietnam. Retrieved from http://www.nestle.com/asset-library/documents/media/news-and-features/2013-july/facts-figures-nestle-vietnam-en.pdf.

Nestlé Indonesia (2017). History of Nestlé Indonesia. Retrieved from https://www.nestle.co.id/eng/aboutus/historynestle.

Nestlé Philippines (2017). Nestlé Philippines, Inc., a short history. Retrieved from https://www.nestle.com.ph/aboutus/history.

Nestlé Thailand (2017). Location of operations. Retrieved from https://www.nestle.co.th/en/aboutus/our_locations.

Nestlé Vietnam (2017). Nestlé Vietnam history. Retrieved from https://www.nestle.com.vn/aboutus/nestlevietnam.

Nor, N. A., & Abdul Wahab, M. A. (2016). Exploring the potentials of coffee industry in Malaysia Food and Fertilizer Technology Center for the Asian and Pacific Region. Retrieved from http://ap.fftc.agnet.org/ap_db.php?id=574&print=1.

OECD (2014). Global value chains: Challenges, opportunities, and implications for policy. Organization for Economic Cooperation and Development. Retrieved from https://www.oecd.org/g20/topics/trade-and-investment/gvc_report_g20_july_2014.pdf.

OECD (2017). Capturing the 2030 Agenda in development co-operation results: SDGs, targets, indicators, and 'Leave no one behind'. OECD Development Assistance Committee: Retrieved from https://www.oecd.org/dac/results-development/docs/2030-Agenda-in-results.pdf.

Penarredonda, J. L. (2017, November 6). The disease that could change how we drink coffee. BBC Future. Retrieved from http://www.bbc.com/future/story/20171106-the-disease-that-could-change-how-we-drink-coffee.

Pendergrast, M. (2010). *Uncommon grounds: The history of coffee and how it transformed the world*. New York: Basic Books.

Pham, T. H., & Tien, D. N. (2014). Bitter taste brewing in Vietnamese coffee. East Asia Forum. Retrieved from http://www.eastasiaforum.org/2014/03/15/bitter-taste-brewing-in-vietnamese-coffee/ (1 September 2016).

The Philippine Star (2016, June 26). Nestlé helps secure future of coffee farming. *The Philippine Star*. Retrieved from http://www.philstar.com:8080/agriculture/2016/06/26/1596562/nestle-helps-secure-future-coffee-farming.

Philippine Statistics Authority (2016). Other crops: Volume of production by geolocation, crop and year, coffee 2016. CountrySTAT Philippines. Retrieved from http://countrystat.psa.gov.ph/selection.asp (1 September 2016).

Pieterse, M. T., & Silvis, H. J. (1988). *The world coffee market and the International Coffee Agreement*. Wageningen: Wageningen Agricultural University.

Pritchard, B., & Fagan, R. (1999). Circuits of capital and transnational corporate spatial behaviour: Nestlé in Southeast Asia. *International Journal of Sociology of Agriculture and Food*, 8, 3–20.

Project Café 2016 South East Asia (2015). Booming Southeast Asian coffee shop market shows accelerated growth in 2015. Allegra World Coffee Portal. Retrieved from https://www.worldcoffeeportal.com/Latest/InsightAnalysis/2015/Solid-growth-revealed-for-South-East-Asian-coffee.

Rainforest Alliance (2013). Certified sustainable coffee grows rapidly as more companies commit to sourcing. Rainforest Alliance. Retrieved from http://www.rainforest-alliance.org/press-releases/sustainable-coffee-grows.

Rangan, V. K., Chase, L., & Kahel, S. (2015, February). The truth about CSR. *Harvard Business Review*. Retrieved from https://hbr.org/2015/01/the-truth-about-csr.

Robbins, R. H. (2013). Coffee, fair trade, and the commodification of morality. *Reviews in Anthropology*, 42(4), 243–263.

Sagebien, J., & Whellams, M. (2010). CSR and development: Seeing the forest for the trees. *Canadian Journal of Development Studies*, 31(3–4), 483–510.

Severino, R. (2001). The ASEAN Way and the rule of law. United Nations Public Administration Network. Retrieved from http://www.asean.org/.

Silva, R. O. (2016). Coffee market report December 2016. International Coffee Organization. Retrieved from http://www.ico.org/documents/cy2016-17/cmr-1216-e.pdf.

Slim, H. (1995). What is development? *Development in Practice*, 5(2), 143–148.

Specialty Coffee Association of America (2009). SCAA protocols, grading green coffee. Retrieved from http://www.scaa.org/PDF/resources/grading-green-coffee.pdf.

Specialty Coffee Association of America (2017). Coffee plants of the world. Retrieved from http://www.scaa.org/?page=resources&d=coffee-plants-of-the-world.

Stratfor (2014, July 24). Southeast Asia's quiet international importance. Retrieved from https://worldview.stratfor.com/article/southeast-asias-quiet-international-importance.

Technavio (2016a). Global instant coffee market 2016–2020. Retrieved from https://www.technavio.com/report/global-non-alcoholic-beverages-global-instant-coffee-market-2016-2020.

Technavio (2016b). Global roast and ground coffee market 2016–2020. Retrieved from https://www.technavio.com/report/global-non-alcoholic-beverages-roast-and-ground-coffee-market.

Teuber, R. (2010). Geographical indications of origin as a tool of product differentiation: The case of coffee. *Journal of International Food & Agribusiness Marketing*, 22(3–4), 277–298.

Toledo, V. M. (2012). Coffee and sustainability: The multiple values of traditional shaded coffee. *Journal of Sustainable Agriculture*, 36(3), 353–377.

UN (2015). Transforming our world: The 2030 Agenda for Sustainable Development. New York: United Nations.

UN Comtrade (2017). UN Comtrade database. UN, Department of Economic and Social Affairs, Statistics Division, Trade Statistics. Retrieved from https://comtrade.un.org/data.

UN Global Compact (2011). Partnership fundamentals: A 10-step guide for creating effective UN — Business partnerships. Retrieved from https://business.un.org/documents/resources/partnership_fundamentals.pdf.

UN Global Compact (2013). UN–business partnerships: A handbook. Retrieved from https://business.un.org/documents/resources/un-business-partnership-handbook.pdf.

UN Global Compact (2017a). Farm Mountain Coffee Denmark information page. Retrieved from https://www.unglobalcompact.org/what-is-gc/participants/12245.

UN Global Compact (2017b). Java Mountain Coffee information page. Retrieved from https://www.unglobalcompact.org/what-is-gc/participants/86151.

UN Global Compact (2017c). Sithar Coffee Co. information page. Retrieved from https://www.unglobalcompact.org/what-is-gc/participants/89361.

UN Global Compact (2017d). Starbucks Coffee Co. information page. Retrieved from https://www.unglobalcompact.org/what-is-gc/participants/8771.

UN Global Compact. (2017e). Making global goals local business. Retrieved from https://www.unglobalcompact.org/docs/about_the_gc/MakingGlobalGoalsLocalsBusiness2017.pdf.

UN Global Compact (2017f). UN–Business partnership passport. Retrieved from https://business.un.org/en/passport.

UN Office of Internal Oversight Services (2017). OIOS support to the follow-up and review of the SDGs: An advisory. UNOIOS, Inspection and Evaluation Division. Retrieved from https://oios.un.org/resources/2017/09/FcdUD2er.pdf.

UN Sustainable Development Knowledge Platform (2018). UN Global Compact: Implementation plan for the 2030 Agenda for Sustainable Development. Retrieved from https://sustainabledevelopment.un.org/content/documents/8965Global%20Compact.pdf.

US Bureau of Labor Statistics (2017). Price inflation for roasted coffee since 1939. Retrieved from http://www.in2013dollars.com/Roasted-coffee/price-inflation/1963.

USDA Foreign Agricultural Service (2017). Coffee: World markets and trades, 2017/18 forecast overview. Retrieved from https://apps.fas.usda.gov/psdonline/circulars/coffee.pdf.

Verdin, M. (2016). Commerz ups coffee price forecasts, as Brazil robusta values soar. Retrieved from http://www.agrimoney.com/news/commerz-ups-coffee-price-forecasts-as-brazil-robusta-values-soar--9834.html (1 September 2016).

Wenick, A. (2016). Can coffee become the world's first 100 percent sustainable agricultural product? Public Radio International. Retrieved from https://www.pri.org/stories/2016-03-20/can-coffee-become-world-s-first-100-percent-sustainable-agricultural-product.

World Bank (2017). The World Bank Group's twin goals, the SDGs, and the 2030 Development Agenda. Retrieved from http://www.worldbank.org/en/programs/sdgs-2030-agenda#2.

World Food Summit (2017). Better food for more people. Ministry of Environment and Food, Denmark. Retrieved from http://eu-refresh.org/sites/default/files/WORLD%20FOOD%20SUMMIT%202017%20-%20Presentation%20-%20Better%20Food%20for%20More%20People%20%28version%2020%20April%202017%29.pdf.

Wu, D. (2013, May 22). Rethinking the development gap: ASEAN's inclusive growth imperative. *The Diplomat*. Retrieved from http://thediplomat.com/2013/05/rethinking-the-development-gap-aseans-inclusive-growth-imperative/.

CHAPTER 17

Cooperation of Cooperatives: Partnership towards SDGs

Saikaew Thipakorn

1 Introduction

The Sustainable Development Goals (SDGs) are a set of universal goals based on the desire to end poverty, protect the planet, and ensure prosperity and peace for all. There are 17 goals and 169 targets. The SDGs build on the successes of the Millennium Development Goals (MDGs), extended to include additional issues such as climate change, peace, and justice. The 17 goals are overlapping and interconnected: in order to fully achieve one goal, it is necessary to also achieve associated targets from other goals. In addition, the SDGs require the cooperation of all parties, nations, and peoples, rich and poor. However, no clear directions for how the goals should be achieved have been formulated; neither does the SDG documentation specify how partnerships should be formed.

Cross-border cooperation is normally spearheaded by government agencies or large business enterprises with extensive, often global, networks. There are rules and regulations that govern such cooperation. For grassroots organizations, following the same rules is a much greater challenge. In the case study presented in this chapter, the Japanese cooperatives found operations at the level of the government and large business to be untrustworthy; they therefore decided to use their own principles to govern their international endeavour. The result was a successful cooperation with Thai agricultural cooperatives.

1.1 *Research Question*

This chapter explores the partnership between Japanese and Thai cooperatives. The origin of this partnership was a scheme to ensure food safety for Japanese consumers. Ultimately, this cooperation led to developments in organic farming and capacity building for Thai cooperatives. The research questions to be addressed are therefore: how much does this cooperation contribute to the objectives of the SDGs? What lessons could be learned from this cooperation?

 | DOI:10.1163/9789004391949_018

1.2 *Significance of the Study*

SDG 17 emphasizes international cooperation between various actors: developed and developing countries; public, private, and civil society partnerships; and different forms of North–South, South–South, and triangular regional and international cooperation. It also stresses certain factors such as finance, technology, capacity building, trade, and systemic issues including policy and institutional coherence. The case study in this chapter is an example of North–South cooperation, although this is not a donor–recipient relationship. The study shines a light on a form of partnership between two equal partners: Japanese cooperatives (representing the North) and Thai cooperatives (the South).

A cooperative is "an autonomous association of persons united voluntarily to meet their common economic, social, and cultural needs and aspirations through a jointly owned and democratically controlled enterprise".[1] It is a very particular kind of private organization. While international cooperation between government agencies or business entities must follow specific procedures and rules if it is to succeed, international cooperation between cooperatives does not follow the same rules due to its special, voluntary nature. This chapter explores the pattern of cooperation between cooperatives, its success and challenges. Lessons learned from this case study could be useful in promoting successful partnerships for achieving the SDGs at the grassroots level.

1.3 *Research Method*

In conducting this research, the author visited three Japanese cooperatives, Daiichi-wo-mamoru kai, Seikatsu Club, and JA Zenchu, in 2015. The purpose of the visits was to study how these cooperatives could become strong and sustainable community enterprises. The research identified the teikei system as an important core principle that brought success for the Japanese consumer cooperatives. In Thailand, the export of organic bananas to Japan by agricultural cooperatives in the south of Thailand was the focus of study. Here, the author visited Ban Lat Cooperative and Thayang Cooperative — two cooperatives that were the learning centres and inspiration for many other cooperatives. This research indicated that the teikei system was extended through the international cooperation between the Japanese and Thai cooperatives. The hypothesis of this chapter is, therefore, that the teikei system lies behind the successful cooperation of these cooperatives.

1 See the website of the International Cooperative Alliance: https://www.ica.coop/en/whats-co-op/co-operative-identity-values-principles.

2 Literature Review

This chapter touches on two important issues: North–South partnership and partnership between business organizations in the form of international trade. On the issue of partnership between the North and the South, mainstream research tends to focus on donor–recipient relationships, with government or business conglomerates as the dominant actors in the partnership. Studies on partnership in the context of the SDGs are mainly undertaken by scholars and NGOs who are actors and partners themselves. The private sector and business organizations are encouraged to take part in the attempt to achieve the SDGs: however, how business organizations should cooperate in this endeavour has not been adequately explored.

A study by the Institute for Human Rights and Business (IHRB, 2015) argues that, when economic growth was seen as the only way to provide for the rapidly expanding world population, business and industrial sectors were given the leading role. In the past two decades, however, there has been a shift towards the belief that, in our efforts to provide for the world population, we should not deny future generations the chance to provide for themselves. Irresponsible business practices have been blamed for pollution in soil, air, and water, for creating an ever-widening gap between rich and poor, and for causing inequalities across the globe. In the process of adopting the SDGs, therefore, the role of the business sector was emphasized as an essential partner, with the potential to combat these global challenges. Business leaders, however, remain dubious about this.

The relationship between the SDGs and business is contradictory. It is the nature of (capitalist) business to deliver wealth and prosperity, even at the cost of environmental degradation. Asking businesses to deliver sustainable development is a major challenge. On the government side, it is very difficult for governments to implement large-scale projects that can address all the interrelated issues covered by the SDGs. At the same time, it is also unethical for governments to dilute the goal of sustainable development in order to attract private investment. The IHRB study (2015, pp. 6–7) highlights three main challenges for public–private partnerships:

1) *The right kind of partner*. The profit orientation of business is usually in conflict with the public-good orientation of development. Partnership between business and development should therefore be based on an understanding of responsibilities and agreement around practical contributions. The terms of a partnership must provide "clarity and accountability" (ibid., p. 6).

2) *The right kind of growth.* The SDGs call for "fast and sustained but also green and inclusive" growth (ibid.). These are qualities which are not normally seen as compatible.
3) *The right kind of financing.* Financing mechanisms should ideally enhance the "complementary strengths of public and private sectors", as well as overcoming the contradictions between foreign investment and inclusive and sustainable growth (ibid., p. 7).

The IHRB report focuses on large corporate businesses, suggesting that it views public–private partnership in pursuit of the SDGs as very problematic.

The form of business institution that is best suited to providing both wealth and development is small scale and localized. Cooperatives are in a good position to perform this task. Cooperatives are a kind of social enterprise which can be seen as business entities at the people level. As the cooperative's main responsibility is towards its members, international concerns are not its primary focus. Studies on cooperatives reflect this. According to the International Cooperative Alliance,[2] the core principles of cooperatives are:

1. voluntary and open membership
2. democratic member control
3. members' economic participation
4. autonomy and independence
5. education, training and information
6. cooperation among cooperatives
7. concern for community.

Cooperatives are thus, by nature, a sustainable and participatory form of business, and could provide an alternative model of enterprise which contributes to the realization of the SDGs. A study on cooperatives and the SDGs conducted by the International Labour Organization and the International Cooperative Alliance (ILO and ICA, 2014) found that, although cooperatives were not actively engaging in implementing the SDGs, they were making significant contributions to the realization of the objectives of the goals. In particular, agricultural cooperatives were recognized for their poverty-reduction efforts (SDG 1); they also provided for food security and good nutrition (SDG 2). Agricultural cooperatives solved many challenges that small producers faced in their endeavours to produce food, with services such as group purchasing and group marketing helping farmers to overcome obstacles. The same study mentions the role of cooperatives in bridging the trade barriers between developed and developing

2 See the website of the International Cooperative Alliance: https://www.ica.coop/en/whats-co-op/co-operative-identity-values-principles.

countries. By linking cooperatives in the South to markets in the North, they encourage ethical production and provide markets for products which result from social development projects in local communities in the South (ibid.).

In the case of Thailand and Japan, research was conducted by the Devawongse Varopakarn Institute, Thailand's Ministry of Foreign Affairs, on the Japan–Thailand Economic Partnership (JTEPA) and the export of bananas by Thai agricultural cooperatives (Vichian Tanthamaroj, 2014). Under the JTEPA, signed in 2007, there would be a tax exemption for Thai bananas exported to Japan, beginning from 4,000 tons during the first year. The amount would be gradually increased to 8,000 tons in the fifth year. However, the amount of bananas exported to Japan decreased after the JTEPA took effect, while the amount of bananas sold domestically increased. The study showed that Thai cooperatives did not benefit from the JTEPA tax exemption scheme, and that the scheme did not provide a sufficient incentive for organic banana export to Japan (ibid.). The research focused on cooperative banana exports under the JTEPA, but little is known about the wider context of the partnership between Japanese and Thai cooperatives. It is therefore important to investigate how this partnership operates successfully.

My own study on Japanese cooperatives as a model of social enterprise for Thai organizations revealed that the teikei system is the basis for the success of Japanese consumer cooperatives (Saikaew Thipakorn, 2016). Teikei is closely linked to Japanese organic agriculture, and is the term used by the Japan Organic Agriculture Association (JOAA). The meaning of teikei is "organic human relationship" between producers and consumers. The idea is to create an alternative distribution system to replace conventional markets. In the teikei system, producers and consumers reach consensus and mutual understanding through conversation. Both sides take responsibility for supporting their own delivery system. The Japanese Consumer Cooperative Union (JCCU) uses the word 'Sanchoku' which is a kind of community-supported agricultural philosophy. The terms 'teikei' and 'Sanchoku' share the same three core principles:

- *Traceability*: the origin of the product and identification of the producer are traceable.
- *Standardization*: consumer cooperatives in each region are in direct contact with producers; inspection of products is carried out bilaterally by producers and the cooperatives.
- *Communication*: the direct flow of information and communication between producers and cooperative members is actively encouraged.

The Sanchoku principle and the teikei system were originally a response from Japanese society to industrial and economic expansion after World War II, which caused serious environmental pollution and the chemical contamination of food. They supported the growth of organic agriculture in Japan. The resulting system of consumer cooperatives corresponds well with the objectives of sustainable development. Japanese cooperatives introduced this system to Thai cooperatives. By examining how well Thai cooperatives and Thai society have accepted this system and whether organic agriculture in Thailand has benefited from teikei, we can shed some light on the issue of sustainable development in Thailand.

3 Consumer Cooperatives in Japan

Cooperatives are part of the Japanese culture. Formal agricultural cooperatives were first introduced in Japan in 1843, in the form of credit unions formed by a group of farmers. In 1900, the first cooperative law was implemented, the Industrial Cooperative Law. The law legalized four types of cooperatives: credit cooperatives, marketing cooperatives, purchasing cooperatives, and shared machine cooperatives (Vichian Tanthamaroj, 2014). Nine years later, the credit cooperatives were allowed to expand their operations to include other business, allowing them to transform themselves into all-purpose cooperatives.

Consumer cooperatives in Japan pre-date World War II, but they were disbanded during the War. When Japan was under the control of the Supreme Commander for the Allied Powers (SCAP), cooperatives and trade unions were seen as institutions that could instil democracy into Japanese daily life. Therefore, in 1948, a new Cooperative Law was passed and consumer cooperatives were reborn (Thompson, 2017). They provided the most important distribution system during the post-war period, when Japan faced food shortages and starvation.

Consumer cooperatives fall under the supervision of the Japanese Ministry of Health, Labour, and Welfare, and their main business is to supply food and other daily necessities to their members. They have distinctive characteristics, such as exclusive member purchase and active member participation. Japanese consumer cooperatives are divided into four categories: retail, health, housing, and insurance. In 1951, Toyohigo Kagawa — a leader in the Japanese labour and democratic movements — founded the Japanese Consumer Cooperative Union (JCCU) as the apex organization; in 2011, "590 cooperative organizations were affiliated to JCCU" (ICA, 2011, p. 6). The JCCU represents and serves the interests of all the consumer cooperative societies of Japan. Its activities include:
- the development of cooperative products and their distribution to members
- mail-order business

- the realization and enhancement of food safety governance from consumers' perspective
- activities and businesses concerned with peace
- environment and welfare
- nationwide inter-cooperative business collaboration
- promotion of other regional alignments.[3]

3.1 *Growth of Consumer Cooperatives*

According to the Japanese Cooperative Law of 1948, members were not allowed to join other consumer cooperatives beyond their prefecture. This acted as a limitation on the growth of consumer cooperatives.

During the 1970s and 1980s, as the Japanese economy was on the rise, consumer cooperatives experienced tremendous growth. When the economy experienced a downturn, many cooperatives were down-sized, but did not go out of operation. These days, consumer cooperatives are important to everyday life in Japan. An amendment to the Consumers Livelihood Cooperative Society Law in 2007 allowed primary cooperatives to expand their business to the adjacent prefectures, meaning that consumer cooperatives can now merge across prefectures. This gave them the opportunity to grow. In 1996, 688 primary JCCU food cooperatives had a combined membership of 14 million households (Moen, 1997). In 2011, retail consumer cooperatives had 19 million members, with each member spending an average of \$147 per month at the cooperative. The members also invest heavily in their cooperatives. On average, each member holds \$403 in shares in their cooperative (ICA, 2011). The JCCU itself has more than 28 million members nationwide and has been involved in international activities since its early years. In 1987, it established the Asian Consumer Cooperative Fund.

3.2 *Cooperatives and Organic Farming*

Consumer cooperatives in Japan are closely related to organic farming. Interest in organic agriculture in Japan began in the early 1970s and the Japan Organic Agriculture Association was established in 1971 (JOAA, 1993). As indicated above, environmental contamination as a result of hasty industrialization, including the heavy use of chemicals in agriculture and industrial pollution in natural water sources and soil, became a serious cause for concern in the late 1960s, with the incidence of conditions such as Minamata disease. Consumers, especially in urban areas, began to be very anxious about the safety of food,

3 See the website of the JCCU: http://jccu.coop/eng/jccu/index.php.

and made increasing efforts to obtain uncontaminated food, such as milk and eggs. At the same time, some farmers, who had begun to realize the dangers of chemical additives to food and soil, started to practise organic farming.

The case of Daichi-wo-mamoru kai provides a good example of a consumer cooperative. In 1975, a group of neighbours wanted good quality organic vegetables and contacted a group of farmers directly. They promised to buy, if the farmers promised to produce good quality organic products. The consumers and producers worked together in order to guarantee a truly chemical-free agricultural production process. The newly formed cooperative arranged seminars to introduce new technology to both consumers and producers. It also arranged farm tours as a form of quality verification for consumers and encouragement for farmers to keep up the good work. The cooperative managed its own distribution system, to cut out middlemen, to keep down the cost of transportation, and to ensure that it maintained direct control on the quality of the products. The case of Daichi-wo mamoru kai highlights a strong relationship between consumer cooperative, producers, and organic farming.

Of course, organic farming involves more than growing agricultural produce without chemical inputs. Organic farming must be supported by a group of people with understanding of and commitment to environmental problems, food safety issues, and a 'green' attitude. Organic farming cannot be done on a large scale; it cannot deliver out-of-season products or produce fruits and vegetables of a standardized size and shape. It is vulnerable to climate change and natural disasters. Farmers involved in mainstream marketing systems are reluctant to practise organic farming, because of the risk that their farm produce might not have a market. Consequently, in Japan, a direct relationship between consumers and producers has been developed in order to guarantee that market, creating a win-win situation for both parties.

The essential component of this relationship is teikei, which is a system of community-supported agriculture. The concept of teikei involves small-scale, local, organic farming and volunteer-based, non-profit partnerships between producers and consumers. According to JOAA (1993), the 10 principles of teikei are:

1 Principle of mutual assistance
2 Principle of intended production
3 Principle of accepting the produce
4 Principle of mutual concession in the price decision
5 Principle of deepening friendly relationships
6 Principle of self-distribution
7 Principle of democratic management

8 Principle of learning among each group
9 Principle of maintaining the appropriate group scale
10 Principle of steady development.

Teikei brings together consumers and producers to help each other to accomplish their goals. Interviews at Daichi-wo-mamoru kai and the Seikatsu Club, two of the Japanese cooperatives involved in the research, revealed that meetings between consumers and producers took several forms. For example, the consumer co-ops send representatives to visit farms from which they are intending to buy food and products. The co-ops acknowledge producers' problems, such as drought and natural disaster, and try to help or offer support. The co-ops encourage consumers to write letters and send them to farmers who produce food, so that producers receive feedback about their products directly from consumers. Sometimes, the co-ops arrange special lectures from experts on issues of concern to both consumers and producers, promoting an equal understanding of problems and situations among all parties.

In order to support organic agriculture, consumers themselves have to take an active role. They have to realize that food will not come in an ideal shape and size. They have to adjust their menu to the food available and they have to try to consume all the produce that is provided. Some co-ops organize a 'han' system comprising groups of consumers who live in the same area. Members take turns to transport and distribute the produce from the farms. Some well-established co-ops set up their own centres for chemical and contamination checks, develop their own standard of product inspection, and manage packaging and delivery.

The strength of the Japanese consumer cooperatives is evident in the growth in number of members. In addition, the increased demand from consumers for 100% organic produce has caused a shift in the mode of agricultural production: farmers are more willing to use non-chemical farming techniques, which has a positive impact on the environment. It can thus be argued that the teikei system used by Japanese consumer cooperatives can make a serious contribution to the achievement of SDG 2, which aims to "end hunger, achieve food security and improved nutrition and promote sustainable agriculture". Moreover, as a framework for international trade between Japanese and Thai cooperatives, the teikei system can also contribute to SDG 17: "strengthen the means of implementation and revitalize the global partnership for sustainable development". This is the focus of the remainder of this chapter. Implementing teikei as the basis of international trade helped facilitate this partnership at the people level. In order to explain further, we have to understand the background of Thai cooperatives and organic agriculture in Thailand.

4 Japanese–Thai Cooperation

In the south of Thailand, the weather and soil are suitable for producing the best bananas in the region. Bananas are one of the 10 fruits produced in Thailand for both domestic consumption and export. There are many kinds of bananas, but the one exported to Japan is Hom Thong. The export of bananas to Japan is well established: Japan's demand for bananas is estimated at 8,000 tons/year, with banana exports from Thai cooperatives accounting for 2,000 tons/year. The total value of exports is over 50 million baht (US$1.5 m.) per year.[4] This generates more than 20,000 baht/month (US$625) for one family (Thairath Online, 2015). The Japan–Thailand Economic Partnership Agreement (JTEPA) of 2010 stipulated that Thailand could export 4,000 tons of bananas per year, increasing by 1,000 tons per year up to a maximum of 8,000 tons. However, the success of this cooperation should not only be measured through the volume of exports/imports, the income of members, and the number of cooperative employees involved: it has also introduced the concept and feasibility of organic farming to Thai people, especially in the southern provinces, and has encouraged Thai villagers to work together in the form of cooperatives.

4.1 *Agricultural Cooperatives in Thailand*[5]

Cooperatives in Thailand were originally formed to counter market failures. When Thailand established trading relations with Western countries, there was a shift away from farming for domestic consumption, to farming for export. Agricultural products became vulnerable to price fluctuation on the world market, as well as natural disasters. Villagers frequently got into debt, and bankers who stepped in to offer loans took advantage of their plight. Many villagers lost their own land and faced bankruptcy. Cooperatives started to appear in 1915, during the reign of King Rama V, with the objective of helping poor farmers.

The first cooperative in Thailand was an agricultural cooperative called Wat Chan Cooperative Unlimited Liability, located in the Pitsanulok Province. This province was selected because of its low population density at the time. The founder, Prince Pittayalongkorn, hoped that the cooperative would enable newly migrated farmers from the south to provide for themselves. The purpose

4 Figures from Thailand's National Bureau of Agricultural Commodity and Food Standards (ACFS): Agricultural trend: Organic banana entering Japanese market (October 2011). Retrieved from the ACFS website (in Thai).

5 Data from this and the following sections come from the author's fieldwork.

of the co-op was originally to promote business cooperation among members: members would help each other and help the society. The outcome was a success: farmers did not have to use their land as collateral to borrow money, but instead had access to low-interest loans from the co-op to buy seeds, fertilizer, and equipment. They could earn their own living and hold on to their own land. Moreover, the government organized training and introduced new farming techniques to them. Overall, the quality of their lives improved.

According to the Cooperative Act B.E.2542 of 1999, cooperatives in Thailand are categorized into seven types:

1 agricultural cooperative
2 land settlement cooperative
3 fisheries cooperative
4 consumer cooperative
5 thrift and credit cooperative
6 service cooperative
7 credit union cooperative.

The first three are classed as agricultural cooperatives, and the other four as non-agricultural cooperatives. They fall under the supervision of the Cooperative Promotion Department of the Ministry of Agriculture. In 2016, there were a total of 8,263 cooperatives and cooperative groups in Thailand, of which 4,684 were agricultural cooperatives. Only 13.72% were listed as non-active.

4.2 *Banana Export*

Two groups of Japanese and Thai cooperatives have been involved in the export of organic bananas to Japan. The first group comprises Tohto Consumer Cooperative Society and the agricultural producers' cooperative corporation Wagoen, in cooperation with Thayang Agricultural Cooperative, Petchburi Province. The exporter is Otento (Thailand) Co. Ltd., a subsidiary of Wagoen. Otento's business operation is the sale of Japanese-style farm products, and the production of these in Thailand.

Thayang Agricultural Co-op Ltd. is located in Amphur Thayang, Petchburi Province, where 80% of the population rely on farming for their livelihoods. The main products are rice, different kinds of banana, lemon, rose apple, sugar cane, mango, red bean, and coconut. Thayang Co-op Ltd. was established in 1966 as a credit co-op for production, with 360 families — 12 groups from three Tambons (districts) — as founding members. It became Thayang Agricultural Co-op Ltd. in 1970, and its activities have expanded to include financial businesses, central markets, and product gathering.

In 1991, Mr Surin Cholprasert, then Director General of the Department of Cooperative Promotion, introduced Mr Hiroyuki Yamamoto, then Chairman of Tohto Consumer Cooperative Society, to Thayang Agricultural Cooperative, famous producer of bananas. The Japanese wanted Thayang Cooperative to produce organic bananas (Gros Michel variety) for export to Japan. However, the first shipment of 500 kg of bananas, in July 1992, was rejected, due to lack of experience and knowledge in shipping and transportation. Thayang Co-op turned to local academic scholars for help in improving their shipping and transportation techniques, and a few months later a 6-ton batch of bananas was successfully shipped to Japan. Since then, the amount of exported bananas has increased steadily, with the contract between the two cooperatives being renewed annually.

Tohto Cooperative decided to buy bananas through Wagoen Co-op. Shipping is handled by the Otento company, a subsidiary of Wagoen. Otento is responsible for maintaining contact with Thayang Cooperative, conducting inspections according to Wagoen standards, and managing the export of the bananas. Thayang Co-op not only produces bananas itself, according to Wagoen's standards, but also gathers and collects bananas from co-op members and delivers them to Otento.

Thayang Cooperative has become an outstanding co-op in Thailand. In 2012, its membership included 219 communities — 62 groups living in 12 Tambons. It exported 35 tons of bananas per month. Its channels for selling bananas have been diversified. Bananas which are rejected for export are sold in the domestic market at almost the same price as export. The large convenience store chain, 7-Eleven, sells bananas from Thayang Co-op. It operates with a small loan from the Bank for Agriculture and Agricultural Cooperatives, but in fact it is financially independent: the loan is maintained as a sign of the good relationship between the bank and the cooperative. Interviews by the author with some co-op members revealed member satisfaction: although organic farming yields a smaller amount of produce, the benefits of better health to the consumer, and their own families, is considered more important.

For the second group, the Japanese importer Pan Pacific Foods Corporation acts as middleman, managing and coordinating the cooperation among cooperatives. It imports bananas from Thailand and distributes them to many Japanese consumer cooperatives. This second group includes six main groups of Thai cooperatives in Chumporn Province, Petchburi Province, and Surat Thani Province: details of three of these are given below. In its role as coordinator, Pan Pacific Foods liaises between the Yodogawa co-op in Japan, and Thai cooperatives. Representatives from Pan Pacific visit the banana producers at their farms regularly, give advice on technical details regarding production and management, acknowledge farmers' problems, and arrange regular meetings between representatives of Yodogawa and the Thai co-ops.

The Tungkawat group in Amphur Lamae, Chumporn Province, was established in 1993 when villagers heard that a Japanese team was coming to the area to search for organic bananas. They decided to get together, rent a bus, and go to Petchburi Province to study organic banana farming and banana packaging for export at Thayang Co-op. After the trip they formed their own cooperative with 36 founding members, using their pooled resources to buy young banana shoots from Thayang Co-op. The Pan Pacific Foods Corporation managed details of the plantation, and sent news of the Tungkawat group to Japan. As a result, the Yodogawa Cooperative from Osaka came to visit, and signed a MOU with the Tungkawat group. They made their first shipment of organic bananas in 1994.

According to the members, the Tungkawat group was hard working and strongly motivated to make a success of organic banana farming. Although members had no prior knowledge about organic farming, they worked together and consulted each other frequently in order to find solutions to their common problems, since the techniques learned from Thayang could not all be adopted in their area. They learned that sincerity in their communication with consumers in Japan created trust among consumers and producers. The consumers understood the problems of the villagers and tried to help. For example, when natural disasters occurred in Chumporn, such as drought or floods, Japanese consumers collected donations and sent them to the affected farmers. This endorsement and appreciation of their efforts encouraged the villagers to continue their work. As a result, the group has become strong and self-reliant, with 230 members who live in many other Amphurs (counties) of Chumporn Province.

Ban Lat Agricultural Co-op is located near Thayang Co-op in Petchburi Province. It began when villagers who produced other crops, such as rice, sugar palm, and lemon, were not successful because of troubles with middlemen. In 1940, a group of 28 savings co-ops joined forces to form an agricultural co-op, with four more savings co-ops joining the following year. In 1969, the group became Ban Lat Agricultural Co-op. It merged with Ban Lat Land Co-op and registered as Ban Lat Agricultural Co-op Ltd in 1975. Now the co-op's activities include a product distribution centre, credit business, purchasing business, sale and preservative food business, services, and products gathering business. The co-op also manages training centres for those who are interested in learning about organic banana farming and co-op management.

Organic banana production in Ban Lat Co-op began in the mid-1990s. Bunjerd Somwang, a former cooperative inspector, invited Mr Yamamoto, Chairman of Pan Pacific Foods Corporation, Tokyo, to visit Ban Lat Co-op. They discussed the possibility of exporting organic bananas, which were in high demand in Japan. An agreement was reached for 6 tons/week, and the first shipment took place in 1996. In 1999, Mr Yamamoto introduced two members from Shutoken Co-op (now Pal System Co-op) to Ban Lat Co-op and the export of bananas was increased to 8 tons/week. The Japanese and Thai co-ops signed an indefinite contract in 2000 in which they agreed to respect each other's rights and cooperate in a friendly fashion.

According to my interview with the co-op manager, the quality of life of co-op members has improved, especially in term of their income levels. The co-op's success was not solely due to accessing Japanese market, but through their interaction with Japanese co-ops, Ban Lat members learned some techniques for producing quality bananas. They were invited to go to Japan to study the distribution process by talking to Japanese customers. In addition, Thai local scholars offered significant assistance. Many projects sponsored by the royal family, such as the Pidthong Foundation, provided funding for members to travel, to visit sites of best practice. Networks of local government agencies supported technical knowledge with their own expertise. The cooperative's intention was to create 'smart farmers' who can adapt knowledge to apply to other aspects of life. In addition, it is hoped that co-op members will be able to support groups in other areas. Since the successful formula of Ban Lat Cooperative cannot be fully applied in areas with different climate and soil circumstance, the smart farmer should be able to adjust the technique to suit the locality. The cooperative is also keen to recruit new members from within its own younger generation. In 2014, it began the 'Farmer Heir Programme' for young people who might want to continue working on the farm. The programme tries to instil the pride of being farmers into the young men and women, combined with the incentive of working near their families.

The Ban Nasarn Agricultural Cooperative began the export of organic bananas, through the Pan Pacific Foods Corporation, in 1995 and now exports 12 tons/month, produced by 125 members. Ban Nasarn Co-op is located in Amphur Ban Nasarn, Surat Thani Province. It is primarily an area of rubber plantation, but after suffering from rubber price fluctuation, many villagers decided to shift to organic bananas which gave them a more stable income. The co-op was established mainly for collecting fruit products for sale. Initially, it could gather only small amounts of bananas, while the demand from Japan was much higher. The co-op then decided to set up a group of banana producers; this group now covers the area of many Amphurs in Surat Thani Province and Nakorn Rachasima Province. As well as exporting to Japan, they try to diversify their market by contacting local convenience stores and 7-Eleven shops in the south, in order to get their bananas on the shelves of local shops.

5 Contribution to Sustainable Development

In assessing whether the cooperation between Thai and Japanese cooperatives of the past 20 years has led to sustainable development, the following concepts are particularly relevant to the analysis.

- *Inclusive development*: this means that nobody should be left behind in the process of development; vulnerable groups, including low-income groups and the marginalized, must be included.
- *Universal development*: this implies that not only poor countries, but every country must be held accountable. All nations must join, hand in hand, in order to develop a sustainable world for future generations.
- *Integrated development*: the process of development encompasses many different but inter-related goals and targets. No goal should be the sole focus; rather, related goals must be systematically planned and implemented together.
- *Locally focused development*: also known as bottom-up development. This stresses the importance of implementing development projects at the grassroots level, where the stakeholders are located.
- *Technology-driven development*: in order to ensure that the positive outcomes of development are widely spread, information technology should be utilized.

The examples of cooperation between the Japanese and Thai cooperatives, described above, all illustrate the presence of these concepts. Looking first at bottom-up and inclusive development: as cited at the beginning of this chapter, the definition of a cooperative is "an autonomous association of persons united voluntarily to meet their common economic, social, and cultural needs and aspirations through a jointly owned and democratically controlled enterprise".[6] Any agricultural cooperative member who wishes to join the scheme is welcome. There are many processes and activities, from growing to marketing and delivering bananas, in which members can get involved and share knowledge and benefits. Those who wish to become producers can apply to be inspected in order to make certain that they are ready to grow bananas according to the requirements of the Japanese co-ops. The only pre-requisites are the right attitude towards cooperative business, a commitment to organic agriculture, and suitable land to grow organic bananas. Members who are not farmers may work in other processes, such as cleaning bananas, selection, or packaging. The whole scheme is carried out by cooperative members and managers.

6 From the website of the International Cooperative Alliance: https://www.ica.coop/en/whats-co-op/co-operative-identity-values-principles.

In terms of universal development, the collaboration between cooperatives in Japan and cooperatives in Thailand is an example of hand-in-hand development between a developed and a developing country. This collaboration has sent an important message about development of cooperatives in Thailand. It has been a catalyst for local Thai villagers and scholars to bring together their knowledge to develop local technology and management know-how in order to produce and export bananas to Japan. This experience has also enabled the cooperatives to expand and diversify their market for domestic consumption.

Given the definition of cooperatives as "association[s] of persons", the cooperation between the Japanese and Thai cooperatives is even more valuable than cooperation between governments. When such cooperation occurs between people, every aspect of live is included and affected. When guests from Thailand visited the Japanese cooperatives, they were taken round to see the country, the consumer cooperatives, and the Japanese way of life. They got direct feedback from Japanese consumers themselves. When Japanese cooperative members visited Thailand, they had a chance to visit the banana plantations to see how the bananas are grown. They understood the difficulties which Thai farmers face. This exchange creates goodwill on both sides and strengthens the cooperation. Thus, as we saw, when Thai farmers faced hardships as a result of natural disaster, Japanese cooperatives gathered donations to support them financially. Furthermore, when the price of bananas is negotiated in advance, it is based on the principle that the price must cover the cost of production and provide a livelihood. This gives encouragement for Thai producers to keep up the important work of organic agriculture.

Information technology is also used by both Thai and Japanese co-ops to promote their import–export activities. In Thailand, Facebook is used by exporting companies to promote their products, especially Hom Thong bananas. The cooperatives themselves create their own Facebook pages; information on how to grow Hom Thong bananas and how to join the co-op's export scheme appears on YouTube and other media. The experience of producers is also disseminated, with those who have had the chance to visit Japan sharing their stories online. These experiences encourage others to expand their horizons and join the cooperative.

5.1 *Contribution to Thai Society*

The Japanese demand for bananas is high, suggesting ample opportunities for Thai banana producers. Nevertheless, since 2008, total banana exports from Thailand have fluctuated, but have not grown significantly: see Table 17.1. There are many reasons for this. Firstly, domestic consumption has increased. As already mentioned, exporting bananas to Japan equipped Thai cooperatives with managerial skills: when bananas did not pass the export quality check, the cooperatives sold them on the domestic market. Cooperatives were thus able to expand their market, and their bananas are now also available in supermarkets and at convenience stores in Thailand. Secondly, producing organic bananas for export is not an easy task. Producers must first prepare the soil to be chemical free and must pay attention to many details. Those who were not able to meet the exacting standards decided to turn away from this production for export scheme.

However, the achievements of the Thai–Japanese cooperation cannot be evaluated solely on levels of banana exports. My research suggests that, by their own assessment, Thai cooperatives see the increased and stable income from banana exports as just one of many key indicators of their success. Their membership has expanded from a few hundred members to thousands of members, which has improved the effectiveness and also the status of their network. Cooperative banana groups now extend beyond the one province in which bananas were initially grown. Because of their stable economic position, these households are considered to have a higher socio-economic standing. The cooperatives have also become resource centres for people who wish

TABLE 17.1 Total banana exports from Thailand, 2000–2012 (thousand tonnes)

Year	Quantity
2000	5.8
2001	5.5
2002	5.2
2003	6.3
2004	17.7
2005	41.8
2006	20.3
2007	21.4
2008	22.8
2009	25.6
2010	22.8
2011	25.2
2012	27.6

SOURCE: FAO (2014, P. 12, TABLE 1)

to learn about cooperative affairs. They not only receive visitors, from inside the country and abroad, but also send members to other groups to provide advice on agricultural affairs, financial management, and other issues. The cooperatives' activities have expanded to cover additional agricultural products, as well as services such as savings and credit and centralized marketing for agricultural products. The cooperatives have begun to recruit new members from their own younger generations. They give education scholarships to schools around the area which their children attend, and they establish special funds for young people who wish to continue working on the farm. While farmers are often considered among the poor of society, children of the cooperatives are taught to be proud of being raised by farmer parents. If they work on the farm, they will be able to take care of their family while earning their own income. The cooperatives thus pass on not only their knowledge to the young ones, but also their pride at being successful farmers.

5.2 *Organic Agriculture in Thailand*

In Thailand, organic agriculture made its first appearance in the 1980s. It had not previously been popular because organic farm products did not meet the standards of size and shape demanded by the consumers, and because most farmers believed that using chemicals would give them a better yield. At that time, there was no consideration for their own or consumers' health. However, news and information about organic farming in other countries stimulated interest in non-chemical products among Thai consumers, and with the royal initiative, through governmental promotions and NGO support, some farmers began to shift into organic farming.

Organic farming would not have taken off without the growth of a market for organic produce. Founded in 1989, the Alternative Agriculture Network is a network of farmers and NGOs which promotes sustainable agriculture. A domestic market for organic products began to emerge in the 1990s, and within a few years, its popularity grew. In many retail stores the number of organic products increased. However, as in Japan, the government's new vocabulary of 'safe food' confused consumers and eroded public confidence during the mid to late 1990s. In 2003, the International Federation of Organic Agriculture Movements (IFOAM) and the Food and Agriculture Organization of the United Nations (FAO), together with Green Net and the Earth Net Foundation, organized an international conference on organic farming in Thailand. At the same time, government agencies clarified and differentiated the definitions of 'safe food' and 'organic products'. Big business corporations expanded their branches with organic food sections, while some retailers moved into selling organic products only. This supportive environment promoted the market for organic

TABLE 17.2 Number of certified organic farms in Thailand, 2007–2015

Year	Number of farms
2007	3,924
2008	3,545
2009	5,358
2010	7,405
2011	7,499
2012	7,189
2013	9,281
2014	9,961
2015	13,154

SOURCE: COMPILED FROM DATA AVAILABLE ON GREEN NET WEBSITE (HTTP://WWW.GREENNET.OR.TH/EN/ARTICLE)

products and led to an increase in the number of certified organic farms (see Table 17.2).

There are still issues to be resolved. One problem in Thailand has been the transportation of goods. Producers and consumers cannot manage to transport the produce themselves: consumers are not willing to drive to collect products from farms, and producers are not equipped with proper transportation to make deliveries. Some organic farms are located in remote areas, which makes it difficult for them to participate in the market. Nevertheless, banana export contracts with Japanese cooperatives have played a part in promoting organic agriculture in Thailand. They provide a high-end market for organic bananas which has given value-added to organic produce in general.

5.3 *Cooperation and the SDGs*

Cooperation between consumers and agricultural cooperatives undoubtedly contributes to SDG 1 (no poverty) and SDG 2 (zero hunger). The research reported here cannot provide figures for the extra income generated by banana exports, partly because each family produces many other crops besides bananas, but the assessment of the co-op members themselves is that, since exporting bananas to Japan, their income has steadily increased. The cooperation has other benefits. It provides bananas for Japan, where there is limited land for agricultural use. It has encouraged and promoted organic agriculture in Thailand, where neither the technology nor the market for organic agriculture had been well developed prior to accessing the Japanese market. These days, organic products are available in the general Thai market and knowledge about organic farming is widespread. The agricultural production system in Thailand has also received an impetus in the direction of sustainable production, thanks to logistical studies of how to transport bananas within Thailand. Bananas must be picked at the right time, and go through inspection and packing processes in order to reach Japan before they become over-ripe. Failure to follow procedures resulted in the first shipment of bananas being rejected. With the help of local Thai experts, the cooperatives were able to improve their logistical procedures — an improvement that can also benefit other sectors.

Taking everything into consideration, it can be argued that this cooperation between cooperatives contributes most of all to the realization of SDG 17: partnership for the goals. It represents a form of cooperation between Japan and Thailand — a developed and a developing country. Since cooperatives are by their nature basically local organizations shaped by local values and way of life, any cooperation between foreign cooperatives allows the chance for those values and ways of life to be demonstrated and shared. Demands are directly communicated, and matters related to supply are discussed and understood. Representatives from Thailand who visited Japan told stories about Japanese society and their experience with consumers who gave them feedback about their bananas. Japanese consumers came to realize the difficulties of Thai producers and tried to support them. The system of teikei provides the underlying connections that bind the consumers and producers together.

6 Conclusion

This study on the cooperation of Japanese and Thai cooperatives was undertaken to highlight the role of the private sector which can be a primary agent in pursuing the objectives of the SDGs. If the nature of the economic system does not allow the business sector to fulfil this role, cooperatives — as a form of business institution — may be in the best position to take up the task. The study found that, by definition, cooperatives are participatory and democratic institutions; Japanese cooperatives are an excellent example of this.

Cooperation between Japanese and Thai cooperatives has had a considerable impact on Thai society. As a result of this cooperation, Thailand has become an active partner in the global arena in pursuing the SDGs. Thai cooperatives have grown stronger and become more sustainable; Thai farmers have been encouraged to stay on their land and continue their work in agriculture, rather than quitting farming and moving into the industrial or service sector. In terms of promoting organic farming in Thailand, the impact has been even greater.

Cooperatives are business organizations at the grassroots level. Business at this level is normally vulnerable to environmental change: for agricultural businesses, vulnerability is even higher. To counter this vulnerability, the teikei system plays an important role, strengthening the partnership between Japanese and Thai cooperatives. Although this partnership was not established with the aim of achieving SDG 17, it actually provides the means to accomplish the objectives of that goal.

References

FAO (2014). Banana market review and banana statistics 2012–2013. Rome: Food and Agriculture Organization of the United Nations. Retrieved from http://www.fao.org/docrep/019/i3627e/i3627e.pdf.

ICA (2011). Japan in *ICA Global 300 Report 2011*. ICA Committee on Consumer Cooperation for Asia and The Pacific.

IHRB (2015). Business and Sustainable Development Goals. Mind the gap: Challenges for implementation. IHRB State of Play Series, Volume Four. London: Institute for Human Rights and Business. https://www.ihrb.org/pdf/state-of-play/Business-and-the-SDGs.pdf.

ILO and ICA (2014). Cooperatives and the Sustainable Development Goals. A contribution to the post-2015 development debate. International Labour Organization and the International Cooperative Alliance. https://sustainabledevelopment.un.org/content/documents/1247ilo.pdf.

JOAA (1993). 'Teikei' system, the producer–consumer co-partnership and the movement of the Japan Organic Agriculture Association. Tokyo: Japan Organic Agriculture Association. Retrieved from http://www.joaa.net/english/teikei.htm.

Moen, D. G. (1997). The Japanese organic farming movement: Consumers and farmers united. *Bulletin of Concerned Asian Scholars*, 29(4), 14–16. Retrieved from http://dgmoen.net/blog/scholarship/the-japanese-organic-farming-movement-consumers-and-farmers-united/.

Thipakorn, Saikaew (2016). Japan. Research report series, Studies on successful development models. Bangkok: Institute of Asian Studies, Chulalongkorn University.

Thairath Online (2015, August 5). Banana: New export to Japan. Retrieved from https://www.thairath.co.th/content/516276.

Thompson, D. (2017, August 7). Japan: 'Land of Cooperatives'. Retrieved from http://www.grocer.coop/articles/japan-land-cooperatives.

Tanthamaroj, Vichian (2014). *JTEPA and banana export by Thai agricultural cooperatives*. Bangkok: Devawongse Varopakarn Institute of Foreign Affairs.

Index

www.ingramcontent.com/pod-product-compliance
Lightning Source LLC
LaVergne TN
LVHW021724021026
849161LV00004BA/672
* 9 7 8 9 0 0 4 3 7 8 2 3 0 *